Digital Innovations in Architecture, Engineering and Construction

Series Editors

Diogo Ribeiro⬤, Department of Civil Engineering, Polytechnic Institute of Porto, Porto, Portugal

M. Z. Naser, Glenn Department of Civil Engineering, Clemson University, Clemson, SC, USA

Rudi Stouffs, Department of Architecture, National University of Singapore, Singapore, Singapore

Marzia Bolpagni, Northumbria University, Newcastle-upon-Tyne, UK

The Architecture, Engineering and Construction (AEC) industry is experiencing an unprecedented transformation from conventional labor-intensive activities to automation using innovative digital technologies and processes. This new paradigm also requires systemic changes focused on social, economic and sustainability aspects. Within the scope of Industry 4.0, digital technologies are a key factor in interconnecting information between the physical built environment and the digital virtual ecosystem. The most advanced virtual ecosystems allow to simulate the built to enable a real-time data-driven decision-making. This Book Series promotes and expedites the dissemination of recent research, advances, and applications in the field of digital innovations in the AEC industry. Topics of interest include but are not limited to:

Industrialization: digital fabrication, modularization, cobotics, lean.
Material innovations: bio-inspired, nano and recycled materials.
Reality capture: computer vision, photogrammetry, laser scanning, drones.
Extended reality: augmented, virtual and mixed reality.
Sustainability and circular building economy.
Interoperability: building/city information modeling.
Interactive and adaptive architecture.
Computational design: data-driven, generative and performance-based design.
Simulation and analysis: digital twins, virtual cities.
Data analytics: artificial intelligence, machine/deep learning.
Health and safety: mobile and wearable devices, QR codes, RFID.
Big data: GIS, IoT, sensors, cloud computing.
Smart transactions, cybersecurity, gamification, blockchain.
Quality and project management, business models, legal prospective.
Risk and disaster management.

Sai On Cheung · Liuying Zhu

Editors

Construction Incentivization

Beyond Carrot and Stick

 Springer

Editors
Sai On Cheung 🆔
Construction Dispute Resolution Research
Unit
Department of Architecture and Civil
Engineering
City University of Hong Kong
Hong Kong, China

Liuying Zhu 🆔
School of Management
Shanghai University
Shanghai, China

ISSN 2731-7269 ISSN 2731-7277 (electronic)
Digital Innovations in Architecture, Engineering and Construction
ISBN 978-3-031-28958-3 ISBN 978-3-031-28959-0 (eBook)
https://doi.org/10.1007/978-3-031-28959-0

This Springer imprint is published by the registered company Springer Nature Switzerland AG
The registered company address is: Gewerbestrasse 11, 6330 Cham, Switzerland

To Members and Friends of the
Construction Dispute Resolution Research
Unit
With Sincerest Gratitude

Foreword by Roger Flanagan

This book is important because it fills a gap in the knowledge and understanding of how incentives in the construction industry operate; it does more than just describing the carrot and stick approach. The chapters have been written by a knowledgeable group of authors who challenge traditional thinking by suggesting conventional incentive design may not be sufficient, and that we have reached a tipping point in the design and management of incentivization.

The chapters follow a logical path taking the reader from understanding the basic principles of incentivization to considering the importance of human behaviour and relationships. They focus on issues, such as the potential impact of incentives on health and safety, and the impact of COVID-19 on performance and incentives, and how unrealistic targets can impact construction. It looks at how incentivization can impact dispute resolution whilst considering some contractual issues. A new approach in incorporating behavioural targets for incentives is proposed to meet the much heralded relational paradigm in construction contracting.

Incentives are not new, but the way they are structured and used in today's construction industry is new. It started with incentives for the workers to be more productive and to benefit from enhanced payments for superior performance. Irving Fisher in 1919, an economist at Yale University, referred to finding ways to put real pep into the workers returning from the World War to counteract the monotony of work. He attempted to understand the fundamental motivation of workers beyond self-preservation and earning a living; increased pay may motivate many workers, but often more is sought in job satisfaction and progression. Motivation today includes getting job satisfaction and working for an enterprise that puts sustainability at the heart of everything it does, cares about the future of the planet, and takes social responsibility seriously.

Motivation, efficiency, and conflict led the work-study specialists to focus on performance and financial rewards, but the non-financial rewards have come more to the fore in recent times. Incentivization was made possible by a series of efforts to stretch the boundaries of incentivization until it became the encompassing framework that it is today. Incentivization can also mean threatening a penalty for failing to meet a target. In some cases, incentives are powerful tools to motivate people to take a certain action, but they can backfire, and might decrease motivation instead of increasing it.

Building incentives into any system is now an integral part of delivering construction projects. An early completion bonus clause in a contract is an incentive for the contractor to deliver before the date of practical completion. The liquidated damages clause is also an incentive to ensure completion on time. Design and engineering consultants have incentive clauses inserted into appointments, for example in designing a power plant that meets a minimum output, with the opportunity to share the financial benefits when the targets are exceeded. New incentive targets can also be for the achievement of zero carbon on a project. Clients are offering all kinds of incentives to project delivery teams, such as waste minimization and re-use of demolished materials. Many investors will only invest in projects where it can be demonstrated that green targets are being met.

This means Incentivization spans many areas that hitherto were not considered. How to use innovation in the design of incentivization has become a requirement. To do this, we must understand how incentivization evolved and how it may evolve in the future. The certainty is that tomorrow will not be like yesterday and new ways of working and thinking are needed.

This is an excellent book written by knowledgeable researchers of the Construction Dispute Resolution Research Unit of the City University of Hong Kong. This book prepares the ground for new knowledge and techniques for the planning and implementation of construction incentivization. I congratulate Prof. Sai On Cheung for leading the Construction Dispute Resolution Research Unit and as joint editor with Dr. Liuying Zhu of this intriguing volume. With the contributions of the authors, this book on construction incentivization has offered new perspectives on the designs and applications of construction incentivization. I highly recommend it to you.

December 2022   Prof. Roger Flanagan, M.Sc., Ph.D.,
                                         D.Sc., D.Sc. (Hon.), Dr. Eng. (Hon.) PPCIOB, FRICS,
                                                    FICE, Professor, University of Reading
                                                                              Reading, UK

                                                    Honorary Professor, Tsinghua University
                                                                           Beijing, China

                                         Past President of the Chartered Institute of Building
                                                                            Bracknell, UK

Foreword by Makarand Hastak

In my 30+ years as an educator, researcher, and consultant in the construction industry, I have often wondered about the effectiveness of incentives in influencing productivity and the outcome of a project. Should the incentives be ongoing, periodic, or singular? In what circumstances do incentives work best? Are they sufficient to increase worker morale and overcome other project constraints? These and many other questions intrigue me about incentives and disincentives in a construction project. So when Prof. Cheung approached me with the suggestion to write a foreword for this book *Construction Incentivization: Beyond Carrot and Stick*, I eagerly accepted the opportunity. Both Editors Dr. Sai On Cheung and Dr. Liuying Zhu are accomplished scholars in the area of Construction Dispute Resolution. Dr. Cheung has written two prior books on the subject, whereas Dr. Zhu wrote her dissertation on Construction Incentivization and continues to do research in this area. They along with seven authors have compiled an excellent volume that addresses these and many other questions that you and I as industry professionals, educators, researchers, and students would have about incentivization on a construction project.

My experience suggests that incentives and disincentives might have a different impact on large and small contractors. One might use them to formulate a winning strategy for a bid, as in a bonus for early completion, while another to extract more productivity out of their crew to safeguard their profits on a job, as in liquidated damages. However, as the authors have pointed out, construction incentivization does not always lead to a positive outcome. Construction incentives are often used to extract additional effort from the contractor with an expectation of a better performance outcome. Do rewards always result in better performance? The incentives and disincentives are often unilaterally decided by the owner/client. However, if the target expectations are unrealistic, then the incentive may not be a sufficient motivator to extract extra effort from the contractor. Would the outcome be better if the contractor is engaged in such decisions?

Mega projects are known to have a higher instance of cost overrun and schedule delays. Would conventional incentives and disincentives work on such projects where the complexity is much higher along with the associated risks and rewards? Such projects demand better coordination and integrated efforts between the project stake-holders. Innovative methods such as Integrated Project Delivery (IPD) or Advanced Work Packaging (AWP) require a higher level of collaboration between the project parties. IPD even requires that a single contract be signed by the project participants to share the risks and the rewards. The same principle applies to AWP where, the project parties are expected to coordinate efforts in the very early stages of a project for the desired outcome. What is the role of incentives and disincentives for such methods of project delivery where multiple agents share risks and rewards? Does Construction Incentivization promote better collaboration between project partici-pants? Is it more effective on public projects or private projects? What about Public Private Partnership, does construction incentivization work in a different way for such project partnerships? This book addresses questions such as these and many others in a very thoughtful and methodical way. The various chapters in this edited volume are written by the editors and seven Ph.D. scholars. They are all associ-ated with the Construction Dispute Resolution Research Unit (CDRRU) that was established by Prof. Cheung twenty years ago at the City University of Hong Kong. Some authors have already received their Ph.D.s, while others are nearing comple-tion of their doctoral studies. As such, the chapters are very well researched, and the authors have addressed the important topics with clarity that is backed by theoretical constructs, data, and analysis.

Does construction incentivization lead to innovation? Often, value engineering is associated with an incentive for sharing the cost savings in order to motivate the contractor to explore innovative solutions to the problem. Contractual constraints do not always support innovation in a construction project forcing contractors to use true and tried methods to safeguard their low profit margins (e.g., in Lump Sum/Fixed-price contracts). Do we need to think beyond the conventional rewards and penalties to facilitate innovation and adoption of new technologies and methods in construc-tion? Construction researchers through their engagement in the various commissions of CIB (The International Council for Research and Innovation in Building and Construction) as well as academic leaders through the Global Leadership Forum for CEM programs (GLF-CEM) have been debating means and methods of inculcating innovation in the construction industry. What role can construction incentivization play in promoting innovation? In this book, the authors have addressed this issue by offering thoughts that go "beyond carrot and stick," including the use of construction incentivization for strategic use, building inter-organizational relationships, raising the psychological well-being to motivate workers, etc.

Should construction incentives be built into the contracts or should they be used more strategically to mitigate risks that were not identified during the pre-construction stage? For example, long-term warranties were in vogue a few years back when various owners including the Department of Transportations (DOTs) in the US were exploring the risks and rewards of requiring warranties on their contracts. During our research on warranties, we observed that large contractors were not worried about the risk of warranties because (i) they had large bonding capacities, (ii) their construction practice and experience suggested that nothing major happens to a well-constructed pavement in five years, and (iii) they were happy to take the extra amount the DOTs were willing to pay for the warranties. While on the other hand, small contractors, who survive on DOT projects, were very concerned, as offering a five-year warranty on a pavement project was a very risky proposition for them as it would block their bonding capacity, and the incentive of large payout was not strong enough to offset that risk. Can contractors always benefit from construction incentives? Should they always go hand in hand with penalties? Are they even necessary to boost worker morale or to reduce safety risks on a construction project? The authors have explored such complex issues in this book in the context of using construction incentives to mitigate the impact of COVID-19-related litigations, to boost safety compliance, and for negotiating construction disputes. In my opinion, this book is a must read for construction practitioners, academics, and students alike. An important topic such as construction incentivization has many facets, and it is imperative that we understand the various nuances of incentivization to improve our project outcomes.

With compliments to the authors for compiling this important volume.

December 2022   Prof. Makarand Hastak, Ph.D., PE, CCP, CRIS, MASCE
                                        Dernlan Family Head and Professor of Construction
                                                    Engineering and Management

                                        Professor of Civil Engineering, Purdue University

                                        President of CIB (Int'l Council for Research and
                                                    Innovation in Building and Construction)

                                        Chair of the Advisory Board of the Global Leadership
                                        Forum for CEM Programs, West Lafayette, IN, USA

Preface

Incentivization has been used for quite some time in the construction industry. Mixed results have been reported. Are the current designs of construction incentivization (CI hereafter) meeting the needs of the ever-changing landscape of capital developments? This book aims to provide a comprehensive review of this question. In addition, the chapters offer new ideas for the formulation of the same to meet the challenges of modern construction projects. For these purposes, this book has three parts. Part I is devoted to the conceptualization of incentivization in construction and aims to provide theoretical anchors for the planning of construction incentive. Part II examines various strategic uses of incentive. The applications are directed towards certain classes or sectors within the construction industry such as procurement and housing supply. Part III gives four specific applications of incentive. Examples include upholding safety at work, mitigating impact of COVID-19, and enhancing dispute resolution.

Part I: Conceptualization

Chapter 1 (A Primer of Incentivization in Construction) examines the motivation theories that underpin the formulation of incentive plans. Moreover, the triggering agents may not match the project characteristics. It is advocated that the use of incentive, disincentive, and their combination thereof should be contingent on the procurement approach and the types of target outcomes. To revamp the design of CI, five functions should be directed: Goal Commitment; Expectation Alignment; Information Exchange; Risk Efficiency; and Relationship Investment. Chapter 2 (Construction Incentivization in Perspective) further elaborates on the idea of using integrative incentivization to cater to a variety of target outcomes. The idea of engendering behaviour change through CI is introduced. Guided by the aforementioned two chapters, Chap. 3 (Incentivization or Disincentivisation) uses practical cases to illustrate the circumstances under which positive push (incentive) and deterrence (penalization) should be used. Incentivization is found to be more effective to solicit unique dedicated actions while disincentivization is an economical option for relatively straightforward projects. Unsatisfactory results of many CI can be attributed to the insufficient provisions to match the differing sources of motivation. Chapter 4 (Behavioural Considerations in Construction Incentivization Planning) suggests the inclusion of behaviour-based outcome targets and interim awards to augment the incentive for behaviour change.

Part II: Strategic Uses

Accomplishing planned project objectives would be facilitated by appropriate strategies. In the context of CI, the strategic actions include aligning the stakeholders' interests, fostering interorganizational relationships (IOR), and committing to smart leadership. Can these strategies incentivize the mega project participants, integrated project delivery (IPD) contractual parties, construction workers, and civil servants? This part of the book aims to discuss these strategic uses. Chapter 5 (Incentivizing Relationship Investment for Mega Project Management) advocates the development of IOR via CI to incentivize the performance of mega projects. Developing IOR is proposed to be the operating agent. Empirically, the IOR was found to be a partial mediator between CI and project performance. Chapter 6 (Multi-agent Incentivizing Mechanism for Integrated Project Delivery) introduces a multi-agent risk/reward sharing mechanism to supplement the structuring of incentive plans for developments using integrated project delivery (IPD) procurement. The proposed risk/reward sharing mechanism is capable to effect equitable sharing across a large group of participants with due regard to their input. Chapter 7 (Would Raising Psychological Well-being Incentivize Construction Workers?) focuses on the strategic raise of the psychological well-being of workers to motivate performance. For example, prevention-focused safety leadership can be used to enhance the perception of the well-being of construction workers. Chapter 8 (Revamping Incrementalism to Incentivize the Land and Housing Policy Agendas in Hong Kong) explains in what ways incrementalism can shape policy. The evolution of the Hong Kong land and housing supply policy is used to illustrate the dynamic of incrementalism, intellectual inquiry, and incentivization.

Part III: Specific Uses

The fragile economics and severe competition have exacerbated the adversarial relationships between contracting parties of construction developments. The long-drawn COVID-19 pandemic has made the situation even worse. This part aims to provide the specific use of incentive/disincentive (I/D) to alleviate the resulting challenges. Chapter 9 (Means to Incentivize Safety Compliance at Work) systematically overviews the effectiveness of I/D in motivating worker safety compliance. Non-financial incentives are more effective in stimulating safety compliance at work. The individual attributes of workers and construction site conditions also affect the effectiveness of CI. The workers, for example, with higher education and expectation of equitable income distribution, are more apt to comply with safety compliance. Chapter 10 (The Role of Incentivization to Mitigate the Negative Impact of COVID-Related Disputes) discusses how to attenuate disputes stemming from the COVID pandemic. COVID-19 has generated new project risks that are out of anticipation of the contracting parties. CI is identified to be effective in mitigating the damages arising from COVID-19. The resulting disputes are best addressed through the application of equitable risk sharing engineered by CI that can be installed without amending the existing contract terms. Chapter 11 (Interweaving Incentives and Disincentives for Construction Dispute Negotiation Settlement) investigates whether incentive stimulates, or disincentive impedes construction dispute negotiation settlement. The answer seems to be affirmative according to the correlation between I/D and negotiators' intention to settle. Chapter 12 (Voluntary Participation as an Incentive of Construction Dispute Mediation—A Reality Check) challenges the utility of embracing voluntary participation as one of the characterizing design features of construction mediation. Two noteworthy incompatible contracting conditions, namely power asymmetry between disputants and quasi-imposition, are identified as potential obstacles that would negate willingness to mediate.

Construction Incentivization: Beyond Carrot and Stick is a timely collection to revisit the use of incentives to promote performance given the unprecedented technical and managerial challenges of modern sophisticated construction projects. A novel idea of incorporating behaviour-based outcome targets is proposed. In addition, using task programmability and outcome certainty as control parameters for the selection of incentive arrangements is pioneering. This collection is research-rich and is supported by insightful practical illustrations.

We are in debt to Prof. Roger Flanagan and Prof. Makarand Hastak for their enlightening Forewords. Special thanks go to Dr. Lizzy Ma for the excellent editorial assistance. Finally, this book cannot be possible without the contribution of the authors who all are members of the Construction Dispute Resolution Research Unit that was established by Prof. Sai On Cheung in 2002. It is indeed a great pleasure to publish this volume with the members of the research unit on the 20th anniversary.

December 2022   Prof. Sai On Cheung, D.Sc., Ph.D.
Director, Construction Dispute Resolution Research Unit,
Department of Architecture and Civil Engineering
City University of Hong Kong
Hong Kong, China

Dr. Liuying Zhu, Ph.D., M.Sc.
School of Management
Shanghai University
Shanghai, China

Contents

Part III Specific Uses

Editors and Contributors

About the Editors

Dr. Sai On Cheung is a Professor of the Department of Architecture and Civil Engineering, City University of Hong Kong. In 2002, Prof. Cheung established the Construction Dispute Resolution Research Unit. Since then, the Unit has published widely in the areas of construction dispute resolution and related topics such as trust and incentivization. With the collective efforts of the members of the Unit, two research volumes titled Construction Dispute Research and Construction Dispute Research Expanded were published in 2014 and 2021, respectively. Professor Cheung is a specialty editor (contracting) of the ASCE Journal of Construction Engineering and Construction. Professor Cheung received his D.Sc. for his research on Construction Dispute. e-mail: saion.cheung@cityu.edu.hk

Dr. Liuying Zhu works at the School of Management, Shanghai University. Dr. Zhu received her M.Sc. degree (Distinction) and Ph.D. from the City University of Hong Kong. Her M.Sc. dissertation was awarded the 2016 Best Dissertation (Master category) by the Hong Kong Institute of Surveyors. Her doctoral study on construction incentivization was completed with the Construction Dispute Resolution Research Unit. Her current research focuses on construction project management, construction incentivization, and project dispute avoidance. Dr. Zhu has published articles in international journals and book chapters on these topics. e-mail: zhuliuying@shu.edu.cn

Contributors

Nan Cao Construction Dispute Resolution Research Unit, Department of Architecture and Civil Engineering, City University of Hong Kong, Hong Kong, China

Sai On Cheung Construction Dispute Resolution Research Unit, Department of Architecture and Civil Engineering, City University of Hong Kong, Hong Kong, China

Pui Ting Chow HKU SPACE Po Leung Kuk Stanley Ho Community College, Hong Kong, China

Keyao Li Future of Work Institute, Curtin University, Perth, WA, Australia

Sen Lin Construction Dispute Resolution Research Unit, Department of Architecture and Civil Engineering, City University of Hong Kong, Hong Kong, China

Qiuwen Ma Construction Dispute Resolution Research Unit, Department of Architecture and Civil Engineering, City University of Hong Kong, Hong Kong, China

Peter Shek Pui Wong School of Property, Construction and Project Management, Royal Melbourne Institute of Technology University, Melbourne, VIC, Australia

Tak Wing Yiu School of Built Environment, University of New South Wales, Sydney, NSW, Australia

Liuying Zhu School of Management, Shanghai University, Shanghai, China

Part I
Conceptualization

Chapter 1
A Primer of Incentivization in Construction

Sai On Cheung and Liuying Zhu

Abstract Many construction projects end with cost overrun, delay and defects. These undesirable outcomes are particularly disheartening with mega projects. The construction industry has been seeking ways to improve project performance and inter alia, incentive schemes have been used widely as one of the means to induce extra efforts from contracting organisations. In 2007, the Hong Kong Government announced the construction of ten mega projects. Notwithstanding these mega projects have all incorporated certain forms of incentive, delay, substantial cost overruns and quality issues have been reported. The authors observed the following pattern of use of construction incentivization (CI): (i) Most of the CI have targets set on time, cost, quality, and safety, (ii) No clear pattern of how CI are developed, (iii) 'Carrots' are used far more often than 'sticks', (iv) The use of CI is far more common in public projects than private projects, (v) Most targets are quantitative, and (vi) Choice of CI is rather incidental. Apparently, there are two major shortcomings of the prevailing CI arrangements. First, CI is anchored on motivation theories that are mostly related to individuals; Second, the targets are outcome based and mainly tied with developers' goals. This outcome-based approach is useful for tasks of high programmability with outcome that can be accurately projected. However, construction tasks, especially those that need innovation, are typically of low programmability with loose outcome predictability. To overcome these shortcomings, this primer suggests that incentivization should aim for effort greater than mere competence and go beyond carrot and stick should be used. It is advocated that integrative incentive should be used and have five functions: (1) Goal Commitment; (2) Expectation Alignment; (3) Information Exchangeability; (4) Risk Efficiency; and (5) Relationship Investment.

S. O. Cheung (✉)
Construction Dispute Resolution Research Unit, Department of Architecture and Civil Engineering,, City University of Hong Kong, Hong Kong, China
e-mail: saion.cheung@cityu.edu.hk

L. Zhu
School of Management, Shanghai University, Shanghai, China
e-mail: zhuliuying@shu.edu.cn; zhuliuying@shu.edu.cn

© The Author(s), under exclusive license to Springer Nature Switzerland AG 2023
S. O. Cheung and L. Zhu (eds.), *Construction Incentivization*, Digital Innovations in Architecture, Engineering and Construction,
https://doi.org/10.1007/978-3-031-28959-0_1

3

Keywords Motivators · Carrot and Stick · Task Programmability · Outcome Predictability · Integrative Incentivisation

1 Introduction

Modern construction projects are characterized by high value, long duration, complex design, and technically sophisticated. Flyvbjerg (2017) summarized from his study on major worldwide mega projects and offered his iron law of mega projects as "over budget, over time, under benefits". 70% of the projects he studied recorded cost overrun of no less than 50% of the respective budget. In Hong Kong, similar project problems are found in the large-scale developments, especially the infrastructural projects initiated by the government. The Hong Kong Development Bureau (Hong Kong Development Bureau, 2018) reported that many of these projects are having extensive delay, substantial cost overrun and embarrassing quality issues. Need for improvement is self-evident. Construction incentivisation (CI) is the primary tool used to motivate contracting organisations to perform. CI is used in this study as a collective term to cover the range of incentive schemes that can be used in construction contracting to engender project performance improvement. Most construction projects, especially those identified as mega, are installed with CI. Moreover, the effectiveness of CI in enhancing performance is not as forthcoming as expected. For example, the projects under the 2017 Hong Kong Ten Mega Program (HKSAR, 2017) have installed various forms of CI, still delay, cost overrun, and huge claims have been reported.

A pilot literature review by the authors on the use of incentive found that there is no standard pattern of how construction incentive schemes are organised. Moreover, typically cost, schedule and quality outcomes are used as targets for sanctioning awards (incentive) or activating the penalisation (disincentive). These targets are sometimes linked to form composite incentive for complex tasks. In cost-plus contracts, schedule incentive scheme often goes hand in hand with cost-incentive scheme (Abu-Hijleh & Ibbs, 1989a). Quality, safety, overall productivity performance, or a combination thereof together with behaviour modification is used as an integrative incentive scheme for complex projects. To deepen the understanding of the implementation of CI in Hong Kong, views from construction professionals was conducted (Zhu & Cheung, 2018). The interviewees include senior professionals with immense experience coming from the government, private developer, leading contractor, and consultant. The interviewees provided the following observations that are quite in line with the findings in the pilot literature review:

- The use of CI is far more common in public projects than private projects.
- Most of the CI have targets set on time, cost, quality, and safety.
- Most targets are quantitative.
- 'Carrot' is used far more often than 'stick'.
- Choice of CI is rather incidental.

- No clear pattern of how CI is developed.

It can be summarised that though the use of CI in construction contracting is not new, the unsatisfactory record suggest that there are some notable knowledge gaps to be filled for the planning and implementation of CI. For example, motivation theories that portrait individual's behaviour have been the primary theoretical explanation of the value of CI. However, individual effort may not be transcended to organisational level. Furthermore, prevailingly used CI are mostly outcome-based with targets imposed by the offering party. This arrangement is not conductive to harvest performance enhancement, in particular for tasks that demand efforts beyond mere competence. This study aims to critically review the conceptual bases of CI. It is advocated that CI should be designed contingent to the task characteristics. As afore stated, using metric CI targets facilitates discrete determination of attainment or otherwise of the outcome targets. Nevertheless, this outcome-based arrangement gives no regard to the fact that construction project performance is a matter of team effort and credits should also be allowed for the efforts expended irrespective of the outcome. Furthermore, the fact that the estimation of the target outcome may not be accurate.

Many construction-industry reviews (Latham, 1994; Egan, 1998; CIRC, 2001) have advocated that project teams should work cooperatively for the good of the project. There have been voluminous studies on how to make construction contracting more cooperative. Trust building is the most notable recommendation (Zuppa et al., 2016). Cheung et al. (2011) analysed 163 responses from construction professionals in Hong Kong and found that trusting partners communicate much better. Zuppa et al. (2016) added that trust is the catalyst in promoting leadership, building team and information sharing between construction project participants in the US. In fact, from a case study of a record-breaking project that generated 450 patents, Zhu et al. (2020) found that singular use of monetary reward cannot deal with complex and technically challenging operations. Instead, embracing commitment through enhanced Inter-organizational relationship (IoR) was instrumental in driving exceptional efforts and fostering innovations. This neatly points to the need to have tailored CI for innovation developments. Furthermore, Eisenhardt (1985) highlighted that outcome-based incentive arrangements will only work for highly programmable tasks where targets can be set with reasonable accuracy. While uncertainty remains one of the key challenges faced by major engineering and construction projects, construction activities, especially those requiring extensive on-site execution are not highly programmable. Behaviour-based incentives would provide the stimulation for conducive behaviours to meet with the challenges as the project unfolds. This has been proved invaluable when innovative ideas are solicited. This study proposes the use of task programmability and outcome predictability to guide the mapping of procurement options with incentive arrangement.

2   The Study

CI is used as the vehicle to motivate performance. Motivation is the urge to perform an act, to obtain a certain object, or to produce a desired outcome (Satinoff and Teitelbaum, 1983). Bootzin et al. (1991) described motivation as a process that energizes, maintains, and directs behaviour towards certain goals. The force can drive decisions and behaviours that are consistent with the pursuit of the goal (Baron, 1995). Another common view of CI is its ability to align goals between developers and contractors. Incorporating CI in parallel with the formal contract is believed to inspire, drive, and direct one's resources for the attainment of the project goals. For example, target cost contracts have been used to link the interests of the contracting parties. There is strong support that CI can be used as a quasi-contractual tool to gauge performance (Zhu & Cheung, 2021). To enhance the instrumentality of CI, the functions of construction incentivization are first to be identified in this section. Analogically, CI functions are the deoxyribonucleic acid (DNA) of any incentive arrangements. Macneil (1974) describes performance, risk and dispute resolution are the three pillars of contract planning. As such, all contracts are built through systematic planning of the three pillars. For this study, CI functions are those indispensable elements of successful incentive arrangements. The search for CI functions starts with review of literature on motivation theories, outcome targets and project performance.

2.1   Motivation Theories

Motivation is what prompts individuals, teams, and organizations to act in a certain way, or develop an inclination for specific behaviour (Kast and Rosenzweig, 1985). Most motivation theories are addressing individuals instead of organizations. Construction project participants are seldom individuals but are complex commercial organizations (Bresnen & Marshall, 2000). The literatures on CI almost implicitly treat organizational and individual goals as more or less the same (Arditi & Yasamis, 1998). A review of motivation theories is presented here-follows.

Maslow's Hierarchy of Needs Theory

Maslow's needs hierarchy is often applied as a maturity model that describes how individuals move up the hierarchy as they develop (Pardee, 1990). Maslow et al. (1987) believed that individuals who possess a constantly growing inner drive would have great potential. Five sources of motivation are suggested: Physiological, Safety, Socialisation; Esteem, and Self-actualisation (Maslow et al., 1987). Two major postulates are developed to present the progressive relationships of these five sources. Firstly, motivation comes from unsatisfied needs. Secondly, when the lower-ordered needs are satisfied, the next higher level of needs then becomes significant determinants of behavior (Acquah et al., 2021). Thus management can match their staff development level by addressing the needs that would most motivate their effort.

McClelland's Need Theory

McClelland's need theory is more employee oriented. The theory proposes that when a need is strong, its effect is to motivate the person to use behavior which would satisfy that need (McClelland, 1965). Three core needs are Affiliation, Achievement, and Power (Pardee, 1990). McClelland further developed the descriptive set of factors that reflect the high need of achievement (McClelland and Johnson, 1984): (1) preference for personal responsibility; (2) tendency of taking moderate goals and calculated risks; and (3) expectation of performance feedbacks. For project management, project leaders should position their team members to capitalise on the respective motivating effect.

Incentive Theory

The incentive theory proposes that individuals will practice certain behaviours in response to specific task requirement or for a reward (Killeen, 1981). It has been useful to describe behaviour under the control of concurrent chained schedules of reinforcement (Killeen, 1985). People may display certain behaviours to achieve a specific result, incite a particular action or receive a reward (Locke et al., 1988). The motivators can be reinforcement, recognition, and rewards. Moreover, these motivators need to be meaningful, specific, challenging, and acceptable to those who are attempting to achieve them. Typical rewards in organisational setting include bonus, praise, opportunity, promotion, salary and improved fringe benefits (Rose & Manley, 2011).

Expectancy Theory

Expectancy theory explains why one's choice of behaviour is influenced by his assessment of the outcome (Oliver, 1974). Behaviours will be directed towards those that would achieve the desired outcome (Wigfield & Eccles, 2002). It is therefore imperative that the outcome to be realistic and achievable (Bandura, 1982). Caveat against bias of overconfidence should also not be underestimated. Expectancy theory projects that motivation is a function of three main factors: the subjective value placed on the reward ('valence'); the perceived likelihood that effort will produce an appropriate level of performance ('expectancy'); and the perceived likelihood that this performance will be converted into an appropriate level of reward ('instrumentality') (Vroom et al., 2005). Management strategies are therefore needed to raise at least one of these factors to enhance motivation. Motivation can therefore be expressed as:

$$\text{Motivation} = \text{Valence} * \text{Expectancy} * \text{Instrumentality}.$$

Herzberg's Hygiene Factors

Herzberg's motivation-hygiene theory argues that there are separate sets of mutually exclusive factors in the workplace that either cause job satisfaction or dissatisfaction

(Herzberg, 1966). Hygiene factors are those that remove hazards from the environment (Herzberg, 1970). Hygiene factors cannot motivate, but may cause negative effects if not satisfied. The motivating factors: like achievement, recognition, responsibility, and advancement, are satisfiers and promote further improvements (Herzberg, 1974).

Competence Theory

Competence theory identifies that individuals' behaviours are guided by the motives of displaying their skills, intelligence and abilities (Mulder, 2017). The desire to demonstrate these qualities would motivate them to feel competent in a particular area. The confidence of ability exert control over individual's motivation and behavior (Bandura, 1978). Competence underpins enhancement in productivity and efficiency. Competent staff are more confident and willing to share with their peers in return for greater recognition. Table 1 summarizes the motivators suggested by the afore-listed motivation theories. Individuals/organizations can be motivated should appropriate carrot be offered. Conceptually, these motivators can be materialistic, hygienic, and aspirational. Materialistic motivators are extrinsic and can be in the form of reward or deterrence. Hygienic factors are mostly extrinsic with the aim of embracing a conducive environment for performance. Aspirational motivators are primarily intrinsic with the aim of stimulating the self-determination/drive to excel. When aspirational motivators are at work, the need for monitoring diminishes.

   With reference to the review of motivators presented in the preceding paragraphs, deterrence against non-performance appears not the mainstream performance motivator. Moreover, in construction, penalty for non-performance is extensively used, primarily as a baseline safety net for the principal. In other words, penalty is stipulated against non-compliance of contract requirements. Typical example is the inclusion of liquidated damages clause to deter late completion. In this connection, the main aim is not searching for extra efforts but to keep the project under control.

2.2   Outcome Targets

The main purpose of incentivisation is to derive efforts to attain certain targets (Meng & Gallagher, 2012a). In practice, incentive targets are inevitably related to cost, time, and quality (Herten & Peeters, 1986). Bayliss et al. (2004) described these as hard targets. In a study of a partnering project, Bayliss et al. (2004) found that behaviour-based soft targets were having even greater impact on project performance. The section discusses three forms of outcome targets: hard, soft, and innovation.

2.2.1   Hard Targets

- *Cost*

Table 1 The motivators suggested by motivation theories

	Maslow's need theory	McClelland's need theory	Incentive theory	Expectancy theory	Herzberg's hygiene factors	Competence theory
Motivators	Physiological	Affiliation	Reinforcement	Valence	Achievement	Competence
	Safety	Achievement	Recognition	Desired outcome	Recognition	Confidence
	Socialization	Power	Reward	Expectancy	The task	
	Esteem			Instrumentality	Responsibility	
	Self-actualization				Advancement	

Cost is one of the most significant outcome targets because almost all CI has a cost-saving/minimising motive. The major differences between these incentives are in the payment method and risk allocation. Some examples include: (1) Fixed-Price Incentive (firm target) Contract that allows adjusting profit and establishing the final contract price by application of a formula based on the relationship of total final negotiated cost to total target cost. The final price is also subject to a price ceiling; (2) Cost reimbursable contract, the contractor is reimbursed the actual costs they incur in carrying out the works, plus an additional fee; (3) Cost-Plus Incentive-Fee contract (CPIF) uses a banded calculation of pain/gain share system is accumulated in this type of contract to incentivise contractors to reduce cost; (4) Cost-Plus-fixed fee contract (CPFF), it provides for a fee consisting of an award amount that the contractor may earn for cost saving (Chan et al., 2010; Kwawu & Laryea, 2014; Perry et al., 2000; Savio et al., 2013).

- *Schedule*

Schedule incentive scheme entails a premium being offered to the contractor for the early completion of the project (Abu-Hijleh & Ibbs, 1989b; Richmond-Coggan, 2001). The key motive behind schedule incentive scheme is to reward directly to contractors for early completion of work and, otherwise, to penalize them for late completion. The design and implementation of schedule-based incentivisation are relatively straight forward. Schedule targets can be further divided into: (1) final project completion date; (2) intermediate milestone periods; (3) intermediate physical completion milestones; or (4) a combination of final and milestone assessments (Abu-Hijleh & Ibbs, 1989b).

- *Quality*

Quality performance can be applied to a wide range of non-cost/time targets such as functionality, defects, and safety. Quality incentive is used for achieving zero or minor defects. Safety incentive scheme seeks to ensure compliance with safety rules and standards (Meng & Gallagher, 2012b). Different from cost or schedule incentive schemes, the assessment for technical performance is more complex.

Time, cost, and quality targets are the three most used outcome targets (Zhu & Cheung, 2018). Using these 'hard' targets are often described as outcome-based approach. Moreover, Boukendour and Hughes (2014) pointed out that one of the major and recurring problems in designing cost incentive contracts is the setting of the cost target and the risk sharing ratio. This challenge equally applies to the setting of time and quality targets.

2.2.2   Soft Targets

Apart from having hard outcome targets, Eisenhardt (1988) advocated that behaviour-based criteria that reflect the ways the parties behave should also be used. For example, developing innovative solutions may have more far-reaching effect than reaching pre-determined hard targets. Every project can be a testing ground for both technical and managerial innovations. Examples of improved project performance include the attainment of: (i) outcome (hard) targets; (ii) behavioural (soft) outcomes; (iii) value creation (technical and managerial innovations); and (iv) efficient dispute resolution.

Soft targets aim for behaviour modification. Intrinsic motivators are believed to have greater influence as far as shaping behaviour is concerned. The basic "law of behaviour" is that higher incentives will lead to greater effort and higher performance. Extrinsic incentive has been tried to motivate employees. Behaviour intervention incentives have been tried to improve school attendance. Gneezy et al. (2011) used 'intervention' to describe the effect of these behaviour-based incentive because of the potential conflicts over use of monetary incentive. The authors argues that this form of "crowding out effect" has been considered quite common in principal-agent relation. Essentially, behaviour-based incentives work on the prosocial desire of the subject. When this is installed together with monetary reward, the effect of intrinsic motivator would diminish. Moreover, the illustrations used by Gneezy et al. (2011) are primarily related to education, contributions to public goods, and developing habits. The reputational desired outcomes may not directly bring tangible benefits. For signature projects, this was however found to be an intrinsic motivator (Zhu et al., 2020). Reputational enhancement will also bring future job opportunities.

The use of soft targets has also been found in safety incentive plans (Yeow & Goomas, 2014). Likewise, Sims (2002) summarised ten forms of safety incentives: stock ownership, special assignments, training and education, recognition, time off, advancement, social gatherings, increased autonomy, prizes, and money. Sparer and Dennertien (2013) classified safety incentive programs (SIP) into leading (behaviour-based) and lagging (outcome-based) safety performance metric programs. Leading SIP include metrics that could predict the future safety performance such as percentage of safety adult, inspection and walkthrough compliance. Lagging SIP make use of past safety performance metrics to reward workers. It was argued that using leading SIP is based on the assumption that reward being contingent on future performance result, behaviour modification for the sake of performance can be effected. Outcome based SIP reward workers for their individual safety performance. Apparently, behaviour-based SIP requires collective efforts of the workers. As such, the motivators should have elements common to the workers whereby a sense of interdependency can be induced. Achieving construction project objectives obviously need the concerted efforts of the team members, behaviour-based incentive should be the logical choice. Nevertheless, the contractual arrangements take no account of interdependency as all contracts are stand alone legal instrument. In this regard, mechanism like integrated project delivery may offer the vital vehicle to work with behaviour-based incentive.

Yeow and Goomas (2014) proposed a hybrid model called Outcome-Behaviour Based Safety Incentive Program (OBBSIP). There are two principles of OBBSIP. Principle One: outcome-based approach through tiered incentive awarded when meeting safety outcome. The award can be team based and related to periodical performance record. Principle Two: behaviour-based approach through a set of expected safety precautions and safety behaviours. There is no need to have performance record to support the effectiveness of the behaviour because the behaviours are selected based on their proven effect in the long run.

In a principal-agent relation, Murdock (2002) argued that implicit contracts and intrinsic motivation are complements. The idea behind this argument is that people value and therefore derive utility from characteristics of the output of their work in addition to how much they are paid for work. The importance is what characteristics would drive the utility. It would be logical to link these characteristics to the high-level needs suggested by Maslow (1984). Zhu et al. (2020) found that the successful use of a project reputation evaluation system had engendered 'extra' effort of the contractors in a record-breaking mega project. The very fact that the project attracted worldwide attention was sufficient to drive the commitment because they considered non-performance was 'face-losing'. Thus, the type of project at stake would have deterministic effect of the reward, be it incentive or disincentive. The potential gains do not exist when the contractor responds only to extrinsic incentives. The obvious gains would be the interests of the contractors per se. These can be profits, recognition and/or future business opportunities. This study therefore does not support the dichotomy of extrinsic and intrinsic motivation, instead, if the prospective gain stimulates intrinsic motivation, both motivators can work and complement each other. In practice, separate rewards should be used respective to the two forms of motivation.

2.2.3   Innovation

Wang et al. (2018) explored the antecedents of an organization's absorptive capacity by examining the role of innovation incentive. In essence, it was found that innovation incentive enhances absorptive capacity through promoting employees' learning. The implication on the study of incentive is how to mobilize employees' appetite for innovation. Innovation can be identified as something new or not used before. This would require overcoming risk averse attitude. If penalty is attached to innovative attempts, it would be hard to solicit novel ideas that are deemed untested. In this study, it was found that teamwork will help building absorptive capacity, seemingly because of the collective wisdom as well as peer pressure. Interestingly, this study found that transformative leadership reduces the functionality of innovation incentive on absorptive capacity. Transformational leaders influence employees by developing close and individualized relationships with them (Carter & Armenakis, 2013). The potential of injustice hampers the trust of employees on the leader. In construction, what should be the role of the employer who is often the incentive initiator. Micro-management by employer would signal distrust. It is therefore suggested that autonomy is key to innovation incentive.

The study of Surapto et al. (2016) affirms the significance of owner-contractor collaboration to accomplish project goals. The findings may not be unexpected, moreover, the implication on procurement approach is far-reaching. When relational attitude and teamwork are the keys to success, appropriate contractual framework is paramount. The authors suggested that partnering/alliance contracts are likely to perform better than lump sum and reimbursement contracts. In essence, principals and contractors should move away from the conventional principal-agent relationship. In terms of linking procurement with types of incentive, it can be projected that incentive featuring both outcome and behavior target would suit procurement methods that emphasize teamwork and collective efforts. This argument can be extended to solicitation of innovation that shares the same success prerequisites.

2.3 Performance

Project Performance (PP) means the degree of accomplishment of project goals. Typically, PP is measured by the attainment of time, cost, quality, and safety. For example, quality incentive schemes are used to discourage substandard works (Meng & Gallagher, 2012b). Cost incentive will be accorded should the project being completed within cost target. Likewise, schedule incentive is used to enable acceleration to mitigate delay.

It is not new knowledge that the success of incentive plans depends on the performance measures that are used. Kauhanen and Napari (2012) also found that performance measurement is difficult because it is hard to reliably measure an employee's contribution to the objective of the firm. Thus, boarder measures that may not be individual specific are used. The incentive plan for construction project also faces similar challenge should the incentive targets require the collective efforts of the participating organizations. Kauhanen and Napari (2012) also found that due to the nature of their jobs, incentive plans for blue- and white- collar employees are quite different. White-collar employees' performance is often tied to the organizational objectives with a longer time span for assessment. For blue-collar employees, their incentive must be assessed more frequently and with discrete targets to make award decision ambivalent.

Gibbs et al. (2009) also opined that performance measurement is perhaps the most difficult challenge in the design and implementation of incentive systems. It is acknowledged that performance may be affected by factors beyond the employee's control. As such, performance under an incentive scheme should be qualified by controllable and uncontrollable risk, distort and manipulation.

Likewise, reward can be tied to performance under the respective conditions. The term 'Bonus' was proposed to be used for performance evaluated ex post. There are merits to have composite incentive arrangements so that different aspects of performance can be targeted. Moreover, the giving of extra bonus ex post may well be nullifying the original arrangement.

Pillars of Performance

The overriding goal of contractual incentive is to achieve better project performance (Richmond-Coggan, 2001). Goal commitment between both parties is considered the first and foremost function of any incentive plan (Locke et al., 1988). According to goal-setting theory (Locke and Latham, 1984), goals need to be meaningful, specific, challenging, and acceptable to those who are attempting to achieve them. When incentives and rewards are contingent on goal attainment, a performer's goal acceptance increases in proportion to the perceived benefits of attaining the goal (Locke & Latham, 1990). The function of goal commitment is to iron out "discrepancy", the difference or mismatch between present state and ideal state. Two types of discrepancies exist (Bandura, 1993). The first is discrepancy creation due to intrinsic motivation of pursuing better performance, one party sets a future, higher goal in an ideal state generate in mind. The second is discrepancy reduction which is the effort people may pay to narrow the gap between existing facts and requirements or feedbacks such as results of previous project evaluations (Locke et al., 1988).These two features direct that the goals of incentive schemes are affected by subjective requirements and objective facts. The major reflections in incentive planning are contractual safeguards and value creation. Contractual safeguards are provisions to facilitate accomplishment of project objectives. For example, quality incentive schemes are used for meeting performance targets. A performance bonus arrangement can be applied to a wide range of performance areas such as quality, functionality, and safety (Meng & Gallagher, 2012b). Value creation refers to the extra project value such as cost-saving, innovations and long-term cooperation. In some projects, bonus was set for contractors to generate technical innovations. Some incentive plans also act as the bridge to link contracting parties to engender common interests and in turn provide the platform for long-term collaborative working (Cheung et al., 2018). On this point, it is well noted that cooperation is central to the wellbeing of a project. Can CI be used as a vehicle for this purpose? To tease out the central issues underpinning the use of CI, it is necessary to distinguish between interpersonal relationships and role relationships. Guitot (1963) advocates that the ways in which individuals make attributions about others' intentions and behaviours will vary significantly if the other is viewed as acting within a "role" as opposed to "qua persona." It is advocated that behaviour may change when individuals were behaving in a role context. Even though individuals may rely on trust in their "qua persona" relationships, they may be unable to do so when acting as agents for their organizations. Accordingly, adopting conducive contracting behaviours by the organization is thus fundamental to project performance.

Conducive contracting behaviours (CCB) are therefore performance enabling. Several categories of behaviours have been reported and well proven to have positive influence on project performance. Behaviours exemplifying trust (Cheung et al., 2011; Wong et al., 2008), open communication (Cheung et al., 2013; Wang et al., 2020), best endeavour (Pang et al., 2015; Williamson, 1985), joint effort (Bayliss et al., 2004; Hetemi et al., 2020) and crest for innovation (Cheung & Chan, 2014), are previously used manifestations CCB.

It is advocated that if contracting parties are working at arm's length, the overall project performance would be hampered. Oliver (1990) offered six Interorganizational relationship (IoR) determinants: efficiency, asymmetry, reciprocity, necessity, stability, and legitimacy. These determinants shall be further developed into IoR measurements. Zhu and Cheung (2021) used efficiency, asymmetry, and reciprocity to measure the level of IoR of construction organizations. Based on transaction cost economics theory (Williamson, 1985), the formation of IoR is prompted by an organization's desire to improve efficiency. Working together would result in higher efficiency (Oliver, 1990) and with less contractual safeguards (Mellewigt et al., 2007). Asymmetry between organisations can be expressed by the power or control one organisation has over another (Oliver, 1990). Power asymmetry can be caused by information differential (Holmstrom, 1979). Thus, power asymmetry can indicate both equity gap and IoR but in opposite scale. Exchange theory (Oliver, 1990) projected that reciprocity would engender cooperation through stimulating interdependency. This would command more enduring cooperation through internalisation.

Barriers against performance

Equity Theory (Adams & Freedman, 1976) advocates that comparing input with output is part of human nature. Adams (1963, 1965) further suggested that whether one will abide a contract depends not only on what he gets, but also on whether his counterpart is getting more than him. If the output/input ratios of the contracting parties are far apart, the party with the lower ratio will feel unfairly treated. He would find ways to reduce this imbalance. In construction, Lindenberg (2000) stated that unfair payment packages, power asymmetry and risk differentiation would hamper trust among the contracting parties. These disparities between the developer and the contractor are collectively described as equity gap (EG). Four main elements of equity gap have been proposed by Zhu and Cheung (2021): risk ownership (Cheung et al., 2014), information, expected return and power (Adams, 1965).

2.4 Implication on Design of Incentive in Construction

2.4.1 Conceptual Forms of Motivators

An ideal CI therefore should trigger the motivators to engender performance. This section discusses the three conceptual forms of motivators.

Materialistic Motivators

Materialistic motivators are those that would satisfy the basic needs of the stakeholders. Monetary reward is a classic example. Principals tend to believe the sole concern of contracting organisations is profit. Thus, monetary reward should be the most welcome incentive reward. Moreover, the attainment of same must be a realistic one. In this regard, parties to an incentive arrangement must be involved to agree on the goals and expectations. Post-contract CI arrangement provides a unique opportunity for the contracting parties to establish common goals. The associated expectations can be elaborated so that the CI parties can express their expectations. Another important function of CI is the improvement in information exchange through more conducive environment that has no bearing on the award of the contract.

Hygienic Motivators

The common law principle of non-prevention requires parties to a contract not to do anything that will prevent others from performing their contract. The civil law principle of good faith expects contracting parties to use their best endeavours to perform their contract. Nevertheless, there is no such legal backing for parties to develop conducive environment for the completion of the project goals. The original contract has already set out the responsibilities and rights of the parties. What can be the consideration for 'extra' efforts? CI offers the valuable avenue. Hygiene factors are those that remove hazards from the environment (Herzberg, 1970). Hygiene factors may not directly motivate, but may cause negative effects if not satisfied. The motivating effect may come from factors like achievement, recognition, responsibility and advancement. CI therefore can provide these though delegation of responsibility, freedom to innovate as well as balanced risk ownership. According to Herzberg (1974), the creation of these satisfiers would facilitate performance.

Aspirational Motivators

In can be said that most of the motivators listed in Table 1 belong to this group. The central belief is that one would figure out ways to improve once they have aspiration to do better. Intrinsic motivation is the answer. Recognition is key to derive the self-motivation. Esteem, self-actualisation, and competence are notable examples of aspirational self-motivators. Nonetheless, applications to construction projects are not that straightforward because change of behaviours is needed. It is advocated that CI can be deployed to solicit relationship investment.

2.4.2   Motivators and Functions of Incentivization

Incentive plans can be useful management tool (Herten & Peeters, 1986) to bring out the best of the contractors (Korlen et al., 2017). Hard targets are indispensably used as the meeting of time, cost and quality expectations remains fundamental for every construction project. When extra is desired, moving away from a confrontation mode of contracting may offer the necessary breakthrough. In this connection, Zhu and Cheung (2021) suggested the use of incentive contract ex post to address the inequity created ex ante. Generically, behavior-based incentive would be an appropriate candidate to foster conducive contracting behavior. Where innovations are being solicited, the risk aversion attitude must be removed. Composite form of incentive should be considered. To operationalize the conceptual underpinnings of motivation, it is suggested that CI should have five functions with due regard to the motivators listed in Table 1. The five functions are: (1) Goal commitment; (2) Expectation Alignment; (3) Information Exchange; (4) Risk Efficiency; and (5) Relationship Investment. The conceptual framework for the design of CI is presented in Fig. 1.

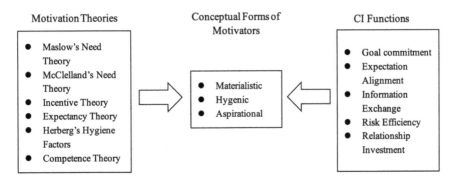

Fig. 1 Motivators and incentivisation

3 A Mapping Framework for Procurement Options and Incentive Arrangement

Most reported studies on incentivisation recommend the need to have aligned goals of the participating parties (Locke et al., 1988). This is not easy to be achieved. In fact, the goals of CI are mostly those of the developers and the interest of the contractors are often secondary (Eisenhardt, 1988). For example, green construction is among the top agenda items of the construction industry across the world. Many governments have used incentive to promote green building practices. Nonetheless, these incentives plans mostly reward or compensatory in nature and participation is voluntary. Saka et al. (2021) also found that apart from incentive that could generate real benefits to the developer like extra construction floor area, there is insufficient evidence to support that developers will adopt green construction to enshrine their reputation. The effect of green labelling on sale enhancement has also yet been demonstrated. This study aptly showed the limitation of pure prosocial approach to incentivize green construction should the interest of the stakeholders are not aligned.

Another challenge is the reliability of the targets. Eisenhardt (1988) argued that realistic target outcomes are only possible for highly programmable tasks. In other words, when the tasks are repetitive and outcome can be predicted with reasonable accuracy like factory production, then using outcome-based incentive is appropriate. However, when the outcome certainty is not high like tasks of low programmability, certain flexibility should be accorded. It is proposed that in formulating a CI, an approach that takes into account of project characteristics and the various forms of outcome targets should be taken. As an illustration, a mapping framework of procurement options with incentive arrangements is proposed (Fig. 2 refers). Since realistic targets are central to the acceptability of a CI, two parameters related CI targets are used.

Task Programmability

Task programmability refers to the extent the tasks can be broken down into discrete work activities for production planning. Task programmability can be assessed by

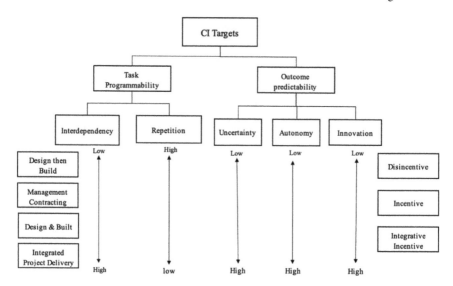

Fig. 2 Mapping Procurement options with Incentive Arrangements

the level of interdependency and repetition. Repetition itself shall warrant ease of programming in terms of scale and learning effect. When tasks are highly interdependent, programming become complex because the uncertainty associated with the outcome.

Outcome Predictability

Outcome predictability is linked somewhat with task programmability. Moreover, predictability is also influenced by the uncertainty associated with the task. Other factors affecting outcome predictability include the degree of contractual autonomy and extent of innovation. Contractual flexibility is typically low when tasks are less complex. For example, in public housing and private residential developments, the tasks are well defined and contractual flexibility is not necessary. Moreover, when innovations are needed to tackle unprecedented challenges, the tasks are undefined, and a flexible contract governance is required to cope with the inevitable ex post adjustments.

Based on the discussion on motivators and task characteristics, it is proposed that the conventional design then build type of construction, disincentive arrangement is appropriate. This form of procurement is used for unsophisticated development where reasonable time is allowed for the design before commencing construction. For more complex projects, detailed design is not possible. On-site decisions are commonly exercised. Response to contingencies must be facilitated for the input of all parties at stake. Incentive is valuable to solicit such inputs. In recent years, the rise in use of building information modelling has led to rethink of procurement approach. To capitalise on the expertise of all contracting parties and advance in information technology. Use of integrative project delivery has gathered momentum.

This form of procurement, however, must be served with collaborative contracting behaviour. It is suggested that integrative incentive would offer the trigger for behaviour modification.

4 Incentivization in Construction

It is well noted that cooperation is central to the wellbeing of a project. Can CI be used as a vehicle for this purpose? To tease out the central issues underpinning the use of CI, it is necessary to distinguish between interpersonal relationships and role relationships. Adams, (1963) advocated that the ways in which individuals make attributions about others' intentions and behaviours will vary significantly if the other is viewed as acting within a "role" as opposed to "qua persona". Behaviour may change when individuals were behaving in a role context. Even though individuals may rely on trust in their "qua persona" relationships, they may be unable to do so when acting as agents for their organizations. Accordingly, adopting conducive contracting behaviours by the organization is thus fundamental to project performance. Back et al., (2013) described incentive plans as predetermined contract strategies that had been designed to motivate project personnel and/or organizations to achieve prescribed project performance objectives. With 90 project data, they found that the level of effectiveness of incentive varied widely. It was suggested that there is no guarantee that incentives will work out as planned. They found using quantified targets for time, cost and quality are less controversial. When qualitative measurement for softer project issues, disagreement over attainment of target is not uncommon because of the subjective nature. Bayliss et al. (2004) proposed the use of longitudinal evaluation to overcome the subjectivity issue of final-shot evaluation. Contingent to the project characteristics, Back et al. (2013) and Ibbs (1991) offered the followings guides for the formulation of construction incentive plans:

- Unilateral versus Negotiated Incentive Plan
- End of project Determination versus Incremental Milestone Determinations
- Quantitative Measures versus Qualitative Measures
- Offsite Determination versus Onsite Determinations
- Win/Lose Bonus
- Carry Over (Retention) Bonus
- Flow Down Structure

Back et al. (2013) also reminded that although there is no empirical evidence to support incentive will bring improved performance. It is still prudent to make sure commitment to perform is in place. In other words, the people involved must be willing to give extra effort. Incentives cannot overcome poor performance of participants due to their inability, unpreparedness, or lack of professional judgment/focus. To summarize, when extra efforts are expected, an integrative approach of incentivization should be adopted. That means on top of using hard targets to keep the baseline performance, behavior-based targets should also be used for behavior modification.

5 Summary

A pragmatic approach is taken for the development of a primer of incentivization in construction. First, motivation theories are reviewed to identify the bases of motivation. It is acknowledged that most of the motivation theories are directed to individuals, their use should therefore be taken with care. In this regard, use of the conceptual forms is proposed. These are: materialistic, hygienic, and aspirational. Five functions of CI are also proposed to trigger these motivators: Goal Commitment, Expectation Alignment, Information Exchange, Risk Efficiency and Relationship Investment. To operationalise this conception, a mapping framework of procurement option with incentive arrangements is used as illustration. The novelty of the framework is the use of task programmability and outcome predictability as the control parameters.

Acknowledgements The work described in this chapter was fully supported by a HKRGC General Research Fund (project number 11202722).

References

Abu-Hijleh, S. F., & Ibbs, C. W. (1989a). Schedule-based construction incentives. *Journal of Construction Engineering and Management, 115*(3), 430–443.https://doi.org/10.1061/(ASCE)0733-9364(1989)115:3(430)

Acquah, A., Nsiah, T. K., Antie, E. N. A., & Otoo, B. (2021). Literature review on theories of motivation. *EPRA International Journal of Economic and Business Review, 9*(5), 25–29.

Adams, J. S. (1963). Toward an understanding of inequity. *Journal of Abnormal and Social Psychology, 67*(5), 422–436. https://doi.org/10.1093/past/96.1.22

Adams, J. S. (1965). Inequity in social exchange. *Advances in Experimental Social Psychology, 2,* 267–299.

Adams, J. S., & Freedman, S. (1976). Equity theory revisited: Comments and annotated bibliography. *Advances in Experimental Social Psychology, 9*(C), 43–90. https://doi.org/10.1016/S0065-2601(08)60058-1

Arditi, D., & Yasamis, F. (1998). Incentive/disincentive contracts: Perceptions of owners and contractors. *Journal of Construction Engineering and Management, 124*(5), 361–373. https://doi.org/10.1061/(ASCE)0733-9364(1998)124:5(361)

Back, W. E., Grau, D., & Mejia-Aguilar, G. (2013). Effectiveness evaluation of contract incentives on project performance. *International Journal of Construction Education and Research, 9*(4), 288–306. https://doi.org/10.1080/15578771.2012.729551

Bandura, A. (1978). The self system in reciprocal determinism. In *American Psychologist* (Vol. 33, Issue 4, pp. 344–358). https://doi.org/10.1037/0003-066X.33.4.344

Bandura, A. (1982). Self-efficacy mechanism in human agency. *American Psychologist, 37*(2), 122–147.

Bandura, A. (1993). Perceived self-efficacy-in-cognitive-development-and-functioning. *Educational Psychologist, 28*(2), 117–148.

Baron, D. P. (1995). Integrated strategy: Market and nonmarket components. *California Management Review, 37*(2), 47–65.

Bayliss, R., Cheung, S. O., Suen, H. C. H., & Wong, S. P. (2004). Effective partnering tools in construction: A case study on MTRC TKE contract 604 in Hong Kong. *International Journal of Project Management, 22*(3), 253–263. https://doi.org/10.1016/S0263-7863(03)00069-3

Bootzin, R. R., Epstein, D., & Wood, J. M. (1991). Stimulus control instructions. In *Case studies in insomnia* (pp. 19–28). Springer.

Boukendour, S., & Hughes, W. (2014). Collaborative incentive contracts: Stimulating competitive behaviour without competition. *Construction Management and Economics, 32*(3), 279–289. https://doi.org/10.1080/01446193.2013.875215

Bresnen, M., & Marshall, N. (2000). Motivation, commitment and the use of incentives in partnerships and alliances. *Construction Management and Economics, 18*(5), 587–598. https://doi.org/10.1080/014461900407392

Carter, M., & Armenakis, A. (2013). Transformative leadership, relationship quality, an employee performance during continuous incremental organizational change. *Journal of Organizational Behavior, 34*(7), 942–958.

Chan, D. W. M., Lam, P. T. I., Chan, A. P. C., Wong, J. M. W., & Author, C. (2010). Achieving better performance through target cost contracts—The tale of an underground railway station modification project. *Performance Measurement and Management in Facilities Management, 28*(5/6), 261–277.

Cheung, S. O., & Chan, K. Y. (2014). *Construction Innovations in Hong Kong: A catalogue.* City University of Hong Kong.

Cheung, S. O., Wong, W. K., Yiu, T. W., & Pang, H. Y. (2011). Developing a trust inventory for construction contracting. *International Journal of Project Management, 29*(2), 184–196. https://doi.org/10.1016/j.ijproman.2010.02.007

Cheung, S. O., Yiu, T. W., & Lam, M. C. (2013). Interweaving trust and communication with project performance. *Journal of Construction Engineering and Management, 139*(8), 941–950. https://doi.org/10.1061/(ASCE)CO.1943-7862.0000681

Cheung, S. O., Wong, W. K., Yiu, T. W., & Pang, H. Y. (2014). *Construction dispute research.* Springer. https://doi.org/10.1007/978-3-319-04429-3

Cheung, S. O., Zhu, L., & Lee, K. W. (2018). Incentivization and interdependency in construction contracting. *Journal of Management in Engineering, 34*(3), 1–13. https://doi.org/10.1061/(ASCE)ME.1943-5479.0000601

Eisenhardt, K. M. (1985). Control: Organizational and economic approaches. *Management Science, 31*(2), 134–149.

Eisenhardt, K. M. (1988). Agency and institutional theory explanations: The case of retail sales compensation. *The Academy of Management Journal, 31*(3), 488–511.

Flyvbjerg, B. (2017). The oxford handbook of megaproject management. *Oxford University Press.* https://doi.org/10.1084/jem.20050882

Gibbs, M., Merchant, K., Van Der Stede, W., & Vargus, M. (2009). Performance measure properties and incentive system design. *Industrial Relations, 48*(2), 237–264.

Gneezy, U., Meier, S., & Rey-Biel, P. (2011). When and why incentives (don't) work to modify behavior. *Journal of Economic Perspectives, 25*(4), 191–210. https://doi.org/10.1257/jep.25.4.191

Herten, H. J., & Peeters, W. A. R. (1986). Incentive contracting as a project management tool. In *Herten H. J., Peeters W. A. R.* (Vol. 4, Issue 1, pp. 34–39). https://doi.org/10.1016/0263-7863(86)90060-8

Hetemi, E., Gemünden, H. G., & Meré, J. O. (2020). Embeddedness and actors' behaviors in large-scale project life cycle: lessons learned from a high-speed rail project in Spain. *Journal of Management in Engineering, 36*(6), 05020014. https://doi.org/10.1061/(asce)me.1943-5479.0000849

Herzberg, F. I. (1966). Work and the Nature of Man.

Herzberg, F. I. (1970). Avoiding pain in the organization. Industry Week. Dec, 7.

Herzberg, F. I. (1974). The wise old Turk. *Harvard Business Review, 54*(5), 70–80.

Holmstrom, B. (1979). Moral hazard and observability. *The Bell Journal of Economics, 10*(1), 74. https://doi.org/10.2307/3003320

Hong Kong Development Bureau. (2018). *Construction 2.0: Time to change.*

Ibbs, C. W. (1991). Innovative contract incentive features for construction. *Construction Management & Economics, 9*, 157–169.

Kauhanen, A., & Napari, S. (2012). Performance measurement and incentive plans. *Industrial Relations, 51*(3), 645–669.

Korlen, S., Essen, A., Lindgren, P., Amer-Wahlin, I., & von Thiele Schwarz, U. (2017). Managerial strategies to make incentives meaningful and motivating. *Journal of Health Organization and Management, 31*(2), 126–141.

Kwawu, W., & Laryea, S. (2014). Incentive contracting in construction. In *Proceedings 29th Annual Association of Researchers in Construction Management Conference, ARCOM 2013, September,* pp 729–738.

Lindenberg, S. (2000). It takes both trust and lack of mistrust: The workings of cooperation and relational signaling in contractual relationships. *Journal of Management and Governance, 4*(1–2), 11–33. https://doi.org/10.1023/A:1009985720365

Locke, E. A., & Latham, G. P. (1990). A theory of goal setting & task performance. Prentice-Hall, Inc.

Locke, E. A., Latham, G. P., & Erez, M. (1988). The determinants of goal commitment. *Academy of Management Review, 13*(1), 23–39. https://doi.org/10.5465/AMR.1988.4306771

Macneil, I. R. (1974). *A Primer of contract planning.*

Maslow, A., & Lewis, K. J. (1987). Maslow's hierarchy of needs. *Salenger Incorporated, 14*(17), 987–990.

Meng, X., & Gallagher, B. (2012a). The impact of incentive mechanisms on project performance. *International Journal of Project Management, 30*, 352–362.

Meng, X., & Gallagher, B. (2012b). The impact of incentive mechanisms on project performance. *International Journal of Project Management, 30*(3), 352–362. https://doi.org/10.1016/j.ijproman.2011.08.006

Murdock, K., The, S., Journal, R., Winter, N., & Murdock, K. (2002). Intrinsic motivation and optimal incentive contracts. *The RAND Journal of Economics, 33*(4), 650–671.

Oliver, C. (1990). Determinants of interorganizational relationships: Integration and future directions published by: Academy of management linked references are available on JSTOR for this article: Determinants of interorganizational relationships: Integration and Futu. *Academy of Management Review, 15*(2), 241–265.

Oliver, R. L. (1974). Expectancy theory predictions of salesmen's performance. *Journal of Marketing Research, 11*(3), 243–253.

Pang, H. Y., Cheung, S. O., Choi, M. C., & Chu, S. Y. (2015). Opportunism in construction contracting: Minefield and manifestation. *International Journal of Project Organisation and Management, 7*(1). https://doi.org/10.1504/IJPOM.2015.068004

Pardee, R. L. (1990). Motivation Theories of Maslow, Herzberg, McGregor & McClelland. A Literature Review of Selected Theories Dealing with Job Satisfaction and Motivation. Business, Feb , 24.

Perry, J. G., Barnes, M., Chan, J. H. L., Chan, D. W. M., Lam, P. T. I., Chan, A. P. C., Chan, J. H. L., Chan, D. W. M., Chan, A. P. C., & Lam, P. T. I. (2000). Target cost contracts: An analysis of the interplay between fee, target, share and price. *Engineering, Construction and Architectural Management, 7*(2), 202–208.

Richmond-Coggan, D. (2001). *Construction contract incentive schemes: Lessons from experience.*

Rose, T., & Manley, K. (2011). Motivation toward financial incentive goals on construction projects. *Journal of Business Research, 64*(7), 765–773. https://doi.org/10.1016/j.jbusres.2010.07.003

Savio, R., De Melo, S., Granja, A. D., & Ballard, G. (2013). Collaboration to extend target costing to non-multi-party contracted projects: Evidence from literature. *Contract and Cost Management, January 2016*, 237–246.

Saka, N., Olanipekun, A. O., & Omotayo, T. (2021). Reward and compensation incentives for enhancing green building construction. *Environmental and Sustainability Indicators, 11*, 100138.

Sims, B. (2002). How successful safety incentive programs reduce injuries without injury hiding. Safety Health Management, (8 May 2002).

Sparer, E. H., & Dennerlein, J. T. (2013). Determining safety inspection thresholds for employee incentives programs on construction sites. *Safety Science, 51*(1), 77–84.

Surapto, M., Bakkar, H. L. M., Mooi, H. G., & Hertogh, M. (2016). How do contract types and incentives matter to project performance? *International Journal of Project Management, 34*, 1071–1087.

Wang, G., Lu, H., Hu, W., Gao, X., & Pishdad-Bozorgi, P. (2020). Understanding behavioral logic of information and communication technology adoption in small- and medium-sized construction enterprises: Empirical study from China. *Journal of Management in Engineering, 36*(6), 1–11. https://doi.org/10.1061/(ASCE)ME.1943-5479.0000843

Wang, W., Chen, Y., & Zhang, S. (2018). Contractual complexity in construction projects: Conceptualization, operationalization, and validation. *Project Management Journal, 49*(3), 46–61.

Wigfield, A., & Eccles, J. S. (2002). Expectancy-value theory of achievement motivation. *Journal of Asynchronous Learning Network, 6*(1), 68–81. https://doi.org/10.1006/ceps.1999.1015

Williamson, O. E. (1985). The economic institutions of capitalism. *In China Social Sciences Publishing House Chengcheng Books Ltd*. https://doi.org/10.5465/AMR.1987.4308003

Wong, W. K., Cheung, S. O., Yiu, T. W., & Pang, H. Y. (2008). A framework for trust in construction contracting. *International Journal of Project Management, 26*(8), 821–829. https://doi.org/10.1016/j.ijproman.2007.11.004

Yeow, P. H. P., & Goomas, D. T. (2014). Outcome-and-behavior-based safety incentive program to reduce accidents: A case study of a fluid manufacturing plant. *Safety Science, 70*, 429–437. https://doi.org/10.1016/j.ssci.2014.07.016

Zhu, L., & Cheung, S. O. (2018). The implementation of incentive schemes in construction. In *11th International Cost Engineering Council (ICEC)World Congress & the 22nd Annual Pacific Association of Quantity Surveyors Conference in 2018*, pp. 1–9.

Zhu, L., & Cheung, S. O. (2021). Towards an Equity-based analysis of construction incentivization. *Journal of Construction Engineering and Management*.

Zhu, L., Cheung, S. O., Gao, X., Li, Q., & Liu, G. (2020). Success DNA of a record-breaking megaproject. *Journal of Construction Engineering and Management, 146*(8), 05020009. https://doi.org/10.1061/(asce)co.1943-7862.0001878

Zuppa, D., Olbina, S., & Issa, R. (2016). Perceptions of trust in the US construction industry. *Engineering, Construction and Architectural Management, 23*(2), 211–236. https://doi.org/10.1108/ECAM-05-2015-0081

Sai On Cheung is a professor of the Department of Architecture and Civil Engineering, City University of Hong Kong. In 2002, Professor Cheung established the Construction Dispute Resolution Research Unit. Since then, the Unit has published widely in the areas of construction dispute resolution and related topics such as trust and incentivization. With the collective efforts of the members of the Unit, two research volumes titled Construction Dispute Research and Construction Dispute Research Expanded were published in 2014 and 2021 respectively. Professor Cheung is a specialty editor (contracting) of the ASCE Journal of Construction Engineering and Construction. Professor Cheung received a DSc for his research in Construction Dispute. Contact email: saion.cheung@cityu.edu.hk.

Liuying Zhu works at the School of Management, Shanghai University. Dr. Zhu received her MSc degree (Distinction) and PhD from the City University of Hong Kong. Her MSc dissertation was awarded the 2016 Best Dissertation (Master category) by the Hong Kong Institute of Surveyors. Her doctoral study on construction incentivization was completed with the Construction Dispute Resolution Research Unit. Her current research focuses on construction project management, construction incentivization, and project dispute avoidance. Dr. Zhu has published articles in international journals and book chapters in these topics. Contact email: zhuliuying@shu.edu.cn.

Chapter 2
Construction Incentivization in Perspective

Sai On Cheung

Abstract Construction incentivization in this book is used as a collective term for all forms of incentive arrangement that aim to engender extra effort of the contracting parties for the improvement of project performance. It is quite often assumed that all enterprises are seeking continual performance. In this regard, incentives in various forms have been used as performance motivator. In construction projects, incentive schemes have also been used to engender performance. Typically, incentive arrangements in construction involve setting cost, schedule, and outcome performance targets. Moreover, the success of incentive schemes is not guaranteed. It had also been found that many projects with incentives still end with project overruns, huge claims, and embarrassing defects. This study identified several design assumptions of conventional incentive that may not suit the ever-increasing complex projects. First, the targets for incentives are often set without consultation with the ultimate project performer. Second, the targets are quantified thus are outcome based. Third, no consideration is given to the behavioral aspect of the incentive. Fourth, there is no appropriate arrangement to solicit superior performance. With reference to the commonly used theoretical underpinnings of incentive arrangements, it is suggested that to have effective construction incentivization, it is necessary to have the scope jointly formulated by the major stakeholders. In this connection, the outcome targets must be agreed. Ideally, risk allocation can be much enhanced should construction incentivization can be used ex post to address ex ante unidentified risks. To bring about superior outcome, incentivization should embrace elements of behavioral performance.

Keywords Construction incentivization · Design assumptions · Competence or beyond

S. O. Cheung (✉)
Construction Dispute Resolution Research Unit, Department of Architecture and Civil Engineering, City University of Hong Kong, Hong Kong, China
e-mail: saion.cheung@cityu.edu.hk

25

1 Introduction

All enterprises are seeking continual performance. In this regard, incentives in various forms have been used as performance motivator. Herten and Peeters (1986) reported the wide use of incentive schemes in many manufacturing sectors such as military developments and aerospace contracts. In construction, incentive schemes have also been used to engender project performance. Likewise, Ibbs (1991) suggested that construction incentive plans can be valuable contract administration tools to enhance project success.

Typically, incentive arrangements in construction involve setting cost, schedule, and quality performance targets (Zhu & Cheung, 2021). That means final project outcomes determine if award will be accorded. Suprapto et al. (2016) analyzed 113 capital projects and found that projects with incentives are likely to perform better if contracting parties value their relation and work as a team. Partnering/alliance contracting approach has also been advocated because of the attempt in developing relational attitude. Ibbs (1991) also recommended that, inter alia, incentive schemes must be fair, and interest balanced.

Nonetheless, the record of the incentives used in construction projects is unconvincing, especially for complex projects. Zhu et al. (2020) reported that many mega projects with incentive schemes still failed to achieve the project targets. Thus, what are the missing links? Boukendour and Hughes (2014) pinpointed that one of the major and recurring problems in designing cost incentive contracts is the setting of target cost and risk sharing ratio. These are essential because of the fundamental issue of maintaining an equitable sharing of risks and rewards to align the interests of the contracting parties, and so to eliminate the adversarial nature of their relationships. The authors further added that an equitable risk-sharing formula would foster trust and cooperation. To this ends, Chapman et al. (2008) highlighted the importance of having a balanced incentive, meaning that incentives should align the interests of client and contractor. These studies also suggest that although incentives do not always work, there are certain design parameters that should be observed. This chapter aims to identify the common issues in the formulation of construction incentivization (CI). CI is used as a collective term that covers all forms of incentive arrangement that seek to improve project performance (Zhu & Cheung, 2022). This study covers the following research tasks:

- Identify the types of incentive scheme commonly used in construction industry.
- Consolidate observations on conventional practice.
- Review the theoretical bases of incentivisation.
- Suggest alternative perspective on the expectations on construction incentivization.

2 Types of Incentives Commonly Used in Construction

Bower et al. (2002) define incentivization as '**a process by which a provider is motivated to achieve extra 'value—added' services over those specified originally and of material benefit to the user**'. The main purpose of incentivization is to adopt client's objectives as well as maximize its own profits (Meng & Gallagher, 2012). Incentive schemes are related to three categories: cost incentive scheme, schedule incentive scheme and quality incentive scheme (Herten & Peeters, 1986).

Zhu and Cheung (2021) studied the use of incentive schemes in the Hong Kong construction industry, 10 structured interviews were conducted with senior construction professionals. The particulars of the interviewees and the incentives used in their respective projects are summarised in Table 1.

The key findings from these interviews are summarized in Table 2:

3 Overview of the Practice of Construction Incentivisation

The commonly used forms of incentive are related to cost, schedule, and quality.

Table 1 The particulars of the interviewees and the incentive schemes used

No	Organisation	Capacity	Incentive scheme used		
			Cost	Schedule	Quality
1	Government	Government department for public facilities other than public housing	✓		✓
2	Government	Government department for public housing			✓
3	Government	Government department for land planning and infrastructure management	✓		✓
4	Developer	Historical building conservation		✓	
5	Developer	Private developer, listed Hong Kong company	✓		✓
6	Developer	Private developer, Mainland capital		✓	✓
7	Contractor	Main contractor	✓		✓
8	Contractor	Main contractor		✓	✓
9	Contractor	Main contractor	✓	✓	✓
10	Consultant	QS Consultant		✓	✓

Table 2 The key findings from the structured interviews

No	Particulars	Types of incentive scheme		
		Cost	Schedule	Quality
1	Incentive schemes provisions	NEC contract with Option C	Responsive acts to prevent project delay	Performance assessment scoring system Pay for safety scheme
2	Aims of the incentive schemes	Communication tools to enhance collaborative working and attract contractors to come to the negotiation table and drive them to focus on the specific targets of the contracts		
		Collaborative working; Generate innovations to save project cost	Quicker completion	Better project performance
3	Barriers against the implementation of the incentive schemes	The conflicts between the project management style and current organizational managing system for adopting target cost contracts	• The manoeuvrability based on limited labour and resources • The contractor may overly on the rewards, they may lay back only for bonus	The standard may be too strict and combat the enthusiasm of the workers
5	Arrangements to enhance the working of the incentive schemes	Target cost estimation at each stage to evaluate the extent of cost-saving	Set milestones and distribute bonus at each stage	Set detailed assessment standards; hold monthly meetings to adjust targets flexibility
6	Effects of the incentive schemes	For individuals, the effect of the incentives usually comes from the pressure of the senior managers. From organizational basis, all the cooperative behaviour is based on the achievable of their commercial benefits		
7	Positive impact on organizational issues	• Commercial benefits • Organizational relationships	• Commercial benefits • Organizational relationships	• Social reputation • Organizational relationships • Working climate of improving the quality of the project

3.1   Types of Incentive

Cost incentive scheme

Cost is one of the most significant performance indicators. Most cost incentives aim to keep cost down either through saving or minimising expenses. These incentives work a bit differently with the types of contracts used. For example, for fixed price contract, CI can provide profit adjustment for project targets set. For cost reimbursement contract, bonus can be allowed should the cost is below certain benchmarks (Kwawu & Laryea, 2014; Perry et al., 2000).

Schedule incentive scheme

When time is of the essence or the project is experiencing unacceptable delay, schedule incentive scheme is used as the bait to accelerate progress. Typically, the contractor is offered a premium for either early completion of the project or compressing the project programme (Abu-Hijleh and Ibbs, 1989; Richmond-Coggan, 2001). In some cases, non-achievement of the incentive schedule outcome would attract a penalty (Abu-Hijleh and Ibbs, 1989).

Quality incentive scheme

Quality incentive schemes are more difficult to formulate and monitor. Essentially, quality targets should be specified. Moreover, it may not be possible to detail quantitatively the required standards. Thus, project employers that aim for high quality finishes sometimes would instigate more stringent quality requirement like limiting the number of defects. Bearing in mind that many minor defects may fall within compliance level individually, would create unacceptable overall final product (Meng & Gallagher, 2012). Compared with cost and schedule incentive schemes, the assessment for quality performance is more complex and sometimes controversial.

In practice, composite arrangements linking cost, schedule and quality performances are commonly used for complex tasks.

3.2   Other Notable Observations

With reference to the afore-mentioned study by Zhu and Cheung (2021), it was found that the incentive schemes used in Hong Kong are more often being initiated by the project employers in the public sector. For private developments, very often incentive schemes are formulated after the project has encountered certain difficulties. In such circumstance, CI is used as a remedial measure. Other than the board use of CI, the following operating patterns are observed: The first observation is unilateral imposition. Incentives are primarily used to solicit efforts from the contractors to resurrect the problem. From the perspective of classical economics, all profit-oriented commercial organizations will respond to benefits derivable from an incentive. Moreover, if the initiator is the sole beneficiary, the commitment of the contractor is unlikely. The situation is even more tricky when the employer is likely to sustain more harm if the problem is not resurrected. One typical example is when incentive for acceleration when the employer has caused project delay. Thus, it is not uncommon to find contractors perceiving unilaterally imposed CI by the employer is only serving the interests of the initiators.

The second observation is CI reward is determined by the attainment or otherwise of predetermined quantified targets. Most of the CI are related to schedule, cost, and quality targets. Understandably these three outcomes are of most concern to the employer. Two issues arise here. First, are the targets realistic? Second, how about other non-quantifiable targets, especially those visual effect of finishes. It has been well documented that incentive targets must be attainable.

The third observation is the award is solely dependent on the achievement of the targets irrespective of the efforts expanded. In this regard, efforts are directed only for the outcome record. This issue is most apparent when innovative ideas are involved. It is not difficult to realise that all innovative ideas are risk prone. Incentive award that takes no account of efforts is not conducive to innovation.

The fourth observation is the absence of clear performance motivator. There is no expectation on the contractor to raise efficiency beyond mere competence. That means there is no expectation of extra effort that goes beyond what has already contracted for. In this connection, superior performance is unlikely.

In views of these observations, the most cited theoretical anchors of incentivisation are discussed in the next section with the aim of identifying the appropriate design concepts for construction incentivization.

4   Theoretical Anchors of Incentivization

The working of CI is inevitably anchored on the concepts of motivation that involves the urging to perform an act, to obtain a certain object, or to produce a desired outcome (Teitelbaum, 1958). Motivation therefore is a process to energize, maintain, and direct behaviours towards attainment of goals (Bootzin, 1991). The force can be 'drive' or 'pull' depending on the nature of the exchange (Baron, 1995). Motivation at work when incentives provide the tangible target to work for. The overriding goal of contractual incentives is to achieve agreed project goals (Richmond-Coggan, 2001). According to goal-setting theory (Locke and Latham, 1984), goals must be meaningful, clear, and achievable. When rewards are contingent on goal attainment, a motivated performer would derive greater effort should the perceived benefits are material and worthwhile (Locke and Latham, 1990). Bandura (1993) further added that a performer would also consider her own ability to attain the goals. Thus, unrealistic goals would not attract performance.

The following theories have been put forward to explain drivers of performance:

- Utility theory
- Principal Agent theory
- Prospect theory
- Self-efficacy theory
- Self-determination theory

4.1 Utility Theory

Utility theory (UT) is about people's choices and decisions. It is concerned with people's preferences and with judgments of preferability, worth, value, goodness, or other similar concepts (Fishburn, 1968). Interpreting utility theory can take two forms: prediction, and prescription. Predictive approach focuses on using utility to predict choice of actual behavior. On the other hand, prescriptive approach offers decision pointers. Unsurprisingly, psychologists are more interested in the predictive approach in recognition of the fact that one's decision is very often influenced by the decision of the others, especially your negotiating counterpart. When in predictive mode, utility theory is widely known as predictive utility theory (PUT). If accurate prediction is possible, prescription shall become plausible. That means, if it were possible to predict accurately the actions of other people (for example, customers or competitors), then the prescriptive approach would have the necessary conceptual foundation. Decision makers can perform their job utilizing different approaches, including applying heuristics. Nevertheless, maximizing utility has been advocated as the most rational approach by the economists.

Prescriptive utility theory is formulated based on the assumption that perhaps is more well-known as a common-sense guideline for the individual to follow in identifying his preferences with justifications. It is a logic-like criterion that consistency and coherence can be attained if preferences are formulated accordingly. It is further suggested that the preferential choices can pass the transitivity test. There are several interrelated purposes of prescriptive utility theory (PUT):

- PUT can be applied as a normative guide to help decision maker to codify his preferences. If one's preferences do not match with the "rational" order, PT would suggest a re-examination of the preferences to identify inconsistency to restore the rational call.
- PUT has the function of helping a decision maker to identify his preferences among complex options. Given the multidimensionality and uncertainty of the options, making preference among them is beyond intuition.
- PUT offers quantitative structure for judgment based on metrics. It is also possible to deploy optimization algorithm to explore the options. The relative strength and weaknesses of the options can be examined in detail.

Notwithstanding the advantages offered by PUT, it is not free from criticism. For example, Burke et al. (1996) devised an experiment to test if expected utility theory works with monetary incentives- a situation identified as Allais Paradox. In simple terms, monetary incentives do not always drive improved performance.

The experiment by Burke et al. (1996) involved college students as subjects and the findings supported the Allais Paradox. Nonetheless, it was found that violations against expected utility theory are significantly reduced when lotteries are real rather than hypothetical. It can be concluded that utility theory and her propositions are logical deduction of expected return on performance. In a nutshell, it works like a cost-benefits analysis. When net benefits are envisaged, it is fair to predict that corresponding performance would follow. Economic rational individuals are expected to follow this "common sense" logic. Moreover, when other non-economic influencers are in action, the prediction is less robust. The question for construction incentivization is whether the assumptions of the utility theory are applicable in construction contracting businesses. Whether the use of composite incentive arrangements can be a plausible way to overcome the drawback of diminishing marginal utility of reward deserves further research efforts.

4.2 Principal Agent Theory

Classical principal agent theory (PAT) (Eisenhardt, 1989) involves a (risk neutral) principal, employing a (work averse) agent to act on his behalf. The agent possesses private information, e.g., about his effort level, the state of nature etc. that is undisclosed to the principal. Thus, the parties are asymmetric in terms of information. The agent is supposed to act to maximize his utility. Concomitantly, he is also work averse in the sense that other opportunities would tempt him to reallocate his resources so that his 'overall' utility is maximized. Trade-offs across jobs are possible. The combination of information asymmetry and the agent's aversion both to work and risk, steer him away from cooperative behavior.

Sappington (1981) outlined four canonical working settings between principal and agent. The first is symmetry of precontractual beliefs. Essentially, this means that both principal and agency share the beliefs about the tasks such as complexity and difficulty and level of efforts needed. As such, it is likely that they can come to a set of common goals for the contract. The second is the agent is presumed to be risk-neutral. However, the reality is seldom the same. What the principal can do is to adopt an equitable risk sharing principle in the contract. The third is the assumption that the agent can be bound to the terms of the contract at no extra costs. Essentially, this view is rather legalistic which the commercial reality may prove difficult. The fourth is the expectation that the agent's performance is publicly observable. This may be the most problematic. Without conscientious effort on monitoring, it is quite unlikely that the principal would know the 'exact' performance of the agent (Grossman & Hart, 1983). Thus, incentives are often used to maintain the desirable performance settings. Typically, an optimal incentive contract involves a pay-for-performance scheme which ties the agent's reward to performance outcomes.

In sum, in a principal-agent relationship, the principal offers a contract to the agent. Once the contract is signed, it is likely that the agent will choose to take actions that maximize his overall utility that the contract allows. Theoretically, the efficiency loss due to the agent's self-interested behavior is measured by comparing the effective outcome under asymmetric information with a fictitious outcome under symmetric information. The redress is to bridge the information gap. Symmetric information simply allows the principal to prescribe and control the desired action. Moreover, the caveat of aligning the interests between the principal and the agent would nullify motivation because the required actions now serve the interest of all actors. The element of self-interest diminishes. Therefore, to address both conflict of interest and information asymmetry, optimal incentive contracts should support partitioning of decision rights and controlling discretionary behavior.

The implications on construction incentivization are the ability to deal with conflict of interest and informational asymmetries between the parties. Raising performance incentives would raise the agent's productivity when risks are not considered. Ironically, psychological concept of intrinsic motivation suggests the opposite. According to cognitive evaluation theory, performance incentives through state-contingent rewards may diminish an agent's intrinsic motivation (Ryan & Deci, 2000b). Likewise, Kunz and Pfaff (2002) examined whether intrinsic motivation would be diminished with the installation of incentives? In fact, Deci (1975) had long found that reward could stifle intrinsic motivation. The presence of extrinsic reward like incentive induces crowding out of intrinsic motivation (Frey, 1997) which is also termed as hidden cost of reward by Lepper and Greene (1978). Moreover, these constructs remain hypothetical, and their existence have not been empirically proven. Heckhausen (1989) proposed that intrinsic motivation has the following manifestations:

a. Intrinsic motivation is internally driven and does not aim to reduce the drive like thirst and hunger.
b. Motivated acts are carried out like leisure time pursuits.
c. Intrinsically motivated behaviors are determined by the performer.

In a principal-agent relation, the potential negative impact of incentive on intrinsic motivation cannot be overlooked. Whether extrinsic motivation will diminish intrinsic motivation depends on the drivers of intrinsic motivation. One such effect is over-justification: attributing one's behavior because of extrinsic reward may undermine the intrinsic motivator. However, if reward convey positive message about the performer's ability or competence, the performer will assume personal responsibility over his behavior. If rewards promote the acquisition of new skills, the perception of intrinsic interest in that activity is deemed necessary. In construction contracting, conflict of interest and information asymmetry are inevitable in employer-contractor relation. The key to motivate work-averse contractor perhaps lies in how incentive arrangements can successfully engender intrinsic motivation.

4.3 Prospect Theory

Prospect Theory (PT) was initiated by Kahneman and Tversky (1979) as a decision-making model. PT offers explanation of some phenomenon that cannot be explained by the Utility Theory developed by Von Neuman and Morgenstern (1953). In essence, UT does not predict well when decisions must be made on events subjected to risks. Basically, utility maximizing may not be the primary decision criteria of risk-taking or risk-averse decision makers. Edwards (1966) put forward three forms of effect when prospect must be considered:

a. Certainty Effect: There is a tendency to underscore probable outcomes in comparison with outcomes that are certain. This tendency would bring about risk-aversion for options involving gains and risk-seeking for options with loss prediction.
b. Isolation Effect: It is of interest to note that it is often the common elements threading across the options are being ignored. Isolation effect would result in framing of a prospect in a way that favors the choice that the decision-maker generates.
c. Reflection Effect: Very often, choices come in pairs of negative and positive prospect (mirroring).

Edwards (1996) further explained that analyzing prospect comes in two phases. The first phase is editing that aims to organize and reformulate the options so that subsequent evaluation and choice can be simplified. Editing thus involves the application of transforming the outcomes and probabilities associated with the offered prospects. The second phase is evaluation during which the prospects sorted out in the editing phase will be considered. In fact, after editing only attainable options will survive and the prospect with the highest value will be selected. The operation of these two phases is supported by the derivation of value function that is based on an accepted reference. The function for gains (risk-averse) is typically concave and convex for losses (risk-taking). The slope of change is steeper for losses than for gains.

Newman (1980) explained how academicians, practitioners, and policymakers are influenced by the Prospect Theory. He contended that Utility Theory is deductive (based on an explicit set of axioms) whereas PT is inductive (based on observations of behavior). Newman (1980) further added that utility theory and prospect theory predict different values of information. "More" information is not necessarily preferred to an agent who behaves according to PT.

In sum, assuming one will not consider the prospect of attaining the reward is likely oversimplifying the reality. Nonetheless, the tendency of risk averse for gains while risk taking for losses suggest that the amount of information to be rendered through an incentivization scheme would be contingent on the risk attitude of the contractor.

Self-efficacy theory

The self-efficacy theory (SET) was first proposed by psychologist Bandura (1977, 1993). One who has self-efficacy would believe that he has the capacity to carry out a task in a way that will achieve the specific goals. The concept of self-efficacy has been applied in many contexts and it is considered essential for performers of incentive schemes. Notably, Bonner and Sprinkle (2002) examined what matters in a monetary incentive-effort-performance relation and found there are three elements of self-efficacy: skill, task, and environment.

Capability can be affiliated with the direct skill possessed by the task performers. Incentive only works for those having the necessary skill for the job. If they lack the skill needed for a given task, their performance will be invariant irrespective of what incentives are offered. Indirect skill is perceptive and may work in a more subtle manner. For example, when one does not perceive having the skill, one would simply stay away from the job. The task itself is also critical. Task complexity will affect how one perceive whether completing the job is feasible. Faced with complex tasks, providing more details can support realistic assessment of one's ability to perform. Thus, in formulating incentives, the tasks and goals must be clear. Only when the performer is convinced that he has the skill to handle (including developing strategy) the complex tasks, the incentive-effort-performance relationship can be attenuated. The third element is the environment and covers all the conditions, circumstances, and influences surrounding the performer. Obvious examples include time pressure, assigned goals and feedback. To get the performer motivated, raising self-efficacy can be an effective means. Task complexity can be handled with greater efforts to improve the clarity of the details. Formulating targets jointly would accord opportunities to tune the task to a manageable scale. Mutually agreed goals and hence performance targets would positively engender committed efforts (direction, duration, and intensity). Another implication on incentive design is the need to establish feedback mechanism to enable learning.

Self-determination theory

Ryan and Deci (2000a, 2000b) proposed the use of Self-determination theory (SDT) to describe human's innate growth tendency and psychological needs. SDT seeks to explain the motivation of behind one's choices if there is no external influences and distraction. Under SDT, human behaviours are self-motivated and self-determined. It can therefore be said that SDT is a humanistic theory. SDT projects that there are three psychological needs to be satisfied should proper functioning is desired. SDT (Ryan & Deci, 2000a, 2000b, 2017) elaborated that human function depends on satisfaction of three basic psychological needs: autonomy, competence, and relatedness.

Based on a meta-analysis on drivers of performance, Cerasoli and Nassrelgrgawi (2016) found that autonomy, competence, and relatedness are pillars of motivated performance. Autonomy energies performance because it reflects the most basic intrinsic desire of humans to be his own agent of the environment. Autonomy is almost synonymous to self-determination; its satisfaction signifies one has control over his own behavior. The associated sense of freedom of choice is pivotal to commitment to perform. The second pillar is competence. Satisfying the psychological need of competence means one is always in favor of demonstrating one's ability, and hence endorsement. Competence. Under SDT, the drive to satisfy competence need predicts enduring efforts to make sure the tasks are performed. As a matter of fact, demonstrating one's ability is fundamentally satisfying. Motivated individuals would confront challenges and feel proud for the skill he possesses to get the job done. Giving proper and timely feedback from a credible source will positively reinforce competence. Relatedness needs address the affective side of human desire of being emotionally bonded and recognized by other affiliates.

Turning now to performance that is conventionally treated as a homogenous, unidimensional construct. This is rather problematic in construction contracting because performance in construction projects is rarely unidimensional. Construction project tasks can categorically be identified as quality or quantity type. Quality-type tasks are those requiring attention to detail, personalization, and careful craftsmanship. Performance indicators thus include creativity, lack of errors, artistic value, and originality etc. Quantity-type tasks are typically repetitive, depend on rote skill, and tend to require less personal investment. These tasks are not offering high level of autonomy and interpersonal facilitation. Thus, the respective indicators include assembly time, quantified output criteria. Performance of quantity-type tasks can better be predicted by incentives while quality-type tasks are more likely to be predicted by factors such as intrinsic motivation and enjoyment. Conventional construction incentives primarily treat construction works as quantity-type. This may as well one of the major drawbacks because quality-type of tasks have proved to be the real challenge as far as project performance is concerned.

Under SDT, those who perceive the three psychological needs are met will outperform those who perceive otherwise. Need satisfaction is a more proximal outcome of incentives and mediates the relationship between incentives and intrinsic motivation. Moreover, mere presence of incentive has little impact on relatedness need. SDT extends the well-established positive link between incentives and performance by showing that need satisfaction and incentives play a joint role in performance improvement. The mere presence of incentives has little to no impact on the degree to which need satisfaction is addressed. The key is making tasks associated with an incentive to embrace autonomy, competence, and relatedness. In this way, both quality and quantity type of tasks can be covered. Emphasizing ownership is a useful way to promote autonomy. Intervention to bolster the need for competence include enabling individuals to get involved in the setting of goals. The very act of setting, striving for, and attaining a goal has a strong impact on perceptions of competence and self-efficacy; both are supposed to have positive impact on performance. As for relatedness, providing feedbacks makes individuals feel more respected. Furthermore, the 'game' must be fair. Perception of injustice impact organizational commitment, turnover intentions, satisfaction, and well-being.

Table 3 An integrated framework for CI design

Theory	Basis	Implications on CI	CI design
Utility Theory (UT)	Utility Maximising Individuals	Net Gain of real possibility	• Clear goals • Real and tangible benefits • Compensate diminishing returns
Principal-Agent Theory (PAT)	Self-interested Principal and Work-averse Agent	Address conflict of interest and asymmetrical information	• Aligned goals and risk preference • Performance observability
Prospect Theory (PT)	Non-rational agent	Expected utility for gains (risk averse) is less than the same quantum of losses (risk taking)	• Agreed targets and rewards • Input from performers
Self-Efficacy Theory (SET)	Ability to perform	Clear goals and target to effect efficacy	• Clear goals • Feedback on performance
Self-Determination Theory (SDT)	Satisfaction of psychological needs of autonomy, competence, and relatedness	Embracing elements of the three psychological needs	• Autonomy to perform • Ability to perform • Appreciation of performance

5 Construction Incentivization in Perspective

This section consolidates the theoretical suggestions deliberated in Sect. 4. An integrative framework is proposed and then followed by an operationalisation of the framework.

5.1 An Integrative Framework for CI Design

Drawing on the theoretical constructs on performance, the following Table 3 presents an integrated framework for CI design.

5.2 Operationalizing the Integrative Framework

To operationalize the conceptual underpinnings of the incentivization to design parameters Table 4 is prepared. Goal, Risk, Reward and Evaluation have been identified with due reference to the case study on construction incentivization conducted

Table 4 Design specificities respective to theories

Design Parameters	Design Specificities	UT	PAT	PT	SET	SDT
Goal	Clear goals	X			X	
	Aligned Goals and risk preference		X			
	Agreed targets and rewards			X		
	Input from performers			X		
	Autonomy to perform					X
	Ability to perform					X
	Composite arrangements	X				
Risk	Aligned goals and risk preference		X			
	Autonomy to perform				X	X
	Ability to perform				X	X
Reward	Real and tangible benefits	X				
	Compensate diminishing returns	X				
	Agreed targets and rewards			X		
Evaluation	Performance observability		X			
	Feedback on performance				X	
	Appreciation of performance					X

by Zhu et al (2020). Against these design parameters, design specificities suggested by the five theories are arranged. Since there are inevitable overlapping, Table 3 is prepared to illustrate the relationships among design parameters, design specificities and the theories.

With reference to Table 4, a design for CI is proposed. Table 4 gives the design specificities under each of the paraments together with the respective reference to theories. It is noted that there is more than one theoretical contribution to the design parameters.

Goal: Establishing goals is probably the first item to be settled for any incentive arrangement. All incentive schemes must have certain goals in mind. Both UT and SET have pointed to the need to have clear goals to serve as the criterion to weight up options. According to the goal-setting theory (Locke and Latham, 1990), goals must be meaningful, specific, challenging, and acceptable to the participants. These requirements nicely sum up the suggestions on goals by other motivating theories. For example, PAT suggests that it is imperative to have the meaningful goals aligned among the stakeholders. These goals must be attainable, thus conform with the project of ADT that the performers must have the ability to achieve the goals. In this connection, the goals should be translated to unequivocal tangible targets. Notwithstanding, two more considerations are suggested. First, the performers should be accorded the freedom to choose the methods to accomplish the targets. Second, to overcome the issue of diminishing returns on utility against rising rewards, composite arrangements like mingling time, cost, and schedule targets can be used to keep the efficiency of the performers at high level.

Risk: An interesting question about the use of incentive is whether the performer is given reward for what she has already contracted for? Paradoxically, if the incentive targets are just what the original contract requires, an CI is serving the function of adjusting the contract terms. This may not be desirable. However, if extra risks are involved, the adjustment will then be legitimized. Thus, PAT explains well the need to link the goals with the risks. It is most likely that the performers are asked to tackle unanticipated risks. To stimulate them to render extra efforts, the risks must be well articulated with the goals of the CI. In this way, the performer will be able to assess their ability to take on the risks at their own course.

Reward: The third design parameter is the reward for the performer. First and foremost, the reward must be commensurate with the risks to be undertaken. Reward must be genuine and material to the performers. The criterion for the reward should also be clear and the fulfillment or otherwise should not create dispute. All these should not be unilaterally decided. Instead, like targets, rewards should also be developed with input from the stakeholders. It is not uncommon that composite incentive arrangements are used in construction projects. Instead of treating different forms of targets as discrete, thoughtful combination of same may offer a unique way to alleviate the issue of diminishing returns of singular target.

Evaluation: Most CI users are only concerned with targets are met. This short-sighted approach will lose the opportunity to improve the performance observability that is considered vital under PAT to curb opportunism. Furthermore, both interim and final feedback should be incorporated to refine the CI. Interim feedback is suggested by SET to reinforce performers to keep the motivation momentum. Feedback on final achievement offers invaluable learning opportunities to upgrade the CI system as well as strengthening of performers' capacity. Feedback can also be a form of appreciation that would be treasured by believers of SDT.

5.3 Discussion

Whether the four observed conventional practice of CI design meet with the afore-mentioned CI prerequisites has been examined. First, unilaterally determined CI runs the danger that the recipients not fully committed to the goals of the CI. Almost all theories discussed in Sect. 4 point to the need to have goals and targets of CI agreed with the stakeholders. Ideally, the goals should be discussed with the aim of developing mutually accepted targets. Open discussion over targets also accords the opportunity in exploring the implications arising from the 'extra' risks to be handled. Another downside of imposition is non-commitment. Sometimes, the CI may have been agreed and signed, but there is no guarantee that the performer will deliver with their best efforts. The commitment issue is also highlighted by PAT.

Second, singular use of quantitative targets for administrative convenience can be problematic. Metric identifications criteria will assist the performers to evaluate if they have the necessary ability to fulfil their promises. Interim feedback can also be facilitated. Thus, there are good reasons to support the use of quantitative targets.

The major critique of the quantitative approach is ignoring the efforts in dealing with the tasks that may be in vain due to uncontrollable circumstances.

Third, recognising effort for reward can be controversial because of the difficulty in evaluating effort. Most project participants would consider they have put in utmost efforts irrespective of the outcome. In other words, it is quite unlikely for contracting parties to admit that they have not directed efforts to perform. Moreover, in high-risk ventures and when innovations are the key, efforts beyond mere competence are needed. The courage in taking the risk in facing potential loss of resources should the anticipated innovation does not materialise must be carefully crafted in a CI. Otherwise, it is very unlikely CI participants would put in the necessary resources.

Fourth, the conventional CI packages are not based on recognised performance motivators. Section Four listed five theories that make valuable suggestions on what would motivate or discourage performance. It is also a fact that there is no universally applicable CI package. Every CI should cater for the need of the project concerned. Moreover, there are certain fundamental issues like the four design parameters listed in Table 4 that every CI designer should go through in formulating an incentive package.

Accordingly, the followings are suggested for the planning of CI:

- The scope of the CI should be jointly formulated by the major stakeholders.
- The CI targets should be agreed by the initiator and the performers.
- Both 'carrot' and 'stick' can be used as deemed appropriate.
- CI can be used ex post to address ex ante unidentified risks.
- CI should embrace elements of behavioral performance.

CI can be an invaluable instrument to review what have not been contemplated ex ante. Under those circumstances, the contractor is required to go beyond what has been contracted for. It is suggested that this would mean CI is asking for something more than that have already contracted for. In fact, extra effort beyond mere competence should be aimed for. In this respect, Meng and Gallagher (2012) conducted a questionnaire survey in the United Kingdom and the Republic of Ireland to analyse the relationship between the use of incentives and the performance of a project. In general, improvements in time and quality could be tracked for projects with incentive schemes incorporated. Moreover, it was also found that 'extra' efforts were the real ultimate element of success.

6   Summary

This first chapter of the volume seeks to put construction incentivization in perspective. In this respect, five theoretical bases of construction incentivization are examined. These are utility theory, principal-agent theory, prospect theory, self-efficacy theory and self-determination theory. Accordingly, design specificities are suggested. In addition, typical incentive arrangements used in Hong Kong were studied. Four key observations were obtained: (1) unilateral formulation by the initiator; (2) only quantified outcome targets are used; (3) only final outcomes count; and (4) no clear motivator can be identified. It is suggested that effective CI should give due consideration of the design specificities suggested by the afore-mentioned theories. This study conceptualises these findings by proposing four key CI design parameters: Goal, Risk, Reward and Evaluation. Goals of CI should be clear and genuinely agreed by the stake holders. CI should not be used to compensate probable under provision for what had been contracted for. Instead, unanticipated ex ante risks are the subject matters of CI. The undertaking of these risks should be within the ability of the performer who should also been given the autonomy over the way to handle the risks. Likewise, the reward must be real and attainable. Positive feedback, both interim and final, will positively reinforces the commitment of the performers to go beyond mere competence in accomplishing the goals.

Acknowledgements The study reported in this chapter was fully supported by a HKSAR RGC GRF project (no. 11202722).

References

Bandura, A. (1977). Self-efficacy: Toward a unifying theory of behavioral change. *Psychological Review., 84*(2), 191–215.

Bandura, A. (1993). Perceived self-efficacy-in-cognitive-development-and-functioning. *Educational Psychologist, 28*(2), 117–148.

Bonner and Sprinkle. (2002). The effects of monetary incentives on effort and task performance: Theories, evidence, and a framework for research. *Accounting, Organizations and Society, 27*, 303–345.

Bower, D., Ashby, G., Gerald, K., & Smyk, W. (2002). Incentive mechanisms for project success. *Journal of Management in Engineering.* https://doi.org/10.1061/ASCE0742-597(2002)18:1(37)

Boukendour, S., & Hughes, W. (2014). Collaborative incentive contracts: Stimulating competitive behavior without competition. *Construction Management and Economics, 32*(3), 270–289.

Burke, M., Carter, J., Gominiak, R., & Ohl, D. (1996). An experimental note on the Allais paradox and monetary incentives. *Empirical Economics, 21*, 617–632.

Cerasoli, N., & Nassrelgrgawi, A. S. (2016). Performance, incentives, and needs for autonomy, competence, ad relatedness: a meta-analysis. *Motivation and Emotion, 40*, 781–813.

Chapman, C., & Ward, S. (2008). Developing and implementing a balanced incentive and risk sharing contract. *Construction Management and Economics, 26*, 659–669.

Deci, E. L. (1975). *Intrinsic motivation.* Plenum.

Edwards, K. D. (1996). Prospect theory: A literature review. *International Review of Financial Analysis, 5*(1), 19–38.

Eisenhardt, K. M. (1989). Agency theory: An assessment and review. *The Academy of Management Review, 14*(1), 57–74.

Fishburn, P. C. (1968). Utility theory. *Management Science, 14*(5), 335–378.

Frey, B. S. (1997). Not just for money. An economic theory of personal motivation, Cheltenham, and Brookfield: Elgar.

Grossman, S. J., & Hart, O. D. (1983). An analysis of the principal-agent problem. *Econometrica, 51*(1), 7–45.

Heckhausen, H. (1989). *Motivation und handeln.* Springer.

Herten, H. J., & Peeters, W. A. R. (1986). *Incentive contracting as a project management tool*, pp. 34–39. https://doi.org/10.1016/0263-7863(86)90060-8

Ibbs, C. W. (1991). Innovative contract incentive features for construction. *Construction Management and Economics, 9*, 157–169.

Kahneman, D., & Tversky, A. (1979). Prospect Theory, an analysis of decision under risk. *Econometrica, 47*(2), 264–291.

Kwawu, W., & Laryea, S. (2014). Incentive contracting in construction. In *Proceedings 29th Annual Association of Researchers in Construction Management Conference*, ARCOM 2013, pp. 729–738.

Kunz, A. H., & Pfaff, D. (2002). Agency theory, performance evaluation, and the hypothetical construct of intrinsic motivation. *Accounting, Organizations and Society, 27*, 275–295.

Lepper, M. R., & Greene, D. (1978). Overjustification research and beyond: Toward a means-ends analysis of intrinsic and extrinsic motivation. In M. R. Lepper & D. Greene (Eds.), *The hidden costs of reward: New perspectives on the psychology of human motivation* (pp. 109–148). Erlbaum.

Meng, X., & Gallagher, B. (2012). The impact of incentive mechanisms on project performance. *International Journal of Project Management*. Elsevier Ltd and IPMA, *30*(3), 352–362.

Newman, D. P. (1980). Prospect theory, implication for information evaluation. *Accounting Organizations and Society, 5*(2), 217–230.

Perry, J. G., Bames, M., Chan, J., et al. (2000). Target cost contracts: An analysis of the interplay between fee, target, share and price. *Engineering, Construction and Architectural Management, 7*(2), 202–208.

Richmond-Coggan, D. (2001). *Construction contract incentive schemes: lessons from experience.*

Ryan, R. M., & Deci, E. L. (2000a). Self-determining theory and the facilitation of intrinsic motivation, social development and well-being. *American Psychologist, 55*, 68–78.

Ryan, R. M., & Deci, E. L. (2000b). When reward compete with nature: the undermining of intrinsic motivation and self-regulation. In C. Sansone & J. Harackiewicz (eds.), *Intrinsic and extrinsic motivation. The search for optimal motivation and performance* (pp 14–54). Academic Press.

Ryan, R. M., & Deci, E. L. (2017). *Self-determination theory: Basic psychological needs in motivation, development, and wellness.* Guilford Publishing.

Sappington, D. E. M. (1981). Incentives in principal-agent relationships. *Journal of Economic Perspectives, 5*(2), 45–66.

Suprapto, M., Bakker, H., & Mooi, H. (2016). How do contract types and incentives matter to project performance?, *International Journal of Project Management. Elsevier Ltd and Association for Project Management and the International Project Management Association, 34*(6), 1071–1087.

Zhu, L., Cheung, S. O., Gao, X., Li, Q., & Liu, G. (2020). Success DNA of a record-breaking megaproject. *Journal of Construction Engineering and Management, 146*(8).

Zhu, L., & Cheung, S. O. (2021). Toward an equity-based analysis of construction incentivization. *Journal of Construction Engineering and Management, 147*(11).

Zhu, L., & Cheung, S. O. (2022). Equity gap in construction contracting: Identification and ramification. *Engineering, Construction and Architectural Management, 29*(1), 262–286.

Sai On Cheung is a professor of the Department of Architecture and Civil Engineering, City University of Hong Kong. In 2002, Professor Cheung established the Construction Dispute Resolution Research Unit. Since then, the Unit has published widely in the areas of construction dispute resolution and related topics such as trust and incentivization. With the collective efforts of the members of the Unit, two research volumes titled Construction Dispute Research and Construction Dispute Research Expanded were published in 2014 and 2021 respectively. Professor Cheung is a specialty editor (contracting) of the ASCE Journal of Construction Engineering and Construction. Professor Cheung received a DSc for his research in Construction Dispute. Contact email: saion.cheung@cityu.edu.hk.

Chapter 3
Incentivization or Disincentivisation

Liuying Zhu

Abstract Construction Incentives and Disincentives (I/D hereafter) arrangements are common project control measures. This study aims to investigate the attributes and scope of application of I/D. Comparatively, incentivisation is more about encouragement of performance improvement by reward provisions. It is objective-driven and generates pressure toward smaller and elite actions. Disincentivisation is less costly and can function well when monetary reward is not the sole performance motivator. It takes effect by penalties to force contractor to comply with their requirements. The analysis of these two mechanisms is further conducted based on two set of case studies in construction industry. It was found that some incentive strategies are attractive for contractor for further negotiation. Moreover, some financial rewards balance the unequal risks and encourage innovation. The proposition of disincentivisation is discussed and illustrated through a case study on the Hong Kong Zhuhai Macau Bridge (HZMB) project (Hong Kong Zhuhai Macau Bridge Authority (2009). Through focus group discussions, it is found that disincentivisation is successful for mega project controlling and unanimous cooperation for multi-agents. The importance of maintaining reputation a signature that disincentivisation is a less costly and viable project control measure.

Keywords Incentivization · Disincentivisation · Project Control · Mega project

1 Introduction

Construction is project-based and needs to be carried out by teams comprised organizations with different specifications. Pursuit of business goals in the form of alliances involves more risks than a single organisation go-it-alone (Das & Teng, 1998). Each organisation is a separate entity that has its own interests and outcome expectations. Project participants would likely put their own interest ahead of the overall project

L. Zhu (✉)
School of Management, Shanghai University, Shanghai, China
e-mail: zhuliuying@shu.edu.cn

goals. Coordinating of these organizations to achieve overall project goals is challenging (Suprapto et al., 2016). Furthermore, opportunistic behaviour would happen when team members seek to maximize their own benefits at the expense of the other members (Lui & Ngo, 2004). Based on transaction theory, Incentives and disincentives (I/D hereafter) arrangements are regularly used as project control measures to alleviate opportunism (Williamson, 1985).

Use of Incentives is based on motivation theories and has been well recognized as catalyst for performance. Behaviour modification theory further introduce the enhancement of I/D on performance (Skinner, 1961). Meng (2015) demonstrated that incentives are key motivator for "better performance". The term "better performance" is explained as (Richmond-Coggan, 2001): (1) Either the developer and/or contractor trying to save a project that is running into difficulties, or (2) Either the developer and/or contractor see additional value by proposing a change. I/D are then developed to meet with the higher expectation based on reality conditions. For I/D, incentives aim to enhance performance through rewards whereas disincentives penalize performance below expected project outcomes (Baker, 1992; Bresnen & Marshall, 2000).

Based on years of research, it is found that valid cases and general guidelines are needed to substantiate how I/D is operated and what purposes it can also achieve. This study therefore aims at investigating the attributes and scope of application of I/D for project performance planning. The research objectives are organized as follows:

(1) Identify the application scenarios of I/D.
(2) Investigate the application of I/D in construction industry; and
(3) Summarize the project prerequisites for the use of I/D.

2 Literature Review

Construction Incentives or Disincentives (I/D) are classical management tool to achieve better project performance (Meng, 2015). It is believed that motivation can be derived from incentivisation, disincentivisation or combination of both (Bubshait, 2003). The primary concern of I/D is to form new cooperation agreements by project participants. 'Carrot or stick' is used to describe developer using I/D arrangements to reward or penalize the contractor for above or under performance respectively (Bubshait, 2003). Specifically, incentives are often used to motivate contractor for excellent performance when disincentives aim to discourage contract violations. With that, both contracting parties do their utmost to enhance project performance. Incentives and disincentives can be used either separately or together. For the combination of incentivisation and disincentivisation, 'Pain gain share system' is commonly used in construction projects (Bresnen & Marshall, 2000). For example, schedule incentive planning is classical as a combination of I&D (Jaraiedi et al., 1995). Developer needs to set financial rewards for early completion and liquidated damages for project delay.

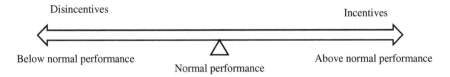

Fig. 1 Incentives and disincentives on a spectrum (Adapted from Meng (2015))

The incentive arrangement aligns specific project goals and contracting parties do their utmost to enhance project value (Bubshait, 2003). Project targets are highly correlated with incentive planning. The selection of I/D also reflects the different expectations of project outcome. For cost management, the anticipation of rewarding cost saving, or penalizing cost overrun reflect different expectations and confidence of project success.

From the perspective of project performance, I&D are both possible contractual tools to serve project control purposes (Hughes et al., 2007). Figure 1 shows the spectrum of I/D.

Oliver (1980) analysed the dynamics of I/D through comparing these two strategies. It is found that incentives are effective for small group of co-operators. Disincentives, comparatively, is costless and have its value when dealing with multiple agents (Hosseinian, 2016). In an ideal situation, if everyone cooperates, the only cost of disincentives is that of threatening to use it (Oliver, 1980). Meng and Gallagher (2012) further analysed the role of disincentives in project monitoring. As disincentivisation is commonly incorporated with project control system, developer's controlling power is enhanced. The penalty also enhances the developer's dominance position in the middle stage of the project and disincentivisation plays a driving role in encouraging best practice and ensuring project success.

To present a clear view of the functions of disincentivisation, the comparison of I&D is summarized in Table 1.

3 Using Incentives as Sweetener for contractor's Negotiation Participation

To further investigate the similarities and differences of I/D, case study approach is conducted to firstly investigate the application status of incentives in Hong Kong construction industry. Case study is a good way to capture collective viewpoints from different project participants across variety management levels (Bryman, 1989). In this section, five construction projects with incentives were studied. Table 2 gives the outlines of the five cases.

The case details are as follows.

Table 1 Comparisons of incentives and disincentives in construction project

No	Content	Incentives	Disincentives	Reference
1	Objective	Motivate better performance apart from the contract	Demotivate underperformance to safeguard the contract	Meng (2015)
2	Manifestation	Financial bonus	The penalty of fine	Chan et al. (2010)
3	Advantage	Attractive	Costless	Oliver (1980)
4	Expectation	Additional value apart from the contract	Contractual safeguards	Hauck et al. (2004)
5	Essence	Encourage contractor to finish their expectations	Force contractor to comply with their requirements	Meng (2015)
6	Function	Motivate small numbers of co-operators and generate pressures toward smaller, more "elite" actions	Motivate unanimous cooperation for multi-agents	Oliver (1980)
7	Sphere of application	Objective and interest alignment	Project monitoring	Meng and Gallagher (2012)

Table 2 Particulars of the five projects that used safety incentives

No	Description	Contract value (billion HKD)	Type of incentive scheme used
A	Public infrastructure	1.7	Cost
B	Tunnel	1.6	Cost
C	Historic buildings revitalisation	1.8	Schedule
D	Residential building	3.2	Quality
E	Residential-commercial complex	0.6	Quality

3.1 Case A. Target Cost Contract for Public Water Storage Project

Location A is in low-lying areas and the rainwater pipes were laid around 30 years. In the previous years, severe flooding problem happens and threatens the living condition in that place. The government aims to seek permanent solution to overcome flooding incidents. A storm water storage project is then scheduled. It contains two phases and aims to have a total storage capacity of 60,000m3 of water. The cost plan of this project is around 1.67 billion HKD.

To better incentivize contractor's cost saving behaviour, the developer adopted New Engineering Contract (NEC) with Option C. A 50–50% gain share system was established between the developer and the contractor. Open book accounting is also adopted to ensure the accuracy and transparency of project cost control.

Based on this agreement, an innovative foundation was designed by contractor. Because of that, this project is considered as a success of saving over 60 million HKD and completed ahead of time. The innovative design also won Platinum grade in BEAM Plus Assessment and Innovation Award from Hong Kong Institution of Engineers.

3.2 Case B. Target Cost Contract for Tunnel Building

This project is part of an island line building which starts in 2009 and finished in 2015. It was anticipated by the government that this metro line can significantly help improve public transportation convenience and create new business opportunities for the connected commercial area. This project is about a 3 kms long tunnel building. The overall budget for this tunnel was HK$15.4 billion.

Considering the technical difficulties and unexpected underground project risks, the developer used target-cost-contract with incentives to enhance communication and contractor's work enthusiasm. A stage-wised gain share system was then set in this contract. The percentage of sharing benefits is strongly connected with different target obligations discussed during the tendering process. The contractor is clearly required to submit up-to-date records for accounting and benefit-sharing.

For the project outcomes, the developer acknowledged the contractor's contribution. Although this project is delayed because of unexpected geological problems for tunnel construction. The project cost is controlled within a reasonable range, and the developer distributes the share of cost-saving to the contractor.

3.3 Case C. Schedule Incentives for Accelerating the Completion of Revitalization Project

In October 2007, the Hong Kong Government introduced Revitalising Historic Buildings through Partnership Scheme. It encourages the social enterprises and service providers to join the project as the Government would pay all the initial costs of renovation (Cheung & Chan, 2012). In this context, this restoration and revitalization project earmarked to become a landmark arts and culture centre. It aims to attract top exhibitions and provide education. This project contains a 20 buildings complex with a history of over 150 years. The original plan for this project is 1.8 billion HKD within 4 years.

Because of the longer-than expected preparation and the dispute of the project in the early stage, the project is delayed for 2 years with double overspend. After several rounds of negotiation, to accelerate the progress of the project, the contract type is changed from lump sum contract to cost reimbursement contract. An incentive scheme was set to motivate the contractor to finish the project with the best endeavour. It contains 5 milestones by dates for physical completion. Financial bonus was also set corresponding to these milestones. The total value of the bonus is around HK$ 3 million which is close to 1% of the contract sum.

During construction, the developer established daily monitoring system for the project procedure. The contractor did significant assessments for packages of additional works. It is shown that the bonus is effective and can compensate the additional labour or materials for earlier completion. This project was completed in time and 2 months earlier than original plan.

3.4 Case D. Safety Schemes in Residential Construction Project

This project locates in the west of Hong Kong. It is a complex of commercial residential buildings. The contract sum is around HK$3.2 billion and the floor area is over 100,000-m square. In response to the call of the government, the managers of this project also pay close attention to safety. The value of the safety incentive was set at HK$5.6 million (0.2% of the contract sum). Two specific safety incentive schemes are set as: 1) Safety Incentive Scheme: the bonus is to reward organizations. It is for good safety performance and disciplinary actions for safety management. The financial rewards are distribute based on achieving specific tasks. 2) Safety Hero Scheme: This Scheme aims to reward individuals. It distributes extra cash bonuses to frontline staffs.

During the project process, developer's safety manager takes responsibility of performance assessment for each quarter for Safety Incentive Scheme. The content is mainly about Effective Safety Management, Safety Performance and Legal Compliance. For Safety Hero Scheme, an average of four Safety Heroes will be selected monthly, cash award of 3.000 HKD will be given to Safety Heroes. For each month, the Safety Manager also nominated ten staffs based on the assessment criteria to The Safety Hero Selection Panel. The panel will review each nominator and decide the final winner. The project passed all the safety examination and won the award by the government for effective safety management.

3.5   Case E. Residential Project with Financial Bonus Scheme

This project is a public housing in the north-east part of Hong Kong. The commercial-residential complex contains 18-storey domestic block providing 620 rental flats, a commercial centre, car parks, a standard primary school and public transport interchange. The technical difficulty is related to the complex geomorphology. The contract value is around HK$ 600 million.

The Bonus Scheme for Building Contracts is set by Hong Kong Housing Authority to encourage high project quality. This scheme consists of Construction Works Bonus at the construction stage and Customer Services Bonus at the maintenance stage to give 1–0.2% of the net contract value, or 7.5 million and 1.5 million whichever is lesser (The Hong Kong Housing Authority, 1998). This scheme aims to award contractors with outstanding performance as an ex-contractual arrangement to motivate contractors to deliver high quality products and services. The main content of the Scheme is about schedule, safety performance, clean disciplinary record and no offence, malpractice, or misconduct causing damage to the developer's image.

All the performance scores were compared with predetermined benchmarks. A contractor would be paid a bonus by the Authority if its final score is above the respective Benchmark Scores, which are reviewed and published quarterly.

Through the investigation of the project, it is found that the assessment system is consistent with the contractor's project internal management style. Based on that, it costs much less effort for the contractor to adopt the award assessment system. The contractor was awarded over 6 million HKD (around 1% of the contract sum) for the Bonus upon contract completion and Customer Service during the project maintenance period.

An investigation was also conducted with other project participants under same Bonus scheme. Compared with this project's contractor, few participants achieve all these tasks. Some of them claimed that the criteria are too harsh. They have little confidence to finish them under high finance pressure. Based on these performance results, this scheme was withdrawn in 2004.

3.6   The Application of Incentives in These Cases

Comparatively, the differences of incentive implementation are based on the different project objectives. For Case A and B, it is found that the cost incentives are incorporated with the management system. The focus of negotiation between developer and contractor is mainly about the defining the scope and sharing ratio of benefits and the adoption of collaborative working. For Case A, the major contribution of the financial incentive scheme is the encouragement of innovation. The design of the new foundation reduced the cost of construction and saved construction time. For Case B, the target cost contract and open book accounting enhanced communication of cost saving. Moreover, the cost saving moderate the risk imbalance between contractual parties to encourage contractor to take risks to accelerate and cooperate with the developer.

For schedule incentives, Case C, D and E all show strong intention of project monitoring and contractual governance. For Case C, specific milestones were both set to further specify the progress targets. As a historic building revitalisation project, it encountered more unforeseeable risks than expected and accident happened. The schedule incentive scheme is therefore set to motivate the contractor to finish this project with the best endeavour. In the end, the completion date is 2 months' earlier than original plan. For Case D and E, specific assessment systems were set to evaluate project performance. Based on that, contractors need to provide performance records in terms of labour training, management meetings and so on. Safety incentives were also used in Case D and contractor's performance was monitored by developer's safety manager. For Case E, the developer also intends to build positive feedback mechanism and try to incorporate project performance into a profound long-term development. As the The Bonus Scheme goes hand in hand with Performance Assessment Scoring System (PASS), the outputs (Works) Assessment and Input (Goals) Assessment were conducted to compare the differences between project goals and outcomes. Differ from other quality incentives, a database of contractors' performance is established and was referred for every tendering process. It takes around 30% of the tendering score. Better performance record is thus necessary for further cooperation.

4 Application of Disincentives in Mega Project Management

As mentioned in Sect. 2, incentives are more often applied to encourage "elite" actions, when disincentives are implemented to motivate unanimous cooperation for multi-agents. Because of that, a mega project involving multiple parties was also investigated to better understand the application of disincentivisation.

4.1 Project Particulars

The Hong Kong-Zhuhai-Macao Bridge (HZMB hereafter) project is planned to promote economic exchanges and cooperation among Hong Kong, Guangdong Province, and Macao. This mega project was started in 2007. It is considered as one of the largest highway projects in China. Also it is considered one of ten mega projects in Hong Kong (Hong Kong Special Administrative Region Government, 2007, 2010). The overall length is 55 km and the main bridge is about 29.6 km. This project also contains a 6.7 km undersea tunnel, two artificial islands and ports in three cities.

This project is completed and opened for use in 2018. Notable accomplishments were reported in terms of time, cost, and innovations. Along with achieving all the project targets, over 400 patents were harvested (Hong Kong Zhuhai Macau Bridge Authority, 2017). Over 400 patents are generated in this project. This project is also considered as a key demonstration project. Moreover, as the one of the longest and biggest bridges in China, this project also greatly promoted the development for construction supply chain development in high-tech material manufacturing, reclamation, and underwater tunnel construction (Hong Kong Zhuhai Macau Bridge Authority, 2017). These technical accomplishments together with the management experience are classical learning materials for future similar oversea tunnels and bridges such as Shen-Zhen and Zhongshan Oversea Tunnel Project.

4.2 The Use of Disincentivisation in HZMB Project

The Hong Kong Zhuhai Macao Bridge Authority (HZMBA hereafter) was established by the Three Governments in 2010. It directly takes the responsibility of construction, operation, and maintenance of the project. HZMBA proposed the following project pledges at the beginning of the project:

(1) Build a world-class cross sea channel.
(2) Provide high quality services for users; and
(3) Become a landmark bridge in China.

Project challenges are also identified at the beginning of the project (Hong Kong Zhuhai Macau Bridge Authority, 2017):

(1) There is no unified construction standard between these three cities.
(2) Multiple risks need to be managed for different components of the bridge, like undersea tunnel, over-sea bridge and artificial islands.
(3) The construction of the project needs to consider the surrounding ecological protection. As the bridge crosses the animal protection area of white dolphins, there are specific requirements for the height, location, and construction method of the bridge.
(4) Multiple project stakeholders are involved in this project. Considering different legal system, it is also difficult to align expectations and manage disputes.

Based on project objectives and challenges, HZMBA developed management system for this project. All these requirements and challenges are transferred into contractual language and developed the Reputation Evaluation System (the System hereafter). Disincentives are also incorporated within the System. The System has the following three parts:

(1) Goal commitment

This System for project management contains 6 goals:

a. Quality management: A quality requirement higher than the industry standard level is set, and the project requires a 120 years' life span.
b. HSE: Health, safety, and environment management.
c. Procedure management: The project needs to be finished within 8 years.
d. Cost: The project should not have excessive cost overruns.
e. Information management: The openness of the
f. Maintain the openness of the system and to adopt industrial standards to promote the interoperability of data exchangeability.
g. Innovation: Cultivate a series of excellent scientific and technological innovations on technology and management.

(2) Monitoring method

Evaluation committee is established to conduct comprehensive evaluation is carried out quarterly by the HZMBA. The committee members are coming from different department. For every 3 months, a quarterly assessment reports of all contractors would be sent correspondingly. A meeting would also be held to discuss all the problems occurred.

(3) Disincentives and performance assessment

2% of contract value was set aside as funding to support the System. The evaluation committee conducted independent evaluation on contractors according to the implementation rules and grading standards. The total score is 100 and the score distribution is concluded in Table 3.

The evaluation is based on mark deduction according to a pre-set scale. There will be significant points when the evaluation committee observe misbehaviours. There is also possibility that the points would be deducted to 0 when contractor make significant errors or major deviations. Table 4 shows the stage-wised payment ratio based on different evaluation scores.

Contractor who receives consistent "D" grade will be counted as breach of contract. HZMBA can thus terminate the contract and change partners.

The detailed particular of the System is presented in Table 5.

Table 3 The score ratio of six project goals

Item	Quality	HSE	Procedure	Cost	Information	Innovation
Score %	35	35	15	5	5	5

Table 4   The evaluation level and corresponding payment ratio of the fund

Comprehensive evaluation score: L	Evaluation level	Bonus payment ratio
$L \geq 90$	AA	100%
$85 \leq L < 90$	A	90%
$80 \leq L < 85$	B	70%
$75 \leq L < 80$	C	50%
$L < 75$ or the qualification is cancelled	D	0

4.3   The Project Control Functions of the System

To further investigate the effect of disincentives applied in this mega project, a focus group discussion was conducted. Ten senior managers from HZMBA, contractor, and supplier who participated in the evaluation of the System were invited. Consensus is reached mainly in the following aspects:

(1) **The System is effective for project controlling.** Specific project goals were set as clear guidance for contractor. The System provides an overall view of all contractors' performance. The disincentives incorporated in the System discourage all the misbehaviours can be detected from different discipline and have been clearly shown through quarterly reports.

(2) **The System is effective in reaching unanimous cooperation.** In the beginning of the project, due to different organizational management style conflicts, deductions help demonstrate the intentions from HZMBA and attract the contractor to come to the negotiation table. As the score rankings for all the project participants were announced in each quarter, it was also found that the System is instrumental for communication and benign competition.

5   The Project Prerequisites for the Use of I/D

5.1   The Project Prerequisites for the Use of Incentivisation

Through the investigations of Hong Kong construction project cases, comparatively, the project prerequisites of applying incentives can be drawn as follows:

(1) *The incentives are highly objective-driven and designed for specific contrac-*

tors All these incentive schemes are rooted in different project targets and organised to fulfil objectives specific to the project. Three different types of incentive schemes (cost, schedule, and quality) are all implemented in Hong Kong construction industry. Moreover, most of incentives investigated are specific tasks for single project participant. It is used to fulfil one specific goal. Based on such attributes, it is also necessary to pay more attention to contractors'™

Table 5 The particulars of the system

Name	Goals commitment	Responsibility allocation		Monitoring method	Assessments	
		Developer's responsibility	Contractor		Content	Ratio
Quality	1. **120 years' project life span** 2. Strictly meets the requirements of national engineering standard system 3. Acceptance rate is **100%**	1. Develop the quality management measures; 2. Set quality policy and goals and establish quality management system; 3. Verify project quality management plan and other documents; 4. Evaluate the performance of different project participants; 5. Investigate major quality accidents; 6. Organize regular quality improvement meetings	1. Directly responsible for construction quality; 2. Strictly enforce the laws and regulations on quality, safety, environmental protection 3. Set up site management 4. Set up site test lab 5. Strictly follow the construction drawings and construction specifications after verification 6. Report and assist in accident investigation. Be responsible for engineering quality accidents caused by construction reasons 7. Carry out the system of personnel training 8. Do the collection, sorting and filing of quality and technical data according to the regulations	1. Unscheduled inspection of the quality condition on site 2. Monthly quality inspection 3. Quarterly basis inspection and yearly evaluation for project performance	Quality management Technology management; Geological prospecting work Design management; Construction supervision; Experiment examination; File management and so on	35%

(continued)

Table 5 (continued)

Name	Goals commitment	Responsibility allocation		Monitoring method	Assessments	
		Developer's responsibility	Contractor		Content	Ratio
HSE	Pursue **zero injury, zero pollution and zero accidents** and reach the advanced level of the international industry	1. Provide resources to improve HSE performance 2. Set annual plans and working targets 3. Prepare a major accident emergency rescue plan and exercise regularly 4. Hold safety meetings on a regular basis 5. Organize HSE accidents investigation	1. Establish and improve the HSE accident emergency system according to the requirements of the administration 2. Provide emergency rescue personnel and necessary emergency supplies 3. Organize regular safety trainings	1. HSE work should be included in daily management work 2. The administration implements regular, irregular, special HSE supervision and regular examination 3. The Contractor shall organize HSE supervision and inspection at least once every half a month 4. The supervisor shall implement supervision 4. The authority carries out quarterly HSE comprehensive evaluation for supervisors and contractors	Control and management of high risks; operation process; Marine animal protection; Waste Management and so on	35%

(continued)

Table 5 (continued)

Name	Goals commitment	Responsibility allocation		Monitoring method	Assessments	
		Developer's responsibility	Contractor		Content	Ratio
Schedule	1. Finish the project **in time** 2. Maximize the utilization of resources under the premise of time limits 3. Have an integrated plan which is closely linked with project management requirements such as quality control and investment control	1. Set the plan, milestones and the total goals 2. Preparation of annual / quarterly / monthly plan 3. Adjust plans 4. Track the actual progress of the site, responsible for the management of the overall progress of the project and analyse the reasons of delay / advance of schedule	1. Reported on time for the phase plan and resource allocation plan of the working out construction project 2. Invest in appropriate personnel and equipment, report on time and truthfully for the progress of the project 3. Carefully analyse the reasons of delay, adjust it in time, put forward reasonable solutions in the report, and submit it to the authority for approval	1. Unscheduled inspection of the process condition on site by Authority; 2 Supervisors should regularly hold site meetings and weekly and monthly progress coordination meetings as required 3. The Authority will examine the schedule management of the participating units annually based on relevant assessment terms of the contract 4. The Contractor shall submit monthly report for examination and approval	Schedule management	15%

(continued)

Table 5 (continued)

Name	Goals commitment	Responsibility allocation		Monitoring method	Assessments	
		Developer's responsibility	Contractor		Content	Ratio
Cost	1. The cost management shall base on **minimum life cycle cost** and **value management**; 2. The preliminary design estimate shall be controlled within the range of floating (**± 10%**) 3. The total cost should be controlled **within** the preliminary design budgetary estimate approved by the Ministry of transportation 4. Improve the efficiency of fund utilization	Supervise and manage the whole process of investment control	1. Carry out and complete the project and repair any defects in the project according to the contract 2. Evaluate the valuations correctly, overestimate or not measuring and pricing in accordance with the actual project progress are prohibited	1. The investment control committee is responsible for the daily assessment and investment control 2. The Authority examines the cost control performance annually based on relevant assessment terms of the contract 3. The contractor is directly responsible for the quality and safety of the engineering construction and shall report in time if there are differences between drawings and actual site 4. The designer shall take responsibility if there are added cost because of the design problems; the supervisor shall take responsibility if there are unnecessary cost because the work is not supervised	Follow the relevant contract process about measurement, VO and payment	5%

(continued)

Table 5 (continued)

Name	Goals commitment	Responsibility allocation		Monitoring method	Assessment	
		Developer's responsibility	Contractor		Content	Ratio
Information	1. Meet the information needs of the project items by three governments and the public; 2. Able to support the management of this project and meet the requirements of the specific joint supervision of three governments and adapt to the multi-level management	1. Set application manager who is responsible for the business implementation and professional management of the related modules of the Department	Each participant should be equipped with computer professionals and a system administrator to be responsible for the daily management and maintenance work	The authority implements quarterly examination and annual comprehensive evaluation to assess the application and management of information system	Integrated information management	5%
Innovation	1. Establish innovation system to ensure **high standard, high quality and high efficiency**; 2. Foster a number of scientific and technological achievements for this project;	Implement Totally Quality Control (TQC) to give full play to the main role of "human for innovation in management and technology	All project participants should cooperate with the innovation management system by administration development, truthfully and timely report the relevant information and data. Reporting should be timely, accurate, not false and concealed	Monthly summary, quarterly inspection and annual comprehensive evaluation are carried out by Authority	Assist the completion of the project acceptance, evaluation, appraisal of the results, and the declaration of awards	5%

concerns. The downside of using financial incentive schemes is the effect of
over-reliance. For example, if the schedule incentives are set, the contractor
may overly focus on the rewards. For some performance incentives, setting
unrealistic goals may destroy the creditability of the incentivisation arrangement. Though there are no extra punishment, unattainable targets make the
incentive scheme meaningless.

(2) *The effectiveness of incentives relies on recipient's confidence of achievements*

It is common that for projects, the contractor needs to achieve a certain level
of performance based on the contract. Higher standard of project performance
requires extra effort. Thus, although incentive schemes are in general welcome
by the contractors, its attractiveness also depends on to how much extra efforts
are needed. As a matter of fact, large organizations are more willing to accept the
incentives because they have sufficient labour/resources to achieve the targets.
The effectiveness of incentives is more related to confidence of goal achievement and goal compatibility. For incentives to work, it is important to create a
comfortable, satisfying working environment and after all attainable targets.

5.2   The Project Prerequisites for the Use of Disincentivisation

The importance of maintaining reputation under a signature project like the HZMB
makes disincentivisation a less costly yet viable option to maintain project performance even for projects are having high risks and facing immense uncertainties. The
followings are necessary actors for the use of disincentivisation:

(1) *Projects can instigate stringent supervision for all project team members* It

is found that disincentivisation works well when stringent project monitoring
is exercised. In HZMB, the System contains basically all major project objectives and provides detailed quarterly performance report, the appraisees are all
members of the project. The scores are indicators of underperformance should
these are below the acceptable norms. Contracting organizations received these
feedbacks from HZMBA and were expected to take necessary action to avoid
penalties.

(2) *Financial reward is not the singular performance motivator* Offering mone-

tary rewards is often used when project conditions are not well defined, risks and
uncertainty are high. Incentives are used to provide buffer for these contingencies. Moreover, for signature projects like the HZMB, maintaining reputation
may well be of the highest priority for the participating contracting organisations. In HZMB, the System served as a performance ranking exercise. The
contracting organisations were very concerned about their positions on the

performance rankings. Keeping face was of vital importance for them. Disincentivisation took effect when performance improvement acts were taken by contracting organisations to save face.

6 Summary

I/D arrangements have been regularly used as project control measures. Incentives are performance motivators due to the embedded monetary rewards. They are attractive to encourage specific contractor to achieve extra project value. Disincentives on the other hand push performance when contracting organisations seek to avoid penalties attached with underperformance. Disincentives thus usually do not involve extra monetary rewards. Through investigating five construction projects with incentivisation, it is further found that incentivisation is objective-driven and applicable for elite actions. For disincentivisaiton, this study found that the HZMB project used a Reputation Evaluation System (the System) to incorporate disincentive arrangements. The HZMB project was a high risk and complex project and prima facie not suitable for use of disincentives. The System worked well to control the performance of the contracting organisations. Performance rankings were taken as records of achievement and contributions in the making of a record-breaking project. The desire to be part of the record making team turned out to the most influential performance motivator.

Acknowledgements The empirical work of this chapter has been reported in a paper entitled "Project control through disincentivisation: A case study of Hong Kong-Zhuhai-Macau Bridge Project" of Collaboration and Integration in Construction, Engineering, Management and Technology: Proceedings of the 11th International Conference on Construction in the 21st Century, London 2019. The support of Contract Planning Department, Hong Kong-Zhuhai-Macao Bridge Authority is duly acknowledged.

References

Baker, G. P. (1992). Incentive contracts and performance measurement. *Journal of Political Economy, 100*(3), 468–505. https://doi.org/10.1086/261831

Bresnen, M., & Marshall, N. (2000). Motivation, commitment and the use of incentives in partnerships and alliances. *Construction Management and Economics, 18*(5), 587–598. https://doi.org/10.1080/014461900407392

Bubshait, A. A. (2003). Incentive/disincentive contracts and its effects on industrial projects. *International Journal of Project Management, 21*(21), 63–70.

Chan, D. W. M., Lam, P. T. I., Chan, A. P. C., Wong, J. M. W., & Author, C. (2010). Achieving better performance through target cost contracts—The tale of an underground railway station modification project. *Performance Measurement and Management in Facilities Management, 28*(5/6), 261–277.

Cheung, E., & Chan, A. P. C. (2012). Revitalising historic buildings through partnership scheme: A case study of the Mei Ho house in Hong Kong. *Property Management, 30*(2), 176–189. https://doi.org/10.1108/EL-01-2014-0022

Das, T. K., & Teng, B.-S. (1998). Between trust and control: Developing confidence in partner cooperation in alliances. *The Academy of Management Review, 23*(3), 491–512.

Flyvbjerg, B. (2017). The Oxford handbook of megaproject management. *Oxford University Press.* https://doi.org/10.1084/jem.20050882

Hauck, A. J., Walker, D. H. T., Hampson, K. D., & Peters, R. J. (2004). *Project alliancing at national museum of Australia—Collaborative process, 130*(February), 143–152.

Hong Kong Special Administrative Region Government. (2007). *The 2007–08 Policy Address Hong Kong.*

Hong Kong Special Administrative Region Government. (2010). *Ten Major Infrastructure Projects.* 2010–11 Policy Address. www.policyaddress.gov.hk/10-11/eng/pdf/projects.pdf

Hong Kong Zhuhai Macau Bridge Authority. (2009). *Introduction of HZMB project.* http://www.hzmb.org/cn/bencandy.asp?id=2

Hong Kong Zhuhai Macau Bridge Authority. (2017). *The research challenges of Hong Kong-Zhuhai-Macau Bridge.*

Hosseinian, S. M. (2016). An optimal time incentive/disincentive-based compensation in contracts with multiple agents. *Construction Economics and Building, 16*(4), 35–53.

Hughes, W., Yohannes, I., & Hillig, J. (2007). Incentives in construction contracts: Should we pay for performance? *CIB World Building Congress,* 2272–2283.

Jaraiedi, B. M., Plummer, I. R. W., & Aber, M. S. (1995). Incentive/disincentive guidelines for highway construction contracts. *121*(1), 112–120.

Lui, S. S., & Ngo, H. (2004). The role of trust and contractual safeguards on cooperation in non-equity alliances. *Journal of Management, 30*(4), 471–485. https://doi.org/10.1016/j.jm.2004.02.002

Meng, X. (2015). Incentive mechanisms and their impact on project performance. In *Handbook on Project Management and Scheduling* (Vol. 2). https://doi.org/10.1007/978-3-319-05915-0_17

Meng, X., & Gallagher, B. (2012). The impact of incentive mechanisms on project performance. *International Journal of Project Management, 30,* 352–362.

Oliver, P. (1980). Rewards and punishments as selective incentives for collective action: theoretical investigations. *The University of Chicago Press Journals, 85*(6), 1356–1375.

Richmond-Coggan, D. (2001). *Construction contract incentive schemes: Lessons from experience.*

Skinner, B. F. (1961). *Analysis of behaviour.* McGraw-Hill.

Suprapto, M., Bakker, H. L. M., Mooi, H. G., & Hertogh, M. J. C. M. (2016). How do contract types and incentives matter to project performance? *International Journal of Project Management, 34*(6), 1071–1087. https://doi.org/10.1016/j.ijproman.2015.08.003

The Hong Kong Housing Authority. (1998). *Preferential tender award system and bonus scheme (BC 112/99).*

Williamson, O. E. (1985). The economic institutions of capitalism. In *China social sciences publishing house chengcheng books ltd.* https://doi.org/10.5465/AMR.1987.4308003

Liuying Zhu works at the School of Management, Shanghai University. Dr. Zhu received her MSc degree (Distinction) and Ph.D. from the City University of Hong Kong. Her M.Sc. dissertation was awarded the 2016 Best Dissertation (Master category) by the Hong Kong Institute of Surveyors. Her doctoral study on construction incentivization was completed with the Construction Dispute Resolution Research Unit. Her current research focuses on construction project management, construction incentivization, and project dispute avoidance. Dr. Zhu has published articles in international journals and book chapters in these topics. Contact email: zhuliuying@shu.edu.cn.

Chapter 4
Behavioural Considerations in Construction Incentivization Planning

Liuying Zhu and Sai On Cheung

Abstract It is quite often assumed that all enterprises seek continual performance. In this regard, incentives in various forms have been used as performance motivators. Typically, incentive arrangements in construction involve setting cost, schedule, and outcome performance targets. Moreover, the success of incentive schemes is not guaranteed. Many projects with incentives still end with project overruns, huge claims, and embarrassing defects. It is advocated that defective design is one of the key causes of the nonfunctioning of incentive arrangements. This study reminds us that there are certain norms to be followed in the planning of construction incentivization. The characteristics of three well-known normative principles are introduced. In addition, this study advocates that construction incentivization should also be planned to engender the commitment of the contracting parties. In this respect, managing behaviours between the parties should be one of the planning norms of construction incentivization. Empirical support is also provided.

Keywords Motivation · Performance · Behavioural outcomes · Planning norms

1 Introduction

Various forms of incentive arrangements have been reported in the preceding chapters. It can be said that incentives are a versatile project management tool when continued performance is pursued. Herten and Peeters (1986) reported the successful use of incentive schemes in manufacturing military products and developing aerospace projects. In construction contracts, incentive schemes have also

L. Zhu (✉)
School of Management, Shanghai University, Shanghai, China
e-mail: zhuliuying@shu.edu.cn

S. O. Cheung
Construction Dispute Resolution Research Unit, Department of Architecture and Civil Engineering, City University of Hong Kong, Hong Kong, China
e-mail: saion.cheung@cityu.edu.hk

been widely used as a contract administration tool to enhance performance, especially from contracting organizations (Ibbs, 1991). Typically, incentive arrangements in construction involve setting rewards for the accomplishment of cost, schedule, and quality outcome targets (Zhu & Cheung, 2021). Effectively, this means that several project outcome aspects are used to determine if a reward can be accorded. Suprapto et al. (2016) analysed 113 capital projects and found that projects with incentives are likely to perform better if contracting parties value their relation and work as a team. Adopting a partnering/alliance contracting approach is considered appropriate because of the emphasis and investment in the relation. Ibbs (1991) further added that, inter alia, incentive schemes must be fair and interest balanced.

Nonetheless, the outcome record of incentive-equipped mega construction projects deploying incentives is far from exciting. Zhu et al., (2020a, 2020b) reported that many large-scale projects with incentive schemes failed to achieve their targets. Thus, why are the incentives not working? Boukendour and Hughes (2014) pointed out that one of the major and recurring problems in designing cost incentive contracts is setting the target cost and risk sharing ratio. These are essential because of the fundamental issue of maintaining an equitable sharing of risks and rewards while aligning the interests of the contracting parties. Minimizing adversity among parties with differing interests is also a long-standing challenge in construction contracting. Serious attempts have been made to suggest quantitative models for risk-sharing formulas (Ma et al., 2021). To this end, Chapman and Ward (2008) highlighted the importance of having a balanced incentive, meaning that the incentives should align with the interests of both the client and the contractor. Thus, these studies suggested that incentives must be thoughtfully planned to achieve the intended objectives. In this regard, the four design parameters identified from relevant theories and reported in Chap. 1 are planning pointers of construction incentivization. This chapter further operationalizes incentive design parameters by examining the incorporation of behaviours as part of construction incentivization (CI hereafter) normative planning.

2   Examples of Normative Principles

Heuristics and norms have played a significant role in human decisions. Both are largely intuition based and developed from the collective wisdom and experience of relevant participating groups. The golden rule may well be the classic example of the normative principle. Three sets of well-recognized normative principles are introduced in this section to illustrate their characteristics.

Table 1 Innovation development principles (Terninko & Zusman, 1998)

– Segmentation	– Convert harm into benefit	– Use strong oxidizers	– Inert environment
– Extraction			– Combining
– Replacement of a mechanical system	– Inexperience short-lived object instead of an expensive durable one	– Copying	– Dynamicity
		– Use of porous material	– Continuity of useful action
– Prior action			
– Transformation of physical and chemical states of an object		– Rejecting and regenerating parts	– Self-service
	– Changing the colour	– Asymmetry	– Flexible film or thin membranes
	– Thermal expansion	– Prior counteraction	
– Cushion in advance	– Local quality	– Spheroidicity	– Universality
– Partial or overdone action	– Counterweight	– Periodic action	– Composite material
	– Inversion	– Mediator	– Rushing through
	– Mechanical vibration	– Use of a pneumatic or hydraulic construction	– Phase transition
– Nesting			
– Equipotentiality	– Feedback		
– Moving to a new dimension	– Homogeneity		

2.1 Principles of Innovation (TRIZ Methodology)

Terninko and Zusman (1998) reported the work of Genrich Altshuller, who developed the TRIZ framework to understand innovation. TRIZ stands for the Theory of Inventive Problem Solving. In essence, it is a method used to systemically analyse the manners in which innovations can be understood. The 40 principles of innovation are shown in Table 1.

Moreover, these principles were developed sixty years ago; thus, with the development of IT and many advanced technologies and crests for sustainability and carbon emission reduction, these principles need updating. The lesson for this study is the way the principles are developed. Essentially, the principles display the pattern of how the innovations were harvested or their characterizing features.

2.2 Principles of Contract Planning

The second set of normative principles was suggested by MacNeil (1974) and is related to the planning of economic exchanges. There are six principles that enshrine the expected functions of commercial contracts.

i. *Permitting and encouraging exchange behaviour*

It is advocated that contracts are tools to record commercial transactions. Guided by freedom of contract, the first principle is to honour the agreements between the contracting parties should they have opted to sign on the dotted line. This makes both legal and business sense when stating the intentions of the parties by way of a written contract. This makes good commercial sense, as contracts are supposed to record the intentions of the parties; seemingly, the performance of a contract is not meant to be prevented.

ii. *Reciprocity*

By its nature, economic exchanges involve reciprocating acts from the contracting parties. It is not difficult to identify rights accompanied by respective obligations in every contract. Thus, reciprocity lies at the heart of every contractual relationship.

iii. *Role effectuation*

Specific roles of the contracting parties or their agents are delineated in a contract. Technically, this empowering act is necessary, especially for their agents, e.g., architects, engineers, and surveyors, as they are not parties to a contract. Thus, it is imperative for the contract to spell out clearly their respective authority in exercising their roles.

iv. *Effectuation of planning*

Under the common law, the principle of nonprevention underlies the performance of contracts. Effectively, this means that no party should do anything to prevent the other contracting parties from performing their responsibilities. In civil law, the principle of good faith is akin in concept. Thus, contractual provisions to facilitate the performance of what has been planned should be included.

v. *Limited freedom of the exercise of choice*

The contract may well be viewed as having the effect of setting the boundary within which the contracting parties operate. Unilateral changes in the boundary are not possible. The choices of the contracting parties are de facto restricted to those that fall within the ambit of the contract. It is therefore incumbent on the contracting parties to plan for the choices that they would like to exercise before the contract is signed.

vi. *Harmonizing contracts with their internal and external social matrices*

A contract only bounds the contracting parties and the provisions if the contract terms are agreed upon, and the parties are free to conclude the same. Moreover, when there are gaps that have not been addressed, industrial norms can be influential references. Likewise, when implied terms are considered, business efficacy is the key. Inevitably, the expectation of society is a deciding factor. This principle reminds the societal dimension even for commercial endeavours.

2.3   Risk Allocation Principles

The third set of principles is quite well known to the construction communities. Abrahamson (1984) was a leading construction lawyer and had exemplary experience in drafting standard forms of construction contract. With reference to a tunnel project, he proposed the following set of risk allocation principles:

"A party should bear a construction risk where:

i.   *The risk is within the party's control;*
ii. *The party can transfer the risk, e.g., through insurance, and it is most economically beneficial to deal with the risk in this fashion;*
iii. *The preponderant economic benefit of controlling the risk lies with the party in question;*
iv. *To place the risk upon the party in question is in the interests of efficiency, including planning, incentive, and innovation; and*
v.   *If the risk eventuates, the loss falls on that party in the first instance and it is not practicable, or there is no reason under the above principles to cause expense and uncertainty by attempting to transfer the loss to another."*

This set of principles can be regarded as the most quoted in construction risk allocation studies (Cheung, 1997) because it embraces the three key allocation criteria of foreseeability, controllability, and manageability (Llyod, 1996). From the project management perspective, allocating a risk to a party who has no information to make a reasonable assessment of the extent of the risk involved is inequitable. Ideally, the party who can control the occurrence of a risk should be in the best position to minimize the occurrence. When the risk materialises, it is most efficient and effective for the party with the suitable capability to manage it so that the impact can be minimized.

2.4 Characteristics of Normative Principles

The term normative refers to the idea that the principles are regarded as standard whereby the subject matter should follow. It is therefore imperative for the principles to have the following credentials:

1. Universal applications can be expected.
2. The principles should be able to stand over time and contexts.
3. The versability of the principles is supported by empirical evidence.
4. Failing to comply with the principles exposes the subject matter to malfunctionality.

In this study, the design objectives of CI are further examined in light of the findings reported in Chaps. 1, 2 and 3. Specifically, behaviour-based design parameters are examined.

3 Construction Incentivization and Project Performance

In Chap. 1, the objectives of construction incentivization are identified through the conceptual lenses of several theories. Chapter 3 discusses when incentives or disincentives should be used. Chapter 5 introduces the importance of managing interorganizational relationships using construction incentivization. These chapters point to the fact that an effective CI should aim to activate the internal drive of the contracting organization for better performance. For this purpose, the equity gap (EG hereafter) between the contracting parties is introduced. It is advocated that the EG is an endogenous factor that has a fundamental influence on parties' contracting attitudes. For ease of reference and making this a stand-alone study, certain parts of Chap. 2, 3 and 5 are repeated in this chapter.

The use of incentives has a long history in capital work projects (Bayliss et al., 2004; Chan et al., 2010; Hughes et al., 2012). Although some encouraging success stories have been reported, there is also no shortage of failing cases (Alfie, 1993; Zhu & Cheung, 2021). Thus, there is no guarantee that project incentives will bring the desired results if CI has not been planned properly. In this section, the ingredients of effective project incentivization are first reviewed. To understand what effective CI should endeavour to achieve, the elements of project performance (PP hereafter) are introduced. A conceptual CI–PP relationship framework and the associated hypotheses are proposed. The primary purpose of incentivization is to solicit 'value-added' services over and above what has already been contracted for (Bower et al., 2002). Matching the needs of the principal and the performance motivators of the agents is therefore central to an effective incentive scheme. Through a literature review, the key features of effective CI have been summarized (Zhu & Cheung, 2021). These include (1) goal commitment (Locke et al., 1988); (2) expectation alignment (Wigfield & Eccles, 2002); (3) information exchangeability (Laffont & Tirole, 1988); (4) risk efficiency (Boukendour & Hughes, 2014); and (5) relationship investment (Adams, 1963). Table 2 summarizes the key components of effective CI.

Turning now to what constitutes project performance, Richmond-Coggan (2001) describes "better performance" in some situations as the degree of effort the project participants exert to save a project that is running into difficulties or to seek additional value by proposing a change. Meng (2012) demonstrated that the aim of an incentive mechanism is to "motivate better performance apart from the contract". The prime elements of project performance (PP hereafter) therefore include (a) contractual safeguards and (b) additional value creation (Zhu & Cheung, 2021). Table 3 gives the relevant details regarding PP.

The intention of having ex post incentivization is to prevent potential slippages in performance. The use of incentivization is based on the theory of organizational behaviour modification (Luthans & Kreitner, 1975) and reinforcement theory (Skinner, 1961). The intuitional expression of CI as a "carrot or stick" is also backed by stimulus–response psychology and self-determination theory (Bresnen & Marshall, 2000; Deci & Ryan, 2009). Most CIs act as 'carrots' to attract contractors to boost performance. Case studies conducted in Australia also found that the success of

Table 2 The key components of effective project incentivization (adapted from Zhu and Cheung (2021))

No	Components	Description	Key references
1	Goal Commitment	• Project members perceive the relationship that can achieve goals by working together • A performer is willing to accept a goal regardless of its difficulty and origin, or the credibility of the assigning person • Goals need to be meaningful, specific, challenging, and acceptable to those who are attempting to achieve them	Locke et al. (1988)
2	Expectation alignment	• Motivation is the perceived likelihood that effort will produce an appropriate level of performance ('expectancy') and the perceived likelihood that this performance will be converted into an appropriate level of reward	Williamson (1979), Vroom (1964)
3	Information exchangeability	• A good information sharing system is established for information exchange and behaviour monitoring	Schieg (2008) Oliver (1990)
4	Risk efficiency	• The allocation of risks and responsibilities are more balanced towards project efficiency • Project members have common attitudes towards risks	Zou et al. (2007), Zou and Zhang (2009) Zhang et al. (2016)
5	Relationship investment	• Status recognition: The party with the power advantage makes more motivational and relational investments towards the party with less power through shared relational attitudes, offering mutual support and developing mutual trust	Cook and Emerson (1978), Oliver (1990) Fu et al. (2015) Richmond-Coggan (2001)

Table 3 Elements of project performance

No	Elements		Descriptions	Key references
1	Contractual safeguards	Cost	Incentive initiator aims to make sure that the project can progress smoothly, and the contract can be fulfilled as agreed	Herten and Peeters (1986)
		Quality		
		Schedule		
2	Value creation	Innovation	Promote innovation to generate project and social benefits	Chan et al. (2011)
		Promotion of project performance	Incentive initiator aims to improve project performance/make the project's performance better than expected	Bresnen and Marshall (2000)
		Long-term commitment	Further relationship investment to enhance dependency	Suprapto et al. (2015)

incentive schemes was achieved through a combination of motivational and commercial objectives (Richmond-Coggan, 2001). In fact, commitment to deliver the agreed incentivization is a necessary condition for the successful use of the scheme (Dulaimi et al., 2003). Accordingly, the first hypothesis of the study is as follows:

H1: Effective construction incentivization (CI) improves project performance (PP).

4   The Behavioural Dimensions of Construction Incentivization

The primary purpose of CI is to solicit 'extra effort' from contracting parties to deliver better performance. It should also be noted that the CI should also befit the needs of the contractor. However, this meeting of minds may not be attained because of the singular use of quantitative targets that are unilaterally set by the incentive initiator. Goal commitment and expectation alignment therefore can hardly be achieved (Meng, 2012). Why is outcome-based CI not delivering, as many motivation theories have suggested? Eisenhardt (1988) highlighted that outcome-based incentive arrangements only work for highly programmed tasks where outcome targets can be set with reasonable accuracy. When projects are full of uncertainties, as in the case of complex infrastructure developments, the incentivizing targets are somewhat difficult to project. In this regard, the ability to master unforeseen eventualities and the concerted efforts of the project team are needed. This approach eliminates the need to deploy behaviour-based performance drivers (Meng, 2012). Regarding performance targets, Eisenhardt (1988) also claimed that behaviour-based criteria that reflect the ways the parties behave should be installed. Stack (2006) summarized that behaviour shaping is an effective method of accounting for responsibilities and promoting progress in complex engineering projects.

The evaluation of construction incentivization should therefore not only be confined to the degree of attainment of hard project targets, such as time, cost, and quality. For example, Rose and Manley (2011) found the critical roles of **project relationships** and **equitable contract conditions** in raising the effectiveness of incentivization arrangements in Australian projects. Zhu et al. (2020a, 2020b) also identified the incentivizing function of the behaviour monitoring system that was applied in a record-breaking mega project. This empirical evidence points to the development of relationism as proposed by Suprapto et al. (2015). In essence, project incentivization should foster cooperative contracting behaviours. It is therefore proposed that to optimize the effect of incentivization (Hughes et al., 2007), behaviour-based arrangements cannot be ignored. In this regard, it is necessary to investigate why contracting parties are not making their utmost efforts. Two forms of attitudinal issues are proposed: (i) equity gap (EG) and (ii) interorganizational relationship (IOR) between the contracting parties.

4.1 Equity Gap Between Contracting Parties

Adams (1963) suggested that whether one abides by a contract depends not only on what one gets but also on whether one's counterpart is getting more. Equity theory explains that a person always compares his or her outcomes-to-inputs ratio with that of the counterpart. Unfair treatment is a prime cause of opportunistic behaviours and disputes (ARCADIS, 2018). Lindenberg (2000) stated that unfair payment packages, power asymmetry and risk differentiation hamper trust among contracting parties. These disparities between the developer and the contractor are collectively described as the equity gap. Four main elements of the EG have been summarized by Zhu and Cheung (2022a): information asymmetry, risk differential, power asymmetry and expected return misalignment. Table 4 gives the details of the EG.

Can CI also be used to reduce uncertainties and balance information asymmetry through additional payments for the enhanced observability of the behaviour of the agent? Boukendour and Hughes (2014) found that project participants make an extra effort only if they feel that they are being fairly treated. In this regard, CI can be used to achieve a more equitable allocation of benefits and risks (Fu et al., 2015b). When the reward is commensurate with the risks involved, contractors can be expected to exert greater effort. CI can also be used to reduce uncertainties and balance information asymmetry through additional payments to raise the observability of the behaviour of the agent (Holmstrom, 1979). The second hypothesis of the study is as follows.

Table 4 Elements of EG on project participants (Zhu & Cheung, 2021)

No	Elements	Description	Key references
1	Information asymmetry	Agent behaviour cannot easily be evaluated during the project's duration	Ross (1973), Chen et al. (2020)
2		The principal may withhold information to avoid additional disputes or risks	
3	Risks differential	Environmental risk differential refers to unforeseeable physical conditions and cost fluctuations because of the market. These risks should be shared by both parties as deemed equitable but was shifted by contractual terms	Fang et al. (2004)
4		Behavioural risks related to the unanticipated contracting behaviour of the contracting parties. Examples are delayed payment and delayed instructions by the principals	
6	Power asymmetry	Sanction power asymmetry refers to the unilateral levy of damages and ordering contract changes between two parties	Chang and Ive (2007)
7		Bargaining power asymmetry is commonly exercised during negotiation. One party with a power advantage can deprive the value of the counterparts' belongings	
8	Expected return misalignment	Contracting parties expect equitable sharing based on their contributions. One party's profit may be squeezed, or it may have more unforeseeable losses	Chang and Ive (2007)

H2: Effective project incentivization should address *ex post* the equity gap that was created *ex ante* to improve project performance (PP).

4.2 Interorganizational Relationship

A conducive interorganizational relationship (IOR hereafter) refers to the conditions whereby organizations can pursue mutual interests (Cropper et al., 2008). Based on transaction costs theory (Williamson, 1985), the formation of IORs is prompted by an organization's desire to improve efficiency. In this chapter, Zhu and Cheung (2022b) summarized the core elements of IORs as interdependency, reciprocity, trust, and relationship continuity. Table 5 provides further details of these IOR elements.

How can project incentivization be utilized to develop interorganizational relationships between contracting parties? Several studies have found that IORs can be enhanced by bridging equity gaps to embrace equalizing power (Cook & Emerson, 1978), establishing distributive justice (Rose & Manley, 2011) and harvesting mutual trust (Suprapto et al., 2016). From the psychological point of view, a bridged equity gap relieves the tension between the contracting parties and serves as a lubricant for cooperation (Smyth & Edkins, 2007). Transaction cost theory further highlights that project participants are interdependent (Williamson, 1979). Dependence asymmetry may also give rise to a power differential (Emerson, 1962). The sense of equity should therefore be addressed as commitment to delivery be envisaged. The potential for using incentivization to develop IORs for project performance improvement has also been reported (Cropper et al., 2008; Kwawu & Laryea, 2014; Oliver, 1990). CI is

Table 5 Elements of IORs

No	Elements	Descriptions	Key references
1	Interdependency	Contractual parties thus rely heavily on each other. The termination of contracts or switching of a partner halfway causes great losses to both parties	Williamson (1985), Cheung et al. (2018)
2	Reciprocity	In reference to exchange theory, motivates reciprocity and emphasizes cooperation, collaboration, and coordination among organizations. It is the key point and the basis for interorganizational relationship development	Emerson (1976), Rose and Manley (2011)
3	Trust	For organizations, trust is seen as a substitute for contractual control. It is central to every transaction that demands contributions from the parties involved and has been identified as the key driver in fostering cooperation	Güth et al. (2000)
4	Relationship continuity	It refers to the stability of the relationship and long-term cooperation. The perceptions of a collaborative working environment and a long-term relationship are important for developing an IOR	Bock et al. (2005)

considered a starting point to enhance relationship quality in project management (Jelodar et al., 2016). It is an important way of reinforcing collaboration and building trust between project participants (Ceric, 2013). Rose and Manley (2011) highlighted the use of CI to foster cooperation and enhance communication (Kwawu & Laryea, 2014). The implications of the EG and IOR on the use of CI are presented as H3a and H3b, respectively.

H3a: Effective project incentivization should address *ex post* **the equity gap (EG) that was created** *ex ante* **to develop a conducive interorganizational relationship (IOR).**

H3b: Effective project incentivization should enhance interorganizational relationships (IORs) to improve project performance (PP).

To summarize the conceptual bases and hypotheses derived therefrom, a CI–EG–IOR–PP relationship framework (RF hereafter) is proposed and presented in Fig. 1.

5 Testing of Hypotheses

The RF (Fig. 1) was empirically tested. A data collection questionnaire was developed to solicit input from practising construction professionals in Hong Kong. The questionnaire had 5 parts. Part 1 introduced the personal particulars; Parts 2, 3, 4 and 5 contained questions about CI, EG, IOR and PP, respectively. The measurement items were developed from the theoretical deliberations of the constructs as summarized in Appendix. Respondents were asked to select a rating on a Likert scale (1–7) that was the most indicative of the project happening. Two methods were used to analyse the data: structural equation modelling and importance-performance map analysis.

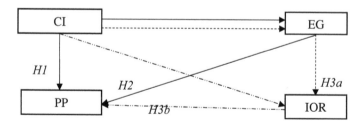

Fig. 1 A CI–EG–IOR–PP relationship framework

Fig. 2 Mediating variable in SEM

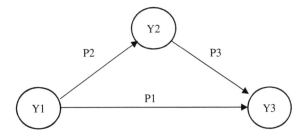

5.1 Structural Equation Modelling

Structural equation modelling (SEM hereafter) was used to examine the structure of the interrelationships expressed in a series of equations, such as a series of multiple regression equations (Hair et al., 2010). For this study, partial least squares SEM (PLS-SEM hereafter) was considered suitable for its ability to analyse complex models (Hair et al., 2010). The software Smart PLS 3 (Ringle et al., 2018) was used.

SEM analysis has two stages. First, the underlying components of each construct need to be verified. All hierarchical component models (HCMs) of CI, EG, IOR and PP were examined (Kuppelwieser & Sarstedt, 2014). Collinearity and redundant variables should be addressed. The second stage is to test the hypotheses. For this part, the mediating analysis in PLS-SEM was also applied. Mediation occurs when a third mediating variable intervenes between two other related constructs (Hair et al., 2014). The general structure of the mediating effect analysis in SEM is shown in Fig. 2.

In Fig. 2, Y1 represents the independent variable, Y2 represents the mediating variable, and Y3 represents the dependent variable. As a result, P1, P2 and P3 are the coefficients between the variables. As presented, P1 shows a direct effect of Y1 on Y3. The mediating effect of Y2 is assessed by P2*P3.

5.2 Importance-Performance Map Analysis

Important performance map analysis (IPMA hereafter) was used to identify key behavioural incentivizing agents. IPMA is an extension of the PLS-SEM analysis. It has been used to study customer services, marketing strategies, information management and better allocation of organizational resources (Magal & Levenburg, 2005). IPMA is a matrix-based technique and evaluates the factors in two dimensions: **importance** and **performance** (Eskildsen & Kristensen, 2006). For this study, the word "performance" in IPMA is like another key construct—project performance (PP). To avoid confusion, the word "**performance**" in the IPMA analysis is replaced by "**satisfaction**". In this way, the importance-performance map analysis is identified as "the importance-satisfaction map analysis" (ISMA).

The ISMA is particularly useful in enriching the interpretation of PLS-SEM results (Hair et al., 2014). It extends the standard reporting of the path coefficient estimates by adding an extra dimension that considers the average values of the latent variable scores (Ringle & Sarstedt, 2016). For the traditional quadrant approach, the total effects represent the predecessor constructs' **importance** in shaping the target construct (e.g., PP), while the average scores collected from respondents represent their **satisfaction** for each factor (Hair et al., 2014). In an ISMA, factors found to have high importance and a low satisfaction score should receive prioritized action by management (Martilla and James, 2019). To avoid the possible discontinuity in the inferred priorities caused by minor changes, the diagonal line approach is further suggested as a supporting approach (Bacon, 2003). The diagonal line approach in essence is a gap analysis where any factor below an upwards sloping 45° line in the ISMA is of high improvement priority.

Moreover, Matzler et al. (2003) found that for some factors, a change in factor satisfaction can be associated with a change in factor importance. The three-factor approach is elaborated on to make up for these defects and to help develop corresponding management strategies in different scenarios. In this study, a three-factor approach was adopted to determine whether the relationship between factor satisfaction and overall satisfaction is linear and symmetric (Matzler et al., 2003). Planning directions for project incentivization are developed as informed by the findings from both the quadrant and diagonal line approaches.

6   Data Analysis

6.1   Data Description

A total of 483 questionnaires were distributed, and 142 valid responses were received. The response rate was 30%, which is considered acceptable, as it is close to the median rate (35.7%) of a survey conducted in the United States with 1,607 organizational academic studies. It is also noted that the response rate of questionnaire surveys for studies conducted in the construction industry usually ranges from 25 to 30% (Easterby-Smith et al., 1991). Table 6 shows the distribution of project nature and type.

In general, the projects by type are quite well represented. Parts 2, 3, 4 and 5 contain questions about CI, EG, IOR and PP, respectively. Respondents were asked to indicate using a Likert scale of 1 (Strongly disagree) to 7 (Strongly agree) how accurate the statement represented the situation of the reference project. The descriptive statistics of the dataset are presented in Appendix.

In Part 2, it can be found that all the factors regarding the use of CI have scores above 4 (midpoint) on a scale of 1 to 7. This suggests that all these arrangements were included in the CI used in the reference projects. Q2.2 and Q2.5 have the highest scores. The standard derivation of these two questions is 0.84 and 0.83, which are

Table 6 The distribution of project nature and type

Q1	Project nature	Num	%
1	Residential	50	35
2	Commercial	27	19
3	Civil/Infrastructure	35	25
4	Composite	30	21
Q2	Project type		
1	Government project	40	28
2	Institutional project	21	15
3	Private project	81	57

also the lowest among the other items in Part 2, indicating that the respondents agreed that achievable common goals were set through CI. Part 3 is about the EG situations of the projects. It was found that all the mean scores were approximately 4 and range from 3 (slightly disagree) to 5 (slightly agree), demonstrating that these gaps may not be notable or might not have been addressed during the construction period. Part 4 and Part 5 show the distributions of IOR and PP. The mean scores for most of the questions regarding IOR are all above 5 (slightly agree), suggesting that basically all the projects that incorporated CI achieved satisfying outcomes regarding IOR. For PP, the most satisfactory result is for project quality (Q5.5). The least satisfied is for project time (Q5.6).

6.2 The Results of PLS-SEM Analysis

To detect collinearity among variables, Pearson's correlation test was conducted (Hair et al., 2010). After analysing the correlation for each part, it was found that Q3.1 is negatively correlated with the other variables listed in Part 3. Q3.1 is about collaborative effort to set common goals for the project. Hair et al. (2014) advised that these types of variables should be removed. The composite reliability and AVE of these factors are summarized in Table 7. For this study, all of the composite reliability indices are above 0.70, indicating inclusion for further analysis (Davcik, 2014). The average variance extracted (AVE) is over 0.4, which is considered adequate when the composite reliability is higher than 0.6 (Fornell & Laecker, 1981).

For the measurement of the structural model, an assessment of collinearity was conducted through a variance inflation factor (VIF) test. The results show that all VIF values are below 5, which indicates that there is no potential collinearity issue (Hair et al., 2014). The R^2, f^2 and Q^2 values of the overall model are also examined and summarized in Table 8.

The R^2 and adjusted R^2 in Table 8 are all greater than 0.10, suggesting an acceptable predictive accuracy of the model (Hair et al., 2014). The effect size (f^2) is also tested to evaluate whether the omitted construct has a substantive impact on the

Table 7 Average variance extracted (AVE)

Factor	Composite reliability	Average variance extracted (AVE)
CI	0.93	0.43
Goal commitment	0.88	0.71
Expectation alignment	0.79	0.57
Risk efficiency	0.85	0.59
Information exchangeability	0.85	0.66
Relationship investment	0.87	0.62
EG	0.89	0.41
Information	0.74	0.42
Power	0.89	0.50
Expected return	0.92	0.78
Risk	0.82	0.46
IOR	0.94	0.50
Interdependency	0.94	0.89
Reciprocity	0.86	0.67
Relationship Continuity	0.89	0.73
Trust	0.92	0.60
PP	0.92	0.48
Contractual Safeguards	0.92	0.61
Value creation	0.84	0.51

endogenous constructs. Table 8 shows that most f^2 values are over 0.02, so they are considered to have significant effects (Cohen, 1988). For the model fit, Stone-Geisser's Q^2 value should also be examined by a blindfolding procedure (Hair et al., 2014). Generally, the PLS-SEM data analysis results fit all these criteria. Figure 3 shows the path coefficients and significance with bootstrapping applied for 5000 samples. All the standard path coefficients are statistically significant.

All the path coefficients (t values for direct effects) of the CI–PP relationship framework are also summarized in Table 9. A negative correlation relationship is obtained between the EG and the other three constructs (CI, IOR and PP). It is also noted that CI is positively correlated with IOR and PP at a 100% significance level.

With the dataset of 142 projects with CI and applying a 5% significance level, the PLS-SEM analysis results support the general framework presented in Fig. 1. The mediating effects of the constructs were further examined in the SEM analysis. Table 10 summarizes the path significance analysis of the relationship framework.

Table 8 R^2, effect size f^2 and Q^2 values of the framework

Factors	R^2	R^2adjusted	Effect size f^2				Q^2 (=1−SSE/SSO)
			CI	EG	IOR	PP	
CI	–	–					
Goal Commitment	0.66	0.66	1.95				0.45
Expectation alignment	0.62	0.61	1.6				0.33
Information exchangeability	0.66	0.65	1.91				0.39
Relationship investment	0.78	0.78	2.64				0.4
Risk efficiency	0.73	0.72	3.53				0.45
Equity gap	0.11	0.10	0.12				0.15
Information	0.34	0.33		0.51			0.18
Power	0.84	0.84		5.33			0.39
Expected return	0.54	0.54		1.17			0.39
Risk	0.52	0.51		1.07			0.19
IOR	0.64	0.63	1.25	0.19			0.3
Interdependency	0.25	0.24			0.33		0.2
Reciprocity	0.66	0.65			4.61		0.41
Relationship continuity	0.82	0.82			1.91		0.56
Trust	0.95	0.95			19.61		0.53
PP	0.59	0.59	0.11	0.04	0.15		0.26
Contractual safeguards	0.92	0.92				11.53	0.52
Value creation	0.72	0.72				2.56	0.34

At a 5% significance level, all the relationships in the structural model are significant. The empirical results support the mediating role of the EG and IOR on PP. To summarize, the relationship between CI and PP is verified (H1). The P values reflect the significance of the indirect effects. The mediating effects are further verified at a 5% significance level. Hypotheses H1, H2, H3a and H3b are thus supported.

7 The Results of ISMA

The ISMA results are used to identify the key behaviour-based performance-incentivizing agent. First, reverse scaling of the EG is applied for interpretation consistency (Ringle & Sarstedt, 2016) (e.g., on a 7-point Likert scale, 7 becomes

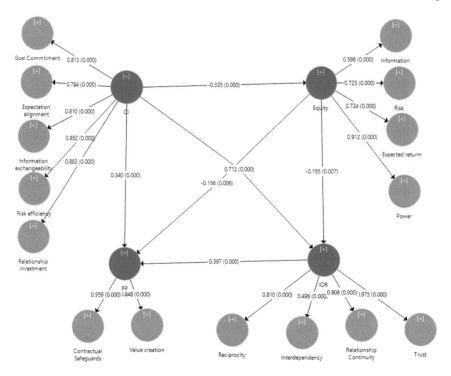

Fig. 3 The PLS-SEM analysis of the relationship framework

Table 9 Path coefficients and significance of the key construct relationships

	Original sample (O)	Sample mean (M)	Standard deviation (STDEV)	T Statistics (|O/STDEV|)	P values
CI -> EG	−0.340	−0.340	0.090	3.720	0.000*
CI -> IOR	0.710	0.710	0.060	12.790	0.000*
CI -> PP	0.340	0.340	0.090	3.750	0.000*
EG -> IOR	−0.190	−0.200	0.070	2.710	0.010*
EG -> PP	−0.160	−0.160	0.060	2.740	0.010*
IOR -> PP	0.400	0.400	0.100	3.990	0.000*

Note "*" denotes significance at the 5% level

1, 6 becomes 2, 5 becomes 3 and 4 remains unchanged). After that, the satisfaction and importance values are computed. The means indicate the respective construct's satisfaction score, with 0 and 100 representing the lowest and the highest satisfaction. For importance values, the total effects (the overall value of the direct and indirect effects) of all the constructs towards the target construct (PP) are calculated. The

Table 10 Significance analysis of the direct and indirect effects for the overall framework

	Original sample (O)	Sample mean (M)	Standard deviation (STDEV)	T Statistics (|O/STDEV|)	P values	Hypothesis
Direct effects						
CI -> IOR	0.710	0.710	0.060	12.790	0.000*	
CI -> PP	0.340	0.340	0.090	3.750	0.000*	H1
Indirect effects						
CI -> EG -> PP	0.050	0.060	0.030	1.960	0.050*	H2
CI -> EG -> IOR	0.070	0.060	0.030	2.540	0.010*	H3a
CI -> IOR -> PP	0.280	0.280	0.080	3.710	0.000*	H3b

Note "*" means significant at the 5% level

three-factor approach is used to estimate the relative impact of each factor for high and low satisfaction. In this regard, the analysis involves the following steps.

(1) The target factor (PP) satisfaction score must be recoded. To distinguish the high/low PP satisfaction score, the mean score is calculated for each sample. As all the questions about PP are measured on a 1–7 Likert scale. An average score lower than 5 (slightly agree) is considered "low satisfaction", while others are considered "high satisfaction". After recording the PP scores, the 142 responses are separated into two groups. Seventy cases have high PP, while the other 72 cases have low PP.

(2) The second step is then to conduct PLS-SEM separately for a heterogeneity assessment (Rigdon et al., 2011). Table 11 shows the importance and satisfaction scores for the two PP groups.

To highlight the differences between high PP and low PP, two separate ISMs are drawn. The quadrant approach and diagonal line approach are both applied for a more holistic analysis. Figures 4 and 5 present the ISM changes to the three main constructs.

Figure 4 shows the location of the three factors in the low PP group. Based on the quadrant approach, grand means have been used to locate the factors in four

Table 11 Heterogeneity assessment results based on different project performance levels

	Low PP		High PP	
	Importance	Satisfaction	Importance	Satisfaction
CI	0.66	58.38	0.55	72.86
EG	0.29	41.79	0.02	51.48
IOR	0.26	67.65	0.46	49.73

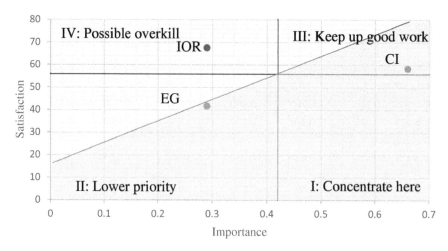

Fig. 4 ISM for the low PP group

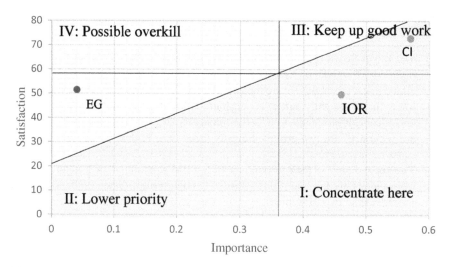

Fig. 5 ISM for the high PP group

quadrants and plot the mean values for the indices on the resulting matrix (Kristensen, 1999). Management should take priority actions for the important factors that affect satisfaction *(Concentrate here)*, followed by "keep up good work" and "less priority". Less attention should be given to quadrant IV, "possible overkill" (Martilla and James, 2019). For this group, CI has the highest importance score and satisfaction score on PP. Developing IORs is found to be less important. Factors falling into the shaded parts of the figure are those requiring management attention according to the diagonal line method. The EG and CI are considered "opportunities", while the IOR is considered "satiated needs". The results are consistent with the quadrant

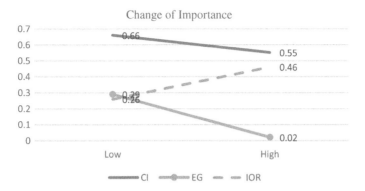

Fig. 6 Changes in importance

approach, suggesting that improving IORs is comparatively less important when project performance is low.

Similarly, Fig. 5 shows the ISMA results for the high PP group. There is a slight change for CI. In contrast, the IOR has moved from Quadrant IV to Quadrant I, "concentrate here," and is considered an "opportunity". This means that for projects that have above average performance scores, improving IORs significantly enhances project performance. There is also a change in the significance of the EG. The management of the EG is much less urgent for the high PP group. To further compare the differences between these two groups, Fig. 6 presents the changes in attribute importance depending on low/high satisfaction of PP:

The observations from Figs. 4, 5 and 6 are listed as follows:

(1) A slight change is found for both the importance and satisfaction scores of CI. Keeping up good work is suggested. According to the third factor approach, these factors are performers that lead to ideal ratings if fulfilled or exceeded all the time. Performers have linear and symmetric relationships with overall satisfaction (Matzler et al., 2003). Thus, CI has been viewed by the respondents as an instrumental tool as far as managing the EG and IOR are concerned.

(2) The impact of the EG on PP differs between the high and low satisfaction score groups. The EG was comparatively higher in the low PP group (higher than the IOR). However, a significant drop occurs when the PP satisfaction score is high. The EG is therefore classified as a basic factor (Matzler et al., 2003) regarded as a prerequisite (Hair et al., 2014). This result shows that the EG causes damage if not bridged (Ceric, 2013; Laffont & Tirole, 1988), and special attention is needed for projects with difficulties.

(3) The IOR can be interpreted as an excitement factor (Matzler et al., 2003) that can raise the overall satisfaction if delivered but does not cause low satisfaction otherwise. In other words, the positive enhancement of these factors has a greater impact on overall satisfaction (Matzler et al., 2003). Promoting IORs through CI is thus important, especially in pursuing exceptional PP.

Based on the aforementioned ISMA results, the key behaviour-based performance incentivizing agents (objective 3) EG and IOR are found to be instrumental for CI planning towards PP enhancement. Furthermore, CI acts as a performer, the EG is a basic factor and the IOR acts as an excitement factor (Matzler et al., 2003).

8 Implications for the Planning of Behaviour-Based CI

The key findings of the study are as follows: (i) effective CI can improve project performance and (ii) the effect of CI can be enhanced by bridging the equity gap to improve interorganizational relationships. This study offers empirical support for the usefulness of having strong IORs to deal with tasks of programmability because of the inherently high level of uncertainty. The following planning directions for CI are suggested:

(1) *Aligning power and expected return*

EG mitigation arrangements could be installed in CI to capitalize on the opportunity ex post when CI is planned. Power can be adjusted in view of the extent of the risks involved (Zhu & Cheung, 2021). To balance the power differential, the ex-ante, more powerful party should share decision-making authority to deal with unforeseen contingencies. The risk–reward reallocation strategy is also instrumental in addressing the return differential (Development Bureau of Hong Kong, 2016). It is important to reward contractors' contributions for their additional work, and setting financial bonuses is commonly suggested as an attractive reward. For mega projects with multiple goals, some nonfinancial rewards, such as early payments and appreciation rewards, can be considered. The promotion of a win–win partnership helps match expectations of return. For projects with high asset specificity, future working opportunities are suggested as incentive rewards to materialize the vision of long-term relationships. This helps both parties change the focus from short-term gain to long-term development. Status recognition is also a suggested method. The weaker party feels better recognized when it is more often engaged in project decisions. In contrast to stringent management styles, greater flexibility should also be given to contractors, especially those in specialist trades.

(2) *Enabling a risk management system*

For risk management, CI can be formulated to (1) prevent excessive risk premiums and (2) develop pain share/gain share working ethos. Traditional thinking suggests that offering risk premiums is a way to restore fairness when contractors assume more risks (Zhang et al., 2016). However, contractors rarely allow for sufficient risk premiums in their tenders due to intense competition. An overly generous risk premium weakens the perception of fairness and thereby hampers interorganizational relationships. Reallocation of risk ex post is therefore suggested when additional information becomes available. The main idea is to manage risks equitably.

Most mega infrastructure projects are complex and full of uncertainties. If risks are identified with the input of the contractor after the award of a contract, the impact analysis can be much more realistic. The situation is more acute for highly nonprogrammable tasks that can only be approached with innovative efforts. In such situations, the inputs of contractors are imperative. This study suggests that CI offers the unique opportunity to tap into the wealth of knowledge and skills of contractors ex post because they are likely to have better information with which to address unanticipated contingencies. Contractors are also more willing to contribute when they are also beneficiaries.

(3) *Aligning goals and expectations*

The effects of any CI depend on the commitments of the parties. Having common goals is the starting point. In addition, these goals should be agreed upon by the contracting parties and with mutual benefits. In addition, the goals must be clearly defined to avoid the possibility of unnecessary disputes. Moreover, this study also points out that aligning contractual parties' expectations of return with respect to the goals is of equal importance. A contractor's motivation can only be maximized when 1) the agent believes that the performance at the desired level is possible; 2) the agent believes that performance improvement efforts will lead to certain positive outcomes; and 3) the outcomes attract the agent. This means that the rewards are attractive enough to engender extra effort (Richmond-Coggan, 2001). Ultimately, the goals of CI must be practically achievable with reasonable effort.

(4) *Promoting interorganizational relationships*

It is further found that when exceptional PP is envisaged, more resources should be devoted to enhancing IORs during construction. Trust and reciprocity are the pivotal IOR drivers (Table 11 refers). A spirit of mutual trust and cooperation would generate interorganizational bounding (Zhang et al., 2021). If a party enters a CI arrangement but believes that his counterpart is going to be self-serving, he is unlikely to conform to the CI. Indeed, CI should remove this scepticism by including trust as a behavioural requirement (Rowlinson, 2012). Recognition is also conducive to fostering trust and upkeeping IORs. The enhancement of collaborative work is the focus of reciprocity.

In sum, it is advocated that CI can be planned to make ex post adjustments to power and risk. Having common goals could foster joint effort. These goals are to be agreed upon ex post and be coupled with behavioural commitments. Accordingly, CI should be planned with both behaviour-based and outcome-based targets.

9 Summary

The project performance of resource-intensive infrastructure developments is of serious concern to investors, be they government or private. The outcome of mega projects has not been satisfactory despite using project incentives that aim to raise performance. It is found that the prevailing use of outcome-based incentive schemes is not effective. The complex physical construction tasks are subject to uncertainties that render a reasonable determination of incentivizing targets. A more reasonable approach is to devise ways that could engender the committed efforts of the whole project team to tackle problems when they arise. The planning of project incentivization should therefore have both outcome-based and behaviour-based components. This study advocates that project incentivization should aim to balance the equity gap (EG) and maintain interorganizational relationships (IORs) to canvass contractors' commitment to raise project performance. Accordingly, a relationship framework is proposed and tested by partial least squares structural equation modelling (PLS-SEM) with 142 sets of project data collected from construction professionals. The study provided empirical support for effective CI offering the unique opportunity ex post to bridge the equity gap to improve interorganizational relationships should improvements in performance be targeted. Furthermore, important-satisfaction map analysis (ISMA) was conducted to confirm that bridging the EG should be an integral part of the management of an IOR. The findings suggested the following planning considerations of CI: (i) aligning power and expected return; (ii) enabling a risk management system; (iii) aligning goals and expectations; and (iv) promoting interorganizational relationships. This study contributes to the planning of CI by proposing behaviour-based components to complement the orthodox outcome-based design.

Acknowledgements The development of equity gap has been quoted from the paper "Toward an Equity-Based Analysis of Construction Incentivization" of the Journal of Construction Engineering and Management. The work described in this chapter was fully supported by an HKSAR RGC project (number 11202722).

Appendix: Data Collection Form and Descriptions

No	Description		Min	Max	Mean	Std	Cronbach(α)
Part 2	**PICI**						0.896
Q2.1	Goal Commitment	Collaborative effort was made between two parties to set common goals for the project	1	7	5.08	1.58	0.490
Q2.2		The incentive plan includes common goals agreed by the contracting parties	4	7	5.76	0.83	
Q2.3		Notable efforts have been directed to fulfil the common goals	3	7	5.72	0.86	

(continued)

(continued)

No	Description		Min	Max	Mean	Std	Cronbach(α)
Q2.4		Extra efforts had been used to fulfil the common goals when confronted with difficulties	3	7	5.70	0.89	
Q2.5	Expectation alignment	The expected performance was achievable for project participants	3	7	5.80	0.84	0.730
Q2.6		Reasonable financial bonus was set to for expected performance	2	7	5.46	1.19	
Q2.7		The performance exceeding expectation led to certain level of rewards	1	7	5.20	1.34	
Q2.8	Information exchangeability	Project information was easier to access than expected under PICI	2	7	5.01	1.08	0.612
Q2.9		Project information was exchanged smoothly under PICI during the whole project	3	7	5.40	0.93	
Q2.10		The project participants' unobserved behaviours were now monitored under of PICI	1	7	4.84	1.11	
Q2.11	Risk efficiency	The tender documents revealed a risk allocation pattern that was more balanced than market norm	1	7	5.00	1.21	0.761
Q2.12		The PICI enabled a risk allocation pattern more equitable than the pattern displaced in the tender documents	2	7	5.03	1.17	
Q2.13		Sufficient resources were provided to promote innovation	2	7	5.04	1.18	
Q2.14		Sufficient resources were provided to prevent project failure	2	7	5.32	1.07	

(continued)

(continued)

No	Description		Min	Max	Mean	Std	Cronbach(α)
Q2.15	Relationship investment	The spiCIrit of partnership was promoted to pursue mutual benefits for the project	2	7	5.57	1.16	0.790
Q2.16		Provisions are included in the construction incentivisation to compensate works due to unforeseen events	2	7	5.26	1.21	
Q2.17		The compensation for item Q3.16 was based on the principle of deriving win–win situation	3	7	5.57	1.04	
Q2.18		The CI focused more on long-term returns instead of short-term gain	1	7	5.25	1.05	
Part 3	**Equity Gap**						0.872
Q3.1.1	Information	At the bidding stage, the developer had an information advantage about the project details	1	6	3.78	1.09	0.561
Q3.1.2		At the bidding stage, the developer had an information disadvantage about the contractor's ability	1	7	3.76	1.32	
Q3.1.3		At the construction stage, the contractor had an information advantage relating to market changes	2	6	4.13	1.02	
Q3.1.4		At the construction stage, the developer could not monitor comprehensively the Contractor's behaviour relating project performance	1	7	3.89	1.20	
Q3.2.1	Risk (Environmental)	Unforeseeable physical conditions	1	7	4.23	1.37	0.792
Q3.2.2		Cost fluctuation (inflation of prices)	1	7	3.76	1.30	

(continued)

(continued)

No	Description		Min	Max	Mean	Std	Cronbach(α)
Q3.2.3		Unforeseeable loss because of adverse climatic conditions	1	7	3.80	1.27	
Q3.2.4	Risk (behaviour risk)	Unforeseeable loss because of defective design	1	7	3.53	1.50	
Q3.2.5		Time for payment	1	6	3.83	1.10	
Q3.2.6		Time for providing information/instructions	1	7	3.83	1.33	
Q3.3.1	Expected return	At the bidding stage, price competition was fully leveraged to drive down contractor's profit	1	6	4.12	1.33	0.859
Q3.3.2		The return for one of the parties was not commensurate to his contribution in resources to the project according to the contract	1	6	3.81	1.09	
Q3.3.3		At the construction stage, return for changes was not commensurate to his contribution in resources to the project	1	6	3.90	1.19	
Q3.4.1	Sanction power	At the construction stage, unilateral termination by the contractor presented greater threat than the developer	1	6	3.81	1.36	0.855
Q3.4.2		Unilateral decision authority over project dispute had been the major weapon used by the developer to achieve his own goals	1	6	3.51	1.39	
Q3.4.3		At the construction stage, the developer was unwilling to cooperate for events which are critical to the contractor	1	6	3.15	1.28	
Q3.4.4		At the construction stage, the contractor was unwilling to cooperate for events which are critical to the developer	1	6	3.18	1.18	

(continued)

(continued)

No	Description		Min	Max	Mean	Std	Cronbach(α)
Q3.4.5	Bargaining power	At the bidding stage, the contractor felt more constrained and sacrificed in negotiating contract terms in relation to compensation for foreseeable losses	1	7	4.20	1.26	
Q3.4.6		At the construction stage, the developer felt more constrained and sacrificed in renegotiation of contract terms in relation to compensation for foreseeable losses or disputes	1	6	3.94	1.29	
Q3.4.7		The developer felt being forced to settle claims below his entitlements for change of work	1	6	3.70	1.13	
Q3.4.8		Making compromise was needed for the developer in view of the time pressure in switching contractor	1	6	4.31	1.24	
Part 4	**IOR**						0.909
Q4.1	Interdependency	The loss of transaction cost was unrecoverable when switching to another counterpart	3	7	5.13	1.08	0.679
Q4.2		The loss of time was unrecoverable when switching to another counterpart	3	7	5.12	1.18	
Q4.3		The loss of project information and data was unrecoverable in switching to another counterpart	1	7	4.30	1.15	
Q4.4	Reciprocity	Shared norms were developed between the two senior management teams	3	7	4.96	1.00	0.762

(continued)

(continued)

No	Description		Min	Max	Mean	Std	Cronbach(α)
Q4.5		Project participants felt being fairly treated when putting efforts towards the attainment of the common goals	1	7	5.05	1.20	
Q4.6		A no-blame culture was established between the two contracting parties	1	7	4.48	1.25	
Q4.7	Trust	A good management system was established to reinforce goal achievement such as continual improvement, profit making and business expanding	3	7	5.27	0.95	0.901
Q4.8		Misunderstandings were avoided by open communication	4	7	5.63	0.95	
Q4.9		Information in the contract document was explained to the affected parties	3	7	5.35	1.01	
Q4.10		Project participants had a good interaction to obtain more information from the other party	3	7	5.55	0.86	
Q4.11		It is believed that one of the parties had confidence to work with the other if they are honest	4	7	5.64	0.75	
Q4.12		Both parties were considerate to understand the other parties' needs and feelings at work	2	7	5.20	0.98	
Q4.13		Being considerate had enhanced the working capacity of the counterpart	3	7	5.26	0.99	
Q4.14		A good inter-organizational relationship was built between two parties	3	7	5.40	0.98	

(continued)

(continued)

No	Description		Min	Max	Mean	Std	Cronbach(α)
Q4.15	Relationship continuity	Both parties perceived that the working environment was collaborative	3	7	5.51	0.96	0.814
Q4.16		Both parties perceived those future working opportunities were likely	3	7	5.44	0.98	
Q4.17		Both parties were willing to accept short-term dislocation believing that it will balance out in the long run	3	7	5.32	0.96	
Part 5	**Project performance**						0.897
Q5.1	Contractual safeguards	The contractor's behaviour could readily be evaluated during the whole project procedure	3	7	5.24	1.07	0.896
Q5.2		Programmable tasks were achieved on each stage during the whole project procedure	3	7	5.09	1.14	
Q5.3		Unexpected situations and difficulties encountered were well-handled	2	7	5.09	1.10	
Q5.4		The project cost was within overall budget	1	7	5.10	1.36	
Q5.5		This project achieved satisfying project quality	3	7	5.41	1.10	
Q5.6		This project finished on time	1	7	4.49	1.65	
Q5.7		The volume of disputes was controlled within a reasonable range	2	7	5.25	1.09	
Q5.8		The amount in dispute was controlled within a reasonable range	2	7	5.20	1.14	
Q5.9	Value creation	Both parties worked together to maximize mutual benefits instead of their own benefits	1	7	5.07	1.24	0.758
Q5.10		Innovations were generated by the developer in this project	1	7	4.53	1.47	

(continued)

(continued)

No	Description		Min	Max	Mean	Std	Cronbach(α)
Q5.11		Innovations were generated by the contractor in this project	2	7	4.63	1.50	
Q5.12		There are promotions of project performance (e.g., cost-saving, shorten the construction period and quality improvements) that are beyond expectations	1	7	4.76	1.42	
Q5.13		Both parties had confirmed a commitment to seek mutual benefits and cooperation in the future	2	7	5.15	1.14	
Valid N		142					

References

Abrahamson, M. W. (1984). Risk management. *International Construction Law Conference*.

Adams, J. S. (1963). Toward an understanding of inequity. *Journal of Abnormal and Social Psychology, 67*(5), 422–436. https://doi.org/10.1093/past/96.1.22

Alfie, K. (1993). Why incentive plans cannot work. *Harvard Business Review, 71*(5), 54.

ARCADIS. (2018). *Global Construction Disputes Report*.

Bacon, D. (2003). A comparison of approaches to importance-performance analysis. *International Journal of Market Research, 45*(1), 1–2.

Bayliss, R., Cheung, S., Suen, H., & Wong, S. (2004). Effective partnering tools in construction: A case study on MTRC TKE contract 604 in Hong Kong. *International Journal of Project Management, 22*(3), 253–263.https://doi.org/10.1016/S0263-7863(03)00069-3

Bock, A. G., Zmud, R., Kim, Y., & Lee, J. (2005). Behavioral intention formation in knowledge sharing: Examining the roles of extrinsic motivators, social-psychological forces, and organizational climate. *MIS Quarterly, 29*(1), 87–111.

Boukendour, S., & Hughes, W. (2014). Collaborative incentive contracts: Stimulating competitive behaviour without competition. *Construction Management and Economics, 32*(3), 279–289. https://doi.org/10.1080/01446193.2013.875215

Bower, D., Ashby, G., Gerald, K., & Smyk, W. (2002). Incentive mechanisms for project success. *Journal of Management in Engineering* [Preprint].

Bresnen, M., & Marshall, N. (2000). Motivation, commitment and the use of incentives in partnerships and alliances. *Construction Management and Economics, 18*(5), 587–598. https://doi.org/10.1080/014461900407392

Ceric, A. (2013). Application of the principal-agent theory to construction management: Literature review. In *Proceedings of the 29th Annual Association of Researchers in Construction Management Conference, ARCOM 2013*, vol. 1, pp. 1071–1081.

Chan, D. W. M., Lam, P., Chan, A., & Wong, J. (2010). Achieving better performance through target cost contracts—The tale of an underground railway station modification project. *Performance Measurement and Management in Facilities Management, 28*(5/6), 261–277.

Chan, D. W. M., Chan, A., & Wong, J. (2011). An empirical survey of the motives and benefits of adopting guaranteed maximum price and target cost contracts in construction. *International Journal of Project Management, 29*(5), 577–590. https://doi.org/10.1016/j.ijproman.2010.04.002

Chang, C. Y., & Ive, G. (2007). Reversal of bargaining power in construction projects: Meaning, existence and implications. *Construction Management and Economics, 25*(8), 845–855. https://doi.org/10.1080/01446190601164113

Chen, Q., Hall, D., Adey, B., & Hass, C. (2020). Identifying enablers for coordination across construction supply chain processes: A systematic literature review. *Engineering, Construction and Architectural Management* [Preprint]. https://doi.org/10.1108/ECAM-05-2020-0299

Cheung, S. O. (1997). Risk allocation: An essential tool for construction project management. *Journal of Construction Procurement, 3*(1), 16–27.

Cheung, S. O., Zhu, L., & Lee, K. W. (2018). Incentivization and interdependency in construction contracting. *Journal of Management in Engineering, 34*(3), 1–13. https://doi.org/10.1061/(ASCE)ME.1943-5479.0000601

Cook, K. S., & Emerson, R. M. (1978). Power, equity and commitment in exchange. *American Sociological Review, 43*(5), 721–739. https://doi.org/10.3102/0034654311405999

Cropper, S., Ebers, M., Huxham, C., & Ring, P. (2008). Introducing inter - organizational relations. *Oxford handbooks online introducing*, pp. 1–23. https://doi.org/10.1093/oxfordhb/9780199282944.003.0001

Davcik, N. S. (2014). The use and misuse of structural equation modeling in management research. *Journal of Advances in Management Research, 11*(1), 47–81. https://doi.org/10.1108/JAMR-07-2013-0043

Deci, E. L., & Ryan, R. M. (2009). Self-determiniation. *The Corsini Encyclopedia of Psychology*, pp. 1–2. https://doi.org/10.1017/CBO9781107415324.004

Development Bureau of Hong Kong. (2016). Practice Notes for New Engineering Contract (NEC)— Engineering and Construction Contract (ECC) for Public Works Projects in Hong Kong. Development Bureau of Hong Kong.

Dulaimi, M. F., Ling, F. Y. Y., & Bajracharya, A. (2003). Organizational motivation and inter-organizational interaction in construction innovation in Singapore. *Construction Management and Economics, 21*(3), 307–318. https://doi.org/10.1080/0144619032000056144

Easterby-Smith, M., Thorpe, R., & Lowe, A. (1991). *Management research: An introduction.* London: SAGE Publications, Inc., p. 2455.

Eisenhardt, K. M. (1988). Agency and institutional theory explanations: The case of retail sales compensation. *The Academy of Management Journal, 31*(3), 488–511.

Emerson, R. M. (1976). Social exchange theory. *Annual Reviews of Sociology, 2*(1976), 335–362.

Eskildsen, J. K., & Kristensen, K. (2006). Enhancing importance-performance analysis. *International Journal of Productivity and Performance Management, 55*(1), 40–60. https://doi.org/10.1108/17410400610635499

Fang, D., Mingen, L., Sik-wah, F., & Liyin, S. (2004). Risks in Chinese construction market— Contractors' perspective. *Journal of Construction Engineering and Management, 130*(2), 235–244. https://doi.org/10.1061/(ASCE)0733-9364(2004)130

Fornell, C., & Laecker, D. F. (1981). *Structural equation models with unobservable variables and measurement error: Algebra and statistics.*

Fu, Y., Chen, Y., Zhang, S., & Wang, W. (2015). Promoting cooperation in construction projects: An integrated approach of contractual incentive and trust. *Construction Management and Economics, 33*(8), 653–670. https://doi.org/10.1080/01446193.2015.1087646

Güth, W., Konigstein, M., Marchand, N., & Nehring, K. (2000). *Trust and Reciprocity in the Investment Game with Indirect Reward.*

Hair, J. F., Black W., Babin, B., & Anderson, R. (2010). *Multivariate data analysis. Seventh, Pearson Education.* https://doi.org/10.1016/j.ijpharm.2011.02.019

Hair, J. F. et al. (2014). *A primer on partial least squares structural equation modelling (PLS-SEM).* SAGE Publications, Inc., p 2455.

Herten, H. J., & Peeters, W. A. R. (1986). Incentive contracting as a project management tool, pp. 34–39. https://doi.org/10.1016/0263-7863(86)90060-8

Holmstrom, B. (1979). Moral hazard and observability. *The Bell Journal of Economics, 10*(1), 74. https://doi.org/10.2307/3003320

Hughes, D., Williams, T., & Ren, Z. (2012). Is incentivisation significant in ensuring successful partnered projects? *Engineering, Construction and Architectural Management, 19*(3), 306–319. https://doi.org/10.1108/09699981211219625

Hughes, W., Yohannes, I., & Hillig, J.-B. (2007). Incentives in construction contracts: Should we pay for performance? *CIB World Building Congress, 2000*, 2272–2283.

Ibbs, C. W. (1991). Innovative contract incentive features for construction. *Construction Management and Economics, 9*, 157–169.

Jelodar, M. B., Yiu, T. W., & Wilkinson, S. (2016). A conceptualisation of relationship quality in construction procurement. *International Journal of Project Management, 34*(6), 997–1011. https://doi.org/10.1016/j.ijproman.2016.03.005

Kuppelwieser, V. G., & Sarstedt, M. (2014). Applying the future time perspective scale to advertising research. *International Journal of Advertising, 33*(1), 113–136. https://doi.org/10.2501/IJA-33-1-113-136

Kwawu, W., & Laryea, S. (2014). Incentive contracting in construction. In *Proceedings of the 29th Annual Association of Researchers in Construction Management Conference, ARCOM 2013,* pp. 729–738.

Laffont, J. -J., & Tirole, J. (1988). The dynamics of incentive contracts, *56*(5), 1153–1175.

Lindenberg, S. (2000). It takes both trust and lack of mistrust: The workings of cooperation and relational signaling in contractual relationships. *Journal of Management and Governance, 4*(1–2), 11–33. https://doi.org/10.1023/A:1009985720365

Locke, E. A., Latham, G. P., & Erez, M. (1988). The determinants of goal commitment. *Academy of Management Review, 13*(1), 23–39. https://doi.org/10.5465/AMR.1988.4306771

Llyod, H. (1996). Prevalent philosophies of risk allocation: And overview. *The Internaional Construction Law Eview, 13*(4), 502–548.

Luthans, F., & Kreitner, R. (1975). *Organisational behavviour modification.* Glenview, IL.

Ma, Q., Li, S., & Cheung, S. O. (2021). Unveiling embedded risks in integrated project delivery. *The ASCE Journal of Construction Engineering and Management* (Available on-line on 25 Oct 2021).

Magal, S. R., & Levenburg, N. M. (2005). Using importance-performance analysis to evaluate e-business strategies among small businesses. In *Proceedings of the Annual Hawaii International Conference on System Sciences, 00*(C), p. 176. https://doi.org/10.1109/hicss.2005.661

Martilla, J. A., & James, J. C. (2019). Importance-performance analysis, *41*(1), 77–79.

Matzler, K., Sauerwein, E., & Heischmidt, K. A. (2003). Importance-performance analysis revisited: The role of the factor structure of customer satisfaction. *Service Industries Journal, 23*(2), 112–129. https://doi.org/10.1080/02642060412331300912

MacNeil, I. R. (1973-1974). The many futures of contracts. *Southern California Law Review, 47.*

Oliver, C. (1990). Determinants of interorganizational relationships: integration and future directions published by: Academy of management linked references are available on JSTOR for this article: Determinants of interorganizational relationships: Integration and Futu. *Academy of Management Review, 15*(2), 241–265.

Richmond-Coggan, D. (2001). *Construction contract incentive schemes: Lessons from experience.*

Rigdon, E. E., Ringle, C., Sarstedt, M., & Gudergan, S. (2011). Assessing heterogeneity in customer satisfaction studies: Across industry similarities and within industry differences. *Advances in International Marketing, 22*(January), 169–194. https://doi.org/10.1108/S1474-7979(2011)0000022011

Ringle, C. M. Ringle, C., Sarstedt, M., & Gudergan, S. (2018). Partial least squares structural equation modeling in HRM research. *International Journal of Human Resource Management*, (January), 1–27. https://doi.org/10.1080/09585192.2017.1416655

Ringle, C. M., & Sarstedt, M. (2016). Gain more insight from your PLS-SEM results the importance-performance map analysis. *Industrial Management and Data Systems, 116*(9), 1865–1886. https://doi.org/10.1108/IMDS-10-2015-0449

Rose, T., & Manley, K. (2011). Motivation toward financial incentive goals on construction projects. *Journal of Business Research, 64*(7), 765–773. https://doi.org/10.1016/j.jbusres.2010.07.003

Ross, S. A. (1973). The economic theory of agency: The Principal' s problem. *The American Economic Review, 63*(2), 134–139.

Rowlinson, M. (2012). A practical guide to the NEC3 professional services contract. *Wiley*. https://doi.org/10.1002/9781118406373

Schieg, M. (2008). Strategies for avoiding asymmetric information in construction project management. *Journal of Business Economics and Management, 9*(1), 47–51. https://doi.org/10.3846/1611-1699.2008.9.47-51

Skinner, B. F. (1961). *Analysis of behaviour*. McGraw-Hill.

Smyth, H., & Edkins, A. (2007). Relationship management in the management of PFI/PPP projects in the UK. *International Journal of Project Management, 25*(3), 232–240. https://doi.org/10.1016/j.ijproman.2006.08.003

Suprapto, M., Bakker, H., Mooi, H., & Hertogh, M. (2016). How do contract types and incentives matter to project performance? *International Journal of Project Management, 34*(6), 1071–1087. https://doi.org/10.1016/j.ijproman.2015.08.003

Suprapto, M., Bakker, H. L. M., & Mooi, H. G. (2015). Relational factors in owner-contractor collaboration: The mediating role of teamworking. *International Journal of Project Management, 33*(6), 1347–1363. https://doi.org/10.1016/j.ijproman.2015.03.015

Terninko, J., & Zusman, A. (1998). *Systematic innovation: An introduction to TRIZ*. CRC Press LLC.

Vroom, V. H. (1964). *Work and motivation*. Wiley.

Wigfield, A., & Eccles, J. S. (2002). Expectancy–value theory of achievement motivation. *Journal of Asynchronous Learning Network, 6*(1), 68–81. https://doi.org/10.1006/ceps.1999.1015

Williamson, O. E. (1979). Transaction-cost economics: the governance of contractual relations. *The Journal of Law and Economics, 22*(2).

Williamson, O. E. (1985) *The economic institutions of capitalism, China social sciences publishing house Cheng books ltd*. https://doi.org/10.5465/AMR.1987.4308003

Zhang, S., Gao, Y., & Ding, X. (2016). Contractual governance: Effects of risk allocation on contractors' cooperative behavior in construction projects. *Journal of Construction Engineering and Management, 142*(6), 1–11. https://doi.org/10.1061/(ASCE)CO.1943-7862.0001111

Zhu, L., Cheung, S. O., Gao, X., Li, Q., & Liu, G. (2020). Success DNA of a record-breaking megaproject. *Journal of Construction Engineering and Management, 146*(8).

Zhu, L., & Cheung, S. O. (2021). Toward an equity-based analysis of construction incentivization, *Journal of Construction Engineering and Management, 147*(11).

Zhu, L., & Cheung, S. O. (2022a). Equity gap in construction contracting: Identification and ramification. *Engineering, Construction and Architectural Management, 29*(1), 262–286.

Zhu, L., & Cheung, S. O. (2022b). Incentivizing relationship investment for project performance improvement, engineering. *Project Management Journal*, (In Press).

Zhu, L., Cheung, S., Gao, X., Li, Q., & Liu, G. (2020b). Success DNA of a record-breaking megaproject. *Journal of Construction Engineering and Management, 146*(8), 05020009. https://doi.org/10.1061/(asce)co.1943-7862.0001878

Zou, P. X. W., & Zhang, G. (2009). Managing risks in construction projects: Life cycle and stakeholder perspectives. *International Journal of Construction Management, 9*(1), 61–77. https://doi.org/10.1080/15623599.2009.10773122

Zou, P. X. W., Zhang, G., & Wang, J. (2007). Understanding the key risks in construction projects in China. *International Journal of Project Management, 25*(6), 601–614. https://doi.org/10.1016/j.ijproman.2007.03.001

Liuying Zhu works at the School of Management, Shanghai University. Dr. Zhu received her MSc degree (Distinction) and PhD from the City University of Hong Kong. Her MSc dissertation was awarded the 2016 Best Dissertation (Master category) by the Hong Kong Institute of Surveyors. Her doctoral study on construction incentivization was completed with the Construction Dispute Resolution Research Unit. Her current research focuses on construction project management, construction incentivization, and project dispute avoidance. Dr. Zhu has published articles in international journals and book chapters in these topics. Contact email: zhuliuying@shu.edu.cn

Sai On Cheung is a professor of the Department of Architecture and Civil Engineering, City University of Hong Kong. In 2002, Professor Cheung established the Construction Dispute Resolution Research Unit. Since then, the Unit has published widely in the areas of construction dispute resolution and related topics such as trust and incentivization. With the collective efforts of the members of the Unit, two research volumes titled Construction Dispute Research and Construction Dispute Research Expanded were published in 2014 and 2021 respectively. Professor Cheung is a specialty editor (contracting) of the ASCE Journal of Construction Engineering and Construction. Professor Cheung received a DSc for his research in Construction Dispute. Contact email: saion.cheung@cityu.edu.hk.

Part II
Strategic Uses

Chapter 5
Incentivizing Relationship Investment for Mega Project Management

Liuying Zhu

Abstract Principal-agent theory (PAT) considers that relational risks for contracting parties are significant and may lead to opportunistic behavior. As mega projects often have high asset specificity and facing great uncertainty, the demand for cooperation between different participants is particularly prominent. Effective moves to enhance interorganizational relationships and alleviate the related bottlenecks are therefore encouraged. Construction incentivization is thus advocated because of its flexibility and high acceptability. This study examines the stimulating effect of construction incentivization on interorganizational relationships for mega projects. A PLS-SEM analysis of 142 projects shows that the interorganizational relationship acts as a mediator between construction incentivization and project performance. Furthermore, developers and contractors have different perceptive views on construction incentivization. It is therefore suggested that construction incentivization should go beyond conventional uses and embrace relationship investment as a goal. Furthermore, there is no substitute for negotiated agreement on incentivization arrangements if mutually aligned interests are pursued.

Keywords Incentivization · Interorganizational relationship · Social exchange theory

1 Introduction

Zeiss (2007) summarized five major challenges facing the construction industry: (1) global climate change; (2) aging infrastructure; (3) shrinking workforce; (4) declining productivity and (5) islands of information. The ability to adapt to the dynamic environment is therefore vitally needed to overcome challenges and to innovate (Flyvberg, 2017; Cheung and Chan, 2014). Comparatively, mega projects have

L. Zhu (✉)
School of Management, Shanghai University, Shanghai, China
e-mail: zhuliuying@shu.edu.cn

© The Author(s), under exclusive license to Springer Nature Switzerland AG 2023   101
S. O. Cheung and L. Zhu (eds.), *Construction Incentivization*, Digital Innovations in Architecture, Engineering and Construction,
https://doi.org/10.1007/978-3-031-28959-0_5

high asset specificity and require multiparty participation. Relational risks in buyer–
seller relationships are recognized by agency theory, which are aggravated by the
complexity and uncertainty of mega projects (Bryde et al., 2019). A noncooperative
attitude is an important factor hindering project performance (PP hereafter). Moreover, the construction project team would dissolve upon completion of the project.
Therefore, long-term benefits are seldom considered by team members (Suprapto
et al., 2016). Opportunistic behavior occurs during the construction stage, which is
not conducive to collaboration and promotes disputes (Zhang et al., 2020a, 2020b).
Effective moves to enhance interorganizational relationships (IORs hereafter) and
alleviate the related bottlenecks are therefore advocated.

What vehicle can be deployed to develop IORs? Williamson (1979) pointed out
that contract incompleteness is unavoidable in complex, long-term transactions.
Therefore, convergent contractual governance is inadequate (Nguyen & Garvin,
2019). The potential use of construction incentivization (CI hereafter) to address
risks identified ex post has been suggested in Chapter 1. In fact, the flexibility and
high acceptability of CI make it important and adequate to address project challenges
(Meng, 2015). Furthermore, the case study of the Hong Kong-Zhuhai-Macao Bridge
Project found that CI can serve the function of IOR maintenance by enhancing information exchange (Zhu et al., 2020). An integrated incentive system was also found to
help the developer obtain additional project updates and enhance interorganizational
communication. Jelodar et al. (2016) further added that incentives are instrumental in
enhancing the quality of project teamwork, as evidenced by team members' commitment and collaboration. Investigating the use of CI on IORs is a valuable organizational study. Accordingly, this chapter reports a study that systematically examines
the use of CI in mega projects to develop IORs for project performance improvement. The findings of this study suggest that the innovative planning of CI should
embrace developing IORs, as put forward by the relevant theories. This study has
the following research objectives:

(1) Identify IORs in mega projects;
(2) Analyze the functions of construction incentivization in mega project manage-
    ment; and
(3) Provide practical recommendations for construction incentivization planning.

2 Interorganizational Relationships in Construction Projects

Interorganizational relationships are the foundation of enduring bonding among organizations (Oliver, 1990). Recent literature focuses mainly on aligning mutual interests
among project participants (Cropper et al., 2008; Manata et al., 2021). The value of
collaboration and cooperation has gradually received attention.

2.1 The Developer-Contractor Interorganizational Relationship

The developer-contractor tensed relationship is commonly observed in construction projects. Based on principal-agent theory (Eisenhardt, 1989), the principal refers to the developer when the agent is the contractor. Cooperation and coordination are usually assumed among project participants. Based on principal-agent theory, different commercial organizations' behavior is driven by their self-interest. In addition, there are distinct aspects of this relationship because of the nature of construction projects. Compared with other projects, the particularity of a construction project is as follows:

(1) Construction project teams are often identified as **temporary organizations** (Cropper et al., 2008). Different from the buyer–seller relationship, construction projects exist for a limited period for prespecified goals. Project participants are commonly unfamiliar and self-interested. Opportunistic behavior may happen during the project.

(2) Mega projects often have **high asset specificity**. Asset specificity refers to durable investments undertaken for transactions. Should the original transaction be prematurely terminated, the opportunity cost incurred for investments is much lower in best alternative uses or by alternative users (Williamson, 1985). If the mega project is not finished, project stage results are irreversible and difficult to utilize. In that case, great loss may result if contract determination happens, especially in the middle or later stage of the project. Transaction cost economics (Williamson, 1979) therefore argues that the specific assets invested in a partnership increase the hazards of opportunism. Relational exchange theory suggests that asset specificity may also enhance trust among contracting partners and lead to more cooperative behavior and higher project performance (Lui et al., 2009). However, in either case, asset specificity affects both the status change and power use of both parties.

2.2 Key Dimensions of the Interorganizational Relationships in Construction

The interorganizational relationship captures the construction project team quality and the dynamic exchange between parties (Song et al., 2020). Zhu and Cheung (2022) identified six dimensions of interorganizational relationships, of which interdependency (Cropper et al., 2008; Fu et al., 2015; Cheung et al., 2018), trust (Cheung et al., 2014), reciprocity (Oliver, 1990) and relationship continuity are considered in this study (Güth et al., 2000; Macneil, 1974).

(1) Interdependency: The three subdimensions of interdependence are uncertainty, asset specificity, and frequency (Williamson, 1985). A 'lock-in' situation occurs when asset-specific investments are made by contractual parties (Williamson,

1979). Interdependence between developers and contractors is also realized when parties perceive that high termination costs are associated with ending the relationship (Sarkar et al., 1998). For construction projects, asset-specific investment substantially increases once projects reach milestones. Project participants thus rely heavily on each other, and the termination of construction contacts or a change in partners may cause significant losses (Guo et al., 2021). Relational exchange theory highlights that interdependency is the pillar of interorganizational cooperation (Kumaraswamy & Anvuur, 2008). After investigating 142 construction projects, Cheung et al. (2018) found that cooperative behavior would be created for contractual parties with high interdependency.

(2) Reciprocity: Reciprocity in construction projects occurs when project participants provide necessary assistance to each other, resulting in a win–win situation. It is one of the bases upon which interorganizational relationships develop (Oliver, 1990). Human altruistic instinct acts as a powerful force to drive people to cooperate rather than confront each other (Fehr & Fischbacher, 2003). Creating a cooperative working environment is also an essential adversarial strategy (Bower et al., 2002). Reciprocity contributes to project collaboration and coordination among project participants (Wang et al., 2019) and is the basis of trust building (Swärd, 2016). A positive relational attitude of reciprocity among team members is beneficial for project efficiency (Suprapto et al., 2016).

(3) Trust: Trust is the foundation of social order (Cheung et al., 2014) and the compensation for contractual control (Zhang et al., 2018). It takes time to develop and maintain mutual trust and major unresolved conflict can destroy trust in a relationship (Ceric, 2016). Mistrust is a potential factor that aggravates speculation and hostility. The evaluation of trust is always a key element of IORs. Cheung et al. (2011) identified three major types of trust in construction contracting: (1) system-based trust; (2) cognition-based trust; and (3) affect-based trust. System-based trust is trust in the performance of systemized open communication. Such arrangements can build trust through strengthened communication among contracting parties. Cognition-based trust develops from confidence in objective knowledge that demonstrates the trustworthiness of the contracting parties. The exchange of such knowledge can be attained through interaction or observation. Affect-based trust develops on a more sentimental platform and involves emotional bonds that connect individuals who value personal attachment.

(4) Relationship Continuity: Relationship stability and continuity are important for IOR long-term development. It has two dimensions: (1) for a specific construction project, the parties involved must be able to fulfill their obligations to ensure the stability of the relationship for a significant period, and (2) both parties must intend to maintain their cooperative relationship over the long term (Bock et al., 2005). This dimension shows that project participants are changing their focus from short-term gain and loss to long-term benefits. In this context, they are also willing to sacrifice short-term interests to obtain more long-term win–win and benefit opportunities. On the other hand, the stability of

their cross-organizational relationship improves. Examples include developing partnerships and creating long-term strategic cooperation opportunities.

3 Relationship Investment from Construction Incentivization

It is proposed that CI can be used to develop IORs to enhance PP. This section first discusses the constructs of CI and PP and then formulates the hypothesized relationships.

3.1 Identification of Construction Incentivization

CI refers to the collective terms of incentive schemes applied in construction projects. The main purpose of incentivization is to motivate project participants and obtain more value than expected (Meng & Gallagher, 2012a, 2012b). CI can be classified based on objective objectives such as cost, schedule, quality, and safety incentive schemes. Based on the nature of the rewards, it can also be divided into financial and nonfinancial incentive schemes (Saka et al., 2021). The underlying needs of the developer and the motivations of the contractor are pivotal and central to CI. To exemplify the four CI design parameters introduced in Chapter One, Zhu and Cheung (2021a) identified that effective CI has the following features: (i) goal commitment (Locke et al., 1988); (ii) expectation alignment (Wigfield & Eccles, 2002); iii) information exchangeability (Bryde et al., 2019; Laffont & Tirole, 1988); iv) risk efficiency (Boukendour & Hughes, 2014); and v) relationship investment (Adams, 1963).

(1) Goal Commitment: The mutual commitment of additional project goals is commonly manifested in CI. It reflects the performer's willingness to cooperate regardless of the difficulty, originality, or credibility of the assigning party (Zhu & Cheung, 2018). For construction projects, CI targets should be agreed upon by contracting parties (Rowlinson, 2012). Extra effort directed toward CI targets for working together should also be clarified (Rose & Manley, 2011). The incentives and rewards are related to the achievable project targets (Locke and Latham, 1990), and extra effort is necessary to fulfill these goals when difficulties arise.

(2) Expectation Alignment: The alignment of goals and expectations is essential in CI planning. Abu-Hijleh and Ibbs (1989) noted that CI targets should be attractive, affordable, and achievable to contractors. For example, financial incentives take effect by compensating the additional effort that a higher return may require. Bridging a project vision can also be a subjective benefit. Bandura's (1982) self-efficacy theory explains that the confidence between two parties underpins the desire for project success. Moreover, the expectation level also influences contracting behavior and the performance of contract commitments

(Blomquist et al., 2016). An appropriate and similar level of confidence should be developed for contractual parties through CI to enhance cooperation and manifest commitment (Das and Teng, 1998).

(3) Information Exchangeability: Information exchangeability holds that an additional information sharing system should be established for CI implementation. For schedule incentives, additional milestones are often set, and rewards are offered. The project procedure is thus more exposed for the developer and helps reduce information asymmetry to solve the agent problem (Schieg, 2008). For mega projects, integrated information sharing systems are established together with performance assessment systems to confer rewards or otherwise (Zhu et al., 2020). Based on the outcome, transaction uncertainty could be reduced. Screening refers to the means for the developer to collect project information for specific tasks (Cropper et al., 2008). As specific tasks are mentioned and additional information sharing platforms are often incorporated, settings relating to communication enrich information exchange, which in turn facilitates project progress and quality control (Hetemi et al., 2020).

(4) Risk Efficiency: Imbalanced risk allocation is a root cause of construction disputes (Zhu and Cheung, 2020). Risk reallocation is a key ammunition of CI (Chapman & Ward, 2008). Risk efficiency refers to the balanced risk toward project efficiency (Zhang et al., 2016) and aligns the risk preferences of stakeholders (Zou & Zhang, 2009). Risk reallocation therefore aims to reduce excessive risk premiums and minimize future construction disputes. Moreover, a fair and efficient risk sharing formula would incentivize contractors by removing suspicion and fostering trust (Boukendour & Hughes, 2014). Innovation is also encouraged when project risks are better allocated and more freedom is allowed (Zou & Zhang, 2009).

(5) Relationship Investment: Relationship investment refers to the motivational and relational move from a power-advantaged party to the invited reciprocation of support and trust. The contracting relationship is promoted to pursue mutual project benefits (Cook & Emerson, 1978). Status recognition is used to offer better recognition of the weaker party and enhance the other party's project engagement (Adams, 1965). Strategic alliances and partnering are also considered incentives for collaboration (Richmond-Coggan, 2001). They both aim to encourage contractors to focus on long-term returns. Their status changes from performance unit to strategic partner, which also improves their trust and participation.

3.2 Project Performance

Project performance (PP hereafter) represents the project outcomes. Multiple dimensions are therefore used due to the many facets of project results (Ahmadi Digehsara et al., 2018). Eisenhardt (1988) argued that performance measured by target outcomes is appropriate for highly programmable tasks only. Moreover, mega projects are often highly complex with low task programmability. Behavior-based criteria are thus necessary to provide a full spectrum of performance. In addition, innovation is also encouraged and cannot be evaluated by programmable tasks (Zhang et al., 2020a, 2020b). The evaluation of project performance thus includes (i) project outcomes in terms of cost, schedule, quality, and safety (Yu et al., 2005); (ii) behavioral outcomes such as joint problem solving and communication (Eisenhardt, 1989; Zhang et al., 2020a, 2020b); and (iii) innovation (technical and managerial) (Dulaimi et al., 2003).

3.3   The Relationships Among CI, IOR and PP

(1)   Effective CI enhances PP improvement

Based on principal-agent theory, the use of CI helps reduce project uncertainty and make more transparent decisions (Zhu & Cheung, 2021b). For example, developers set the incentive of the benefit-sharing ratio to encourage cost savings. For this purpose, an open-book approach is adopted, along with enhanced project information sharing. Observability is therefore increased. Work segregation can also reduce the indeterminacy of other parties (Hosseinian, 2016). Likewise, schedule incentives are set with specific milestones (Wang et al., 2018). Information asymmetry between principal and agent can be reduced by enhancing task measurability (Holmstrom, 1979). In addition, more balanced risks can encourage contracting parties to adopt innovative ideas (Bower et al., 2002).

(2)   IOR mediates the relationship between CI and PP

Apart from the effectiveness of CI based on principal-agent theory, relevant studies point to the multifunction of CI instead of using it solely as financial bait. Rose and Manley (2011) found the importance of providing incentives when cooperation is solicited. IORs thus can be incentivized (Oliver, 1990; Cropper et al., 2008; Kwawu & Laryea, 2014). Incentivization can kickstart IOR development. The different aspects of CI, such as goal commitment, risk allocation and relationship investment, have been found to be essential motivational factors for developing trust (Gunduz & Abdi, 2020). Reallocation of risk perceptions is also beneficial to reinforce trust at the organizational level based on rational pursuit (Yao et al., 2019). With improved IOR, mutual trust can be enhanced with the effect of suppressing opportunistic behavior (Ceric, 2016), raising operational efficiency (Liu et al., 2017) and minimizing construction disputes (Zhu & Cheung, 2020). Enhanced IOR is also instrumental for PP improvement. Collaboration and cooperation are promoted in construction projects, as they are conducive to improving project efficiency (Gunduz & Abdi, 2020). The enhanced relationship reduces the risk premium caused by mistrust during the project procedure and minimizes transaction costs (Kumaraswamy & Anvuur, 2008).

Fig. 1 The conceptual
relationships among CI, IOR
and PP
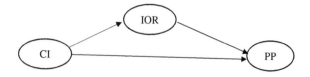

Based on the literature, the relationships of these three factors are like the mediation effect. Figure 1 presents the conceptual relationships of CI, IOR and PP. IOR acts as a mediator between CI and PP:

4 Empirical Study

An online questionnaire was designed to verify the conceptual framework. Construction professionals from the Hong Kong Institute of Architects (HKIA), the Hong Kong Institute of Surveyors (HKIS), the Hong Kong Institute of Construction Managers (HKICM), listed real estate companies and contracting companies located in Hong Kong were invited to participate. The questionnaire focuses on personal particulars (Part 1), the participating project details of CI (Part 2), and the three constructs (Part 3–5). A 7-point Likert scale ranging from 1 (strongly disagree) to 7 (strongly agree) was used to capture the respondents' viewpoints. To obtain valid data, responses with unreasonable filling times were excluded.

The data were analyzed with confirmatory factor analysis (Hair et al., 2010). Partial least squares-structural equation modeling (PLS-SEM hereafter) was applied considering the sample size and the distribution of data (Hair et al., 2014; Henseler et al., 2009). Smart PLS 3 was used to estimate the measurement models and the mediating effect of the key constructs. A hierarchical component model (HCM) is applied for the measurement model of CI, IOR and PP. The mediating effect of IOR was tested based on PLS-SEM. A multigroup analysis (MGA hereafter) was applied. A heterogeneity test was also conducted to check group differences in project roles (the developer/contractor) and the contractual role of CI (CI initiator/recipient).

Over 450 questionnaires were distributed online, and 142 valid responses were obtained. For Part 1, Table 1 presents the personal particulars of these professionals. The table shows that the ratio of management staff and professional staff in this investigation is 1:2. Work experience was basically evenly distributed among these four groups.

Part B investigates the project details incorporating CI. The contractual and organizational roles were investigated. Table 2 presents the cross-check relationship between the organizational role and contractual role of CI:

Most CI was planned and implemented by developers, and contractors were the primary recipients. Among the 73 developer respondents, only 5 have project experience as recipients of CI. To summarize, 79% (68 responses) of the CI projects

Table 1 Personal particulars (Part A)

No	Description	Number	%
1.1	Your position		
1	Management staff	48	34
2	Professional staff	94	66
	Sum	142	100
1.2	Working experience		
1	<5 years	33	23
2	5–10 years	36	25
3	11–20 years	40	28
4	>20 years	33	23
	Sum	142	100

Table 2 The relationship between the organizational role and the contractual role in CI

		The contractual role of CI		Total
		Initiator	Recipient	
Project role	Developer	68	5	73
		93%	7%	100%
	Contractor	18	51	69
		26%	74%	100%
Total		86	56	142
		59%	41%	100.%

investigated were initiated by the developer, and only 21% (18 responses) were initiated by the contractor.

Table 3 presents the details of the projects investigated.

There is a generally even distribution of the project nature, and half of the projects are private projects. Twenty-eight percent of the projects are government projects,

Table 3 Project details

1	Project nature	Num	%
1.1	Residential	50	35
1.2	Commercial	27	19
1.3	Civil/Infrastructure	35	25
1.4	Composite	30	21
2	Project type		
2.1	Government project	40	28
2.2	Institutional project	21	15
2.3	Private project	81	57

and 15% are institutional projects. To obtain a detailed view of the distribution by project nature, a cross check was performed based on these two questions.

Table 4 shows the data for Parts 3–5 of the survey.

The descriptive data for Parts 3–5 are shown in Table 4. For the setting of CI (Part 3), the average scores of most responses are above 4 (neutral), and most of them are higher than 5 (slightly agree). This result shows that these key features are reflected during the project procedure. The highest mean score was obtained for Q3.4 (The expected performance was considered achievable for project participants) (5.80) and Q3.1 (Incentive plans applied common goals set by the contracting parties) (5.76). The standard deviations of these two items are 0.84 and 0.83, respectively. The lowest mean score is Q3.9 (The project participants' unobserved behavior was monitored under CI) (4.84), showing that the CI function of information exposure is comparatively less effective.

The mean scores for most questions regarding IOR are all above 5 (slightly agree). This result shows that a satisfying level of IOR is maintained under CI. The lowest score is Q4.3 (Misunderstandings were avoided through open communication) (4.30). The respondents agreed that IORs were sufficiently maintained in these two areas. Responses with the highest mean scores are related to trust building.

The mean scores of the questions in Part 5 section (Project Performance) are all above 5. Comparatively, all the behavior outcomes have the most satisfying responses, i.e., above 5. For the hard outcome, Q5.7 (This project achieved a satisfactory level of project quality) has the highest mean score. Comparatively, CI created less innovative value for the overall project.

A collinearity test is conducted to identify and eliminate redundant or conflicting variables (Hair et al., 2010). As collinearity impacts the accuracy of the PLS-SEM analysis, redundant or conflicting indicators should be removed based on Pearson's correlation test (Hair et al., 2014). Based on the test result, Cronbach's alpha (α) is also calculated to check internal consistency. A threshold of 0.6 has been proposed (Davcik, 2014).

PLS-SEM Analysis

To evaluate internal consistency and convergent validity, composite reliability tests and average variance extracted (AVE) tests are suggested for PLS-SEM analysis (Davcik, 2014; Hair et al., 2014). An AVE value higher than 0.4 is adequate when the composite reliability level is higher than 0.6 (Fornell & Laecker, 1981). Table 5 shows the composite reliability and AVE of the constructs in this study.

Based on the acceptance of the indices, Fig. 2 shows the PLS-SEM analysis results. Generally, all the coefficients are significant at the 5% level:

Figure 2 presents the analysis results of the empirical study. For each factor, the following is found. (1) For CI, risk efficiency contributes the most (0.870), while information exchangeability contributes the least (0.791) at the 5% significance level. (2) For IOR, trust has the highest contributing value of 0.969, and interdependency has the lowest. 3) For PP, behavior outcome contributes the most (0.939), while innovation contributes the least (0.692).

Table 4 Descriptive statistics for Parts 3–5 of the survey

No	Description	Mean	Std	Cronbach's alpha (α)
Part 3	CI			9.911
Q3.1	Goal Commitment (Locke et al., 1988)			0.800
	The incentive plan includes common goals agreed upon by the contracting parties	5.76	0.83	
Q3.2	Notable efforts were directed to achieve common goals	5.72	0.86	
Q3.3	Extra efforts were used to accomplish common goals when difficulties arose	5.70	0.89	
Q3.4	Expectation Alignment (Willison,1975; Aiken & Hage, 1987; Vroom, 1964)			0.730
	The expected performance was achievable for project participants	5.80	0.84	
Q3.5	A reasonable financial bonus was set to for the expected performance	5.46	1.19	
Q3.6	The performance exceeding expectation led to a certain level of rewards	5.20	1.34	
Q3.7	Information Exchangeability (Schieg, 2008)			0.612
	Project information was easier to access than expected under CI	5.01	1.08	
Q3.8	Project information was exchanged smoothly under CI during the whole project	5.40	0.93	
Q3.9	The project participants' unobserved behavior was monitored under CI	4.84	1.11	

(continued)

Table 4 (continued)

No	Description	Mean	Std	Cronbach's alpha (α)
Q3.10	Risk Efficiency (Zou & Zhang, 2009; Zou et al., 2007) The tender documents revealed a risk allocation pattern that was more balanced than market norms	5.00	1.21	0.761
Q3.11	The CI enabled a risk allocation pattern more equitable than the pattern displaced in the tender documents	5.03	1.17	
Q3.12	Sufficient resources were provided to promote innovation	5.04	1.18	
Q3.13	Sufficient resources were provided to prevent project failure	5.32	1.07	
Q3.14	Relationship Investment (Cook & Emerson, 1978; Oliver, 1990) The spirit of partnership was promoted to pursue mutual project benefits	5.57	1.16	0.790
Q3.15	Provisions are included in the construction incentivization to compensate work due to unforeseen events	5.26	1.21	
Q3.16	Compensation for Item Q3.15 was based on the principle of ensuring a win–win situation	5.57	1.04	
Q3.17	The CI focused more on long-term returns instead of short-term gain	5.25	1.05	
Part 4	IOR			0.909
Q4.1	Interdependency (Cheung et al., 2018) Lost transaction costs were unrecoverable when switching to another counterpart	5.13	1.08	0.679
Q4.2	Lost time was unrecoverable when switching to another counterpart	5.12	1.18	

(continued)

Table 4 (continued)

No	Description	Mean	Std	Cronbach's alpha (α)
Q4.3	Lost project information and data were unrecoverable when switching to another counterpart	4.30	1.15	
Q4.4	Reciprocity (Suprapto et al., 2015) Shared norms developed between the two senior management teams	4.96	1.00	0.762
Q4.5	Project participants sensed being treated fairly when dedicating efforts to the attainment of common goals	5.05	1.20	
Q4.6	A culture free of blame was established between the two contracting parties	4.48	1.25	
Q4.7	Trust (Cheung et al., 2011b) A good management system was established to reinforce goal achievement such as continual improvement, profit generation and business expansion	5.27	0.95	0.901
Q4.8	Misunderstandings were avoided through open communication	5.63	0.95	
Q4.9	Information included in the contract document was explained to the affected parties	5.35	1.01	
Q3.10	Project participants engaged in positive interactions to obtain more information from the other party	5.55	0.86	
Q3.11	Each party had confidence in working with the other because it considered the other party honest	5.64	0.75	

(continued)

Table 4 (continued)

No	Description		Mean	Std	Cronbach's alpha (α)
Q3.12		Both parties understood each other's needs and feelings at work	5.20	0.98	
Q3.13		Considerate behavior enhanced the working capacity of the counterpart	5.26	0.99	
Q3.14		A strong interorganizational relationship developed between the two parties	5.40	0.98	
Q3.15	Relationship Continuity (Bock et al., 2005)	Both parties perceived the working environment as collaborative	5.51	0.96	0.814
Q3.16		Both parties perceived future working opportunities to be likely	5.44	0.98	
Q3.17		Both parties were willing to accept short-term dislocation by considering that it would balance out over the long run	5.32	0.96	
Part 5	PP				0.897
Q5.1	Behavior outcome (Eisenhardt, 1988)	The contractor's behavior could be systematically evaluated during the whole project	5.24	1.07	0.858
Q5.2		Programmable tasks were completed at each stage of the project	5.09	1.14	
Q5.3		Unexpected situations and difficulties were handled effectively by the contracting parties	5.09	1.10	
Q5.4		The contracting parties cooperated with each other to maximize common interests, not just their own interests	5.07	1.24	

(continued)

Table 4 (continued)

No	Description	Mean	Std	Cronbach's alpha (α)
Q5.5	The contracting parties seek cooperation and value cocreation in the future	5.15	1.14	
Q5.6	Hard Outcome (Yu et al., 2005) The project cost fell within overall budget limits	5.10	1.36	0.891
Q5.7	This project achieved a satisfactory level of project quality	5.41	1.10	
Q5.8	This project was completed on time	4.49	1.65	
Q5.9	The volume of disputes was controlled to a reasonable range	5.25	1.09	
Q5.10	The number of disputes was controlled to a reasonable range	5.20	1.14	
Q5.11	Innovation (Dulaimi et al., 2003) Innovations were generated by the developer through the project	4.53	1.47	0.796
Q5.12	Innovations were generated by the contractor through the project	4.63	1.50	
Q5.13	Improvements in project performance (e.g., cost savings, shortened construction periods and quality improvements) surpassed expectations	4.76	1.42	
Valid N	142			

Table 5 Composite
reliability and average
variance extracted (AVE)

	Composite reliability	Average variance extracted (AVE)
CI	0.93	0.43
Goal commitment	0.88	0.71
Expectation alignment	0.79	0.57
Risk efficiency	0.85	0.59
Information exchangeability	0.85	0.66
Relationship investment	0.87	0.62
IOR	0.94	0.51
Interdependency	0.88	0.90
Reciprocity	0.76	0.68
Trust	0.91	0.60
Relationship continuity	0.82	0.73
PP	0.92	0.48
Hard outcome	0.89	0.62
Behavior outcome	0.90	0.64
Innovation	0.88	0.71

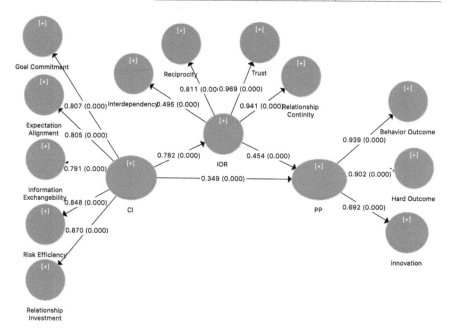

Fig. 2 PLS-SEM analysis result of the framework

Table 6 R^2 value

Factor	R^2	Adjusted R^2
CI	–	–
Goal commitment	0.667	0.664
Expectation alignment	0.650	0.648
Information exchangeability	0.617	0.614
Relationship investment	0.754	0.753
Risk efficiency	0.722	0.720
IOR	0.613	0.610
Interdependency	0.246	0.241
Reciprocity	0.656	0.653
Relationship continuity	0.822	0.821
Trust	0.951	0.951
PP	0.574	0.568
Behavior outcome	0.878	0.877
Hard outcome	0.818	0.817
Innovation	0.477	0.474

The relationships of CI, IOR and PP are also analyzed and validated. Partial mediation means that there is not only a significant relationship between the mediator and the dependent variable but also a direct relationship (e.g., CI and PP). Statistically, the result shows that IOR acts as a partial mediator between CI and PP. The positive relationship between CI and PP is validated, and the coefficient is 0.349. The indirect effect of CI on PP is 0.355 (0.782*0.454 = 0.355), accounting for approximately 50% of the total effect.

SmartPLS3 presents the model fit indices. The R^2 value is the most used measure to evaluate a model's predictive accuracy (Hair et al., 2014). Table 6 shows the R^2 value of the conceptual framework. As R^2 and adjusted R^2 values greater than 0.10 are acceptable (Falk & Miller, 1992),the accuracy of the framework is validated.

Table 7 shows the effect size f^2 and Stone-Geisser's Q^2 values.

In PLS-SEM analysis, the effective size f^2 was examined to evaluate the R^2 values of all endogenous constructs (Hair et al., 2014). For the measurement model, the most effective size f^2 in Table 7 is higher than 0.35, showing that they have large effects (Cohen, 1988). Interdependency has a moderate effect, as the value is higher than 0.15 (Cohen, 1988). The blindfolding procedure is also conducted to assess the Q^2 value. The smaller the difference between the predicted and original values is, the greater the Q^2 value is (Ringle et al., 2018). Table 7 shows that all the Q^2 values are higher than 0.02, which is acceptable, and those higher than 0.35 are considered to have a high effect.

Group differences were also tested by heterogeneity tests to highlight further implications. Views of the developer and contractor, CI initiator and CI recipient were analyzed. Tables 8 and 9 show the group differences.

Table 7 Effect size f^2 and Q^2 values

	Effect size f^2	Q^2 (=1-SSE/SSO)
CI	–	–
Goal commitment	2.001	0.470
Expectation alignment	1.858	0.413
Information exchangeability	1.611	0.338
Risk efficiency	2.593	0.410
Relationship investment	3.073	0.436
IOR	–	0.301
Interdependency	0.327	0.204
Trust	19.572	0.561
Relationship continuity	4.613	0.586
Reciprocity	1.903	0.430
PP	–	0.267
Behavior outcome	7.217	0.550
Hard outcome	4.508	0.489
Innovation	0.914	0.321

Table 8 Group differences between developers and contractors

Description	Path coefficients-diff (developer–contractor)	New p value (developer–contractor)
CI -> Information exchangeability	−0.179	0.035
CI -> Risk efficiency	0.012	0.012
CI -> IOR	0.169	0.003

Table 9 Group differences between CI initiators and recipients

Description	Path coefficients-diff (initiator–recipient)	New p value (initiator–recipient)
CI -> Information Exchangeability	−0.179	0.007
CI -> Expectation Alignment	−0.141	0.003

Table 8 shows that contractors tend to hold a view that CI has a greater effect on information exchange but a slightly lower effect on risk efficiency. Additionally, a stronger connection between CI and IOR is found from the developer's view. Table 9 shows the differences between the CI initiator and the recipient. Similarly, the significance of the difference is also reflected in the contributing value of information exchangeability. Moreover, CI recipients recognize the value more of aligning the expectation of two parties.

5 Discussion and Recommendations

The PLS-SEM analysis empirically validates the hypothesis with 142 responses. Bootstrapping with 5000 samples is adopted, and all the coefficients are significant at the 5% level. It is found that IORs and CI are instrumental for behavior-based project performance improvement. The overall contractual framework also implies that IORs play a mediating role between CI and PP. The results also validate this finding.

The results also show that singular financial rewards are beneficial for project performance enhancement; moreover, relationship investment also improves behavior-based project performance. The focus should be incentivizing relationship investment to engender mutual trust and cooperation. For the heterogeneity test, group differences were detected. Differences were found between developer and contractor. Information exchangeability tends to have a lower contributing value toward CI for developers. As most CI initiators are developers, this difference is also reflected between the CI initiator and recipient. Additionally, the investigation shows that most CI projects are introduced unilaterally. Developers have greater interest in building IORs through CI, which has a less positive effect on recipients in nurturing trust and developing relationship continuity.

Based on both theoretical development and empirical study, recommendations for management are as follows:

(1) *CI should be treated as a stimulator of IOR development.*

Conventional studies of CI have focused mainly on the use of CI to compensate for the extra effort it may cost to improve performance. This study further found that to improve PP, CI should act as a stimulator of IOR development. Different from the traditional concept, relationship investment is found to be the most significant contributor to CI planning, which is less relevant to monetary rewards. Apart from financial incentives, status recognition (partnership) and long-term working opportunities are the sweetener for the contractor to cooperate and maximize project value. Moreover, IOR is the partial mediator between CI and PP. The CI-IOR-PP relationship takes half ($0.782*0.454 = 0355$) of the total effect ($0.355 + 0.349 = 0.704$), representing the key position of IOR in the relationship between CI and PP. For CI design, in the design of incentive mechanisms, the proportion of terms for maintaining IOR deserves project managers' attention.

(2) *Bilateral decisions should be the basis of CI planning*

Another major finding is the differential viewpoints of CI between developer and contractor. The major differences concern the recognition of CI. Developers (most are CI initiators) usually have higher expectations regarding information exchangeability and risk efficiency. However, as the agent, the attitude of the CI recipient is more directly linked to its effect on PP. Bilateral discussion is thus encouraged for the implementation of CI. Negotiating the allocation of risk and expected return promotes the success of CI.

6 Summary of Chapter

Mega projects are classic examples of transactions with high asset specificity and multiparty participation. Relational risks in the buyer–seller relationship is recognized by agency theory. The complexity and uncertainty surrounding mega projects necessitate the use of relationship investment to lubricate the potential working bottlenecks. The flexibility and high acceptability of CI make it a perfect tool to meet project challenges. It is advocated that CI can play a pivotal role in delivering PP through IOR building. This study examines the stimulating effect of CI on IOR development in mega projects. Based on a literature review, the key contributors of IORs are identified as interdependency, trust, reciprocity, and relationship continuity. Goal commitment, risk efficiency, relationship investment, information exchangeability and expectation alignment are essential elements of successful CI. After subjecting 142 project data to PLS-SEM analysis, the IOR was found to be a partial mediator between CI and PP. Accordingly, it is recommended that (1) CI should be treated as a stimulator of IOR development and (2) bilateral decisions should be the basis of CI planning.

Acknowledgements The empirical work of this chapter has been reported in a paper entitled "Incentivizing Relationship Investment for Project Performance Improvement" of Project Management Journal.

References

Adams, J. S. (1963). Toward an understanding of inequity. *Journal of Abnormal and Social Psychology, 67*(5), 422–436. https://doi.org/10.1093/past/96.1.22

Ahmadi Digehsara, A., Rezazadeh, H., & Soleimani, M. (2018). Performance evaluation of project management system based on combination of EFQM and QFD. *Journal of Project Management,* 171–182. https://doi.org/10.5267/j.jpm.2018.4.003

Bandura, A. (1982). Self-efficacy mechanism in human agency. *American Psychologist, 37*(2), 122–147.

Blomquist, T., Farashah, A. D., & Thomas, J. (2016). Project management self-efficacy as a predictor of project performance: Constructing and validating a domain-specific scale. *International Journal of Project Management, 34*(8), 1417–1432. https://doi.org/10.1016/j.ijproman.2016.07.010

Bock, A. G., Zmud, R., Kim, Y., & Lee, J. (2005). Behavioral intention formation in knowledge sharing: Examining the roles of extrinsic motivators, social-psychological forces, and organizational climate. *MIS Quarterly, 29*(1), 87–111.

Boukendour, S., & Hughes, W. (2014). Collaborative incentive contracts: Stimulating competitive behaviour without competition. *Construction Management and Economics, 32*(3), 279–289. https://doi.org/10.1080/01446193.2013.875215

Bower, D., Ashby, G., Gerald, K., & Smyk, W. (2002). Incentive mechanisms for project success. *Journal of Management in Engineering* [Preprint].

Bryde, D. J., Unterhitzenberger, C., & Joby, R. (2019). Resolving agency issues in client–contractor relationships to deliver project success. *Production Planning and Control, 30*(13), 1049–1063. https://doi.org/10.1080/09537287.2018.1557757

Ceric, A. (2016). Trust in Construction Projects. Devon, Taylor & Francis Group.

Chapman, C., & Ward, S. (2008). Developing and implementing a balanced incentive and risk sharing contract. *Construction Management and Economics, 26*(6), 659–669. https://doi.org/10.1080/01446190802014760

Cheung, S. O., Wong, W., Yiu, T., & Pang, H. (2011). Developing a trust inventory for construction contracting. *International Journal of Project Management, 29*(2), 184–196. https://doi.org/10.1016/j.ijproman.2010.02.007

Cheung, S. O., Wong, W., Yiu, T., & Pang, H. (2014). *Construction dispute research, construction dispute research.* Springer.https://doi.org/10.1007/978-3-319-04429-3

Cheung, S. O., Zhu, L., & Lee, K. W. (2018). Incentivization and interdependency in construction contracting. *Journal of Management in Engineering, 34*(3), 1–13. https://doi.org/10.1061/(ASCE)ME.1943-5479.0000601

Cohen, J. (1988). *Statistical power analysis for the behavioral sciences. Second Edi.* Lawrence Erlbaum associates.

Cook, K. S., & Emerson, R. M. (1978). Power, equity and commitment in exchange. *American Sociological Review, 43*(5), 721–739. https://doi.org/10.3102/0034654311405999

Cropper, S., Ebers, M., Huxham, C., & Ring, P. (2008). Introducing Inter-organizational relations. In *Oxford Handbooks Online Introducing,* 1–23. https://doi.org/10.1093/oxfordhb/9780199282944.003.0001

Das, T. K., & Teng, B.-S. (1998). Between trust and control: developing confidence in partner cooperation in alliances. *The Academy of Management Review, 23*(3), 491–512.

Davcik, N. S. (2014). The use and misuse of structural equation modeling in management research. *Journal of Advances in Management Research, 11*(1), 47–81. https://doi.org/10.1108/JAMR-07-2013-0043

Eisenhardt, K. M. (1988). Agency and institutional theory explanations: The case of retail sales compensation. *The Academy of Management Journal, 31*(3), 488–511.

Eisenhardt, K. M. (1989). Agency theory: An assessment and review. *The Academy of Management Review, 14*(1), 57–74.

Eisenhardt, K. M., & Eisenhardt, K. M. (1989). Agency theory: An assessment and review. *The Academy of Management Review, 14*(1), 57–74.

Falk, R. F., & Miller, N. B. (1992). *A primer for soft modelling.* University of Akron Press.

Fehr, E., & Fischbacher, U. (2003). The nature of human altruism—Proximate and evolutionary origins. *Nature, 425*(October), 785–791.

Fornell, C., & Laecker, D. F. (1981). Structural equation models with unobservable variables and measurement error: Algebra and statistics.

Fu, Y., Chen, Y., Zhang, S., & Wang, W. (2015). Promoting cooperation in construction projects: An integrated approach of contractual incentive and trust. *Construction Management and Economics, 33*(8), 653–670. https://doi.org/10.1080/01446193.2015.1087646

Gunduz, M., & Abdi, E. A. (2020). Motivational factors and challenges of cooperative partnerships between contractors in the construction industry. *Journal of Management in Engineering, 36*(4), 1–10. https://doi.org/10.1061/(ASCE)ME.1943-5479.0000773

Guo, W., Lu, W., Hao, L., & Gao, X. (2021). Interdependence and information exchange between conflicting parties: The role of interorganizational trust. *IEEE Transactions on Engineering Management* [Preprint]. https://doi.org/10.1109/TEM.2020.3047499

Güth, W., Königstein, M., Marchand, N., Nehring, K. (2000). Trust and reciprocity in the investment game with indirect reward.

Hair, J. F., Black, W., Babin, B., & Anderson, R. (2010). Multivariate Data Analysis. Seventh, Pearson Education. https://doi.org/10.1016/j.ijpharm.2011.02.019

Hair, F., Jr., J., et al. (2014). Partial least squares structural equation modeling (PLS-SEM). *European Business Review, 26*(2), 106–121. https://doi.org/10.1108/EBR-10-2013-0128

Henseler, J., Ringle, C. M., & Sinkovics, R. R. (2009). The use of partial least squares path modeling in international marketing. *Advances in International Marketing, 20*(May 2014), 277–319.https://doi.org/10.1108/S1474-7979(2009)0000020014

Hetemi, E., Gemünden, H. G., & Meré, J. O. (2020). Embeddedness and actors' BehavIOR in large-scale project life cycle: lessons learned from a high-speed rail project in Spain. *Journal of Management in Engineering, 36*(6), 05020014. https://doi.org/10.1061/(asce)me.1943-5479. 0000849

Holmstrom, B. (1979). Moral hazard and observability. *The Bell Journal of Economics, 10*(1), 74. https://doi.org/10.2307/3003320

Hosseinian, S. M. (2016). An optimal time incentive/disincentive-based compensation in contracts with multiple agents. *Construction Economics and Building, 16*(4), 35–53.

Jelodar, M. B., Yiu, T. W., & Wilkinson, S. (2016). A conceptualisation of relationship quality in construction procurement. *International Journal of Project Management, 34*(6), 997–1011. https://doi.org/10.1016/j.ijproman.2016.03.005

Kumaraswamy, M. M., & Anvuur, A. M. (2008). Selecting sustainable teams for PPP projects. *Building and Environment, 43*(6), 999–1009. https://doi.org/10.1016/j.buildenv.2007.02.001

Kwawu, W., & Laryea, S. (2014). Incentive contracting in construction. In *Proceedings 29th annual association of researchers in construction management conference, ARCOM 2013*, pp. 729–738.

Laffont, J.-J., & Tirole, J. (1988). The dynamics of incentive contracts, *56*(5), 1153–1175.

Liu, J., et al. (2017). Evolutionary game of investors' opportunistic behaviour during the operational period in PPP projects. *Construction Management and Economics, 35*(3), 137–153. https://doi. org/10.1080/01446193.2016.1237033

Locke, E. A., Latham, G. P., & Erez, M. (1988). The determinants of goal commitment. *Academy of Management Review, 13*(1), 23–39. https://doi.org/10.5465/AMR.1988.4306771

Macneil, I. R. (1974). A Primer of contract planning.

Manata, B., et al. (2021). Documenting the interactive effects of project manager and team-level communication behavior in integrated project delivery teams. *Project Management Journal* [Preprint].

Meng, X., & Gallagher, B. (2012a). The impact of incentive mechanisms on project performance. *International Journal of Project Management, 30*, 352–362.

Meng, X., & Gallagher, B. (2012b). The impact of incentive mechanisms on project performance. *International Journal of Project Management, 30*(3), 352–362. https://doi.org/10.1016/j.ijp roman.2011.08.006

Nguyen, D. A., & Garvin, M. J. (2019). Life-cycle contract management strategies in US highway public-private partnerships: Public control or concessionaire empowerment? *Journal of Management in Engineering, 35*(4), 1–13. https://doi.org/10.1061/(ASCE)ME.1943-5479. 0000687

Oliver, C. (1990). Determinants of interorganizational relationships: integration and future directions published by: academy of management linked references are available on JSTOR for this article: Determinants of interorganizational relationships: Integration and future. *Academy of Management Review, 15*(2), 241–265.

Richmond-Coggan, D. (2001). Construction contract incentive schemes: Lessons from experience.

Ringle, C. M., et al. (2018). Partial least squares structural equation modeling in HRM research. *International Journal of Human Resource Management*, (January), 1–27. https://doi.org/10. 1080/09585192.2017.1416655

Rose, T., & Manley, K. (2011). Motivation toward financial incentive goals on construction projects. *Journal of Business Research, 64*(7), 765–773. https://doi.org/10.1016/j.jbusres.2010.07.003

Rowlinson, M. (2012). A practical guide to the nec3 professional services contract. *Wiley*. https://doi.org/10.1002/9781118406373

Saka, N., Olanipekun, A. O., & Omotayo, T. (2021). Reward and compensation incentives for enhancing green building construction. *Environmental and Sustainability Indicators*. Elsevier B.V. https://doi.org/10.1016/j.indic.2021.100138

Schieg, M. (2008). Strategies for avoiding asymmetric information in construction project management. *Journal of Business Economics and Management, 9*(1), 47–51. https://doi.org/10.3846/1611-1699.2008.9.47-51

Song, M., et al. (2020). Testing structural and relational embeddedness in collaboration risk. *Rationality and Society, 32*(1), 67–92. https://doi.org/10.1177/1043463120902279

Suprapto, M., et al. (2016). How do contract types and incentives matter to project performance? *International Journal of Project Management, 34*(6), 1071–1087. https://doi.org/10.1016/j.ijproman.2015.08.003

Suprapto, M., Bakker, H. L. M., & Mooi, H. G. (2015). Relational factors in owner-contractor collaboration: The mediating role of teamworking. *International Journal of Project Management, 33*(6), 1347–1363. https://doi.org/10.1016/j.ijproman.2015.03.015

Swärd, A. (2016). Trust, reciprocity, and actions: The development of trust in temporary inter-organizational relations. *Organization Studies, 37*(12), 1841–1860. https://doi.org/10.1177/0170840616655488

Vroom, V. H. (1964). *Work and motivation*. Wiley.

Wang, T. K., et al. (2018). Causes of delays in the construction phase of Chinese building projects. *Engineering, Construction and Architectural Management, 25*(11), 1534–1551. https://doi.org/10.1108/ECAM-10-2016-0227

Wigfield, A., & Eccles, J. S. (2002). Expectancy–value theory of achievement motivation. *Journal of Asynchronous Learning Network, 6*(1), 68–81. https://doi.org/10.1006/ceps.1999.1015

Williamson, O. E. (1979).Transaction-cost economics: The governance of contractual relations. *The Journal of Law and Economics, 22*(2).

Williamson, O. E. (1985). The Economic Institutions of Capitalism., China social sciences publishing house Chengcheng books ltd. https://doi.org/10.5465/AMR.1987.4308003

Yao, H., Chen, Y., Chen, Y., & Zhu, X. (2019). Mediating role of risk perception of trust and contract enforcement in the construction industry. *Journal of Construction Engineering and Management, 145*(2), 1–13. https://doi.org/10.1061/(ASCE)CO.1943-7862.0001604

Yu, A. G., Flett, P. D., & Bowers, J. A. (2005). Developing a value-centred proposal for assessing project success. *International Journal of Project Management, 23*, 428–436. https://doi.org/10.1016/j.ijproman.2005.01.008

Zhang, L., Huang, S., Tian, C., & Guo, H. (2020a). How do relational contracting norms affect IPD teamwork effectiveness? A social capital perspective. *Project Management Journal, 51*(5), 538–555.

Zhang, M., Lettice, F., Chan, H., & Nguyen, H. (2018). Supplier integration and firm performance: The moderating effects of internal integration and trust. *Production Planning and Control, 29*(10), 802–813. https://doi.org/10.1080/09537287.2018.1474394

Zhang, R., Wang, Z., Tang, Y., & Zhang, Y. (2020b). Collaborative innovation for sustainable construction: The case of an industrial construction project network. *IEEE Access, 8*, 41403–41417. https://doi.org/10.1109/ACCESS.2020.2976563

Zhang, S., Zhang, S., Gao, Y., & Ding, X. (2016). Contractual governance: Effects of risk allocation on contractors' cooperative behavior in construction projects. *Journal of Construction Engineering and Management, 142*(6), 1–11. https://doi.org/10.1061/(ASCE)CO.1943-7862.0001111

Zhu, L., Cheung, S., Gao, X., Li, Q., & Liu, G. (2020). Success DNA of a record-breaking megaproject. *Journal of Construction Engineering and Management, 146*(8), 05020009. https://doi.org/10.1061/(asce)co.1943-7862.0001878

Zhu, L., & Cheung, S. O. (2018). The implementation of incentive schemes in construction. In *11th International Cost Engineering Council (ICEC) world congress and the 22nd annual pacific association of quantity surveyors conference in 2018*, pp. 1–9.

Zhu, L., & Cheung, S. O. (2020). Power of incentivization in construction dispute avoidance. *Journal of Legal Affairs and Dispute Resolution in Engineering and Construction, 12*(2), 1–7. https://doi.org/10.1061/(ASCE)LA.1943-4170.0000368

Zhu, L., & Cheung, S. O. (2021a). Equity gap in construction contracting identification and ramifications. *Engineering, Construction and Architectural Management.* https://doi.org/10.1108/ECAM-09-2020-0725

Zhu, L., & Cheung, S. O. (2021b). Towards an equity-based analysis of construction incentivization. *Journal of Construction Engineering and Management* [In Press]

Zou, P. X. W., & Zhang, G. (2009). Managing risks in construction projects: Life cycle and stakeholder perspectives. *International Journal of Construction Management, 9*(1), 61–77. https://doi.org/10.1080/15623599.2009.10773122

Zou, P. X. W., Zhang, G., & Wang, J. (2007). Understanding the key risks in construction projects in China. *International Journal of Project Management, 25*(6), 601–614. https://doi.org/10.1016/j.ijproman.2007.03.001

Zeiss, G. (2007). Worldwide Challenges Facing the Construction Industry, http://www.geospatial.blogs.com/geospatial/2007/09/convergence.html. Accessed May 24 2018.

Liuying Zhu works at the School of Management, Shanghai University. Dr. Zhu received her M.Sc. degree (Distinction) and Ph.D. from the City University of Hong Kong. Her M.Sc. dissertation was awarded the 2016 Best Dissertation (Master category) by the Hong Kong Institute of Surveyors. Her doctoral study on construction incentivization was completed with the Construction Dispute Resolution Research Unit. Her current research focuses on construction project management, construction incentivization, and project dispute avoidance. Dr. Zhu has published articles in international journals and book chapters in these topics. Contact email: zhuliuying@shu.edu.cn.

Chapter 6
Multi-agent Incentivizing Mechanism for Integrated Project Delivery

Qiuwen Ma

Abstract Integrated project delivery (IPD) has been able to optimize the value for money for the owner by integrating diverse talents from the earliest design stage. The nucleus that contributed to the superior IPD performances is multi-agent risk/reward sharing incentive (RRSI). By overviewing the RRSI, four issues (i.e. setting target cost, incentives for non-cost performances, sharing ratios and caps of risk/reward) are important to IPD participants who involve RRSI. Moreover, a closer inspection of the multi-agent RRSI for IPD revealed that compared to any other procurement strategy, IPD has designed its RRSI with (1) a larger size of risk/reward pool, and (2) a greater amount of incentive pool. This explains why IPD empowers to achieve an effective multidisciplinary integration. However, not all the parties are inclined to join the RRSI, due to risk aversion or the concern of "inequity". To increase the participants' willingness to join the multi-agent RRSI, an optimum sharing model is proposed. With the application of concepts from cooperative game theory and prospect theory, the model can competently incorporate fairness in an optimum sharing of risk/reward. This chapter is helpful for the industry practitioners who are interested in the use of multi-agent RRSI in IPD projects.

Keywords Integrated project delivery · Multi-agent incentivizing mechanism · Risk/reward pool · Incentive pool · Optimum sharing

1 Introduction

Integrated project delivery (IPD) has been structured as an intelligent solution to solve the basic problems that often occurred when using the conventional project delivery methods (Cohen, 2010), i.e. adversarial relations among contract parties, inefficient project delivery, and expensive and low-quality product (Thomsen et al.,

Q. Ma (✉)
Construction Dispute Resolution Research Unit, Department of Architecture and Civil Engineering, City University of Hong Kong, Hong Kong, China
e-mail: qiuwen.ma@my.cityu.edu.hk

© The Author(s), under exclusive license to Springer Nature Switzerland AG 2023
S. O. Cheung and L. Zhu (eds.), *Construction Incentivization*, Digital Innovations in Architecture, Engineering and Construction,
https://doi.org/10.1007/978-3-031-28959-0_6

2009). High level of complexity and uncertainty of the projects would exacerbate these problems (Mesa et al., 2016). Conflicting interests and poor collaboration is considered the most important causes of project failure (Egan, 1998; Latham, 1994). To align the stakeholder interests with project success, IPD provides a contractual incentivization, in which the risk and reward is fairly shared among all contract parties.

In contrast to the traditional project delivery approaches, IPD is characterised by multi-agent risk/reward sharing incentive (RRSI) incorporated in the multi-party agreement (Thomsen et al., 2009). A minimum of owner, architect and main contractor should involve the contractual agreement, jointly setting the target cost and designing a pain/gain incentivization (Ahmed et al., 2021). Aligning the project success with the interests of contract parties via a multi-agent RRSI, real collaboration among the participants can be achieved (Pishdad-Bozorgi, 2017). First of all, RRSI can smartly solve a principal-agent problem wherein the interests of non-owner parties are not always to deliver a quality project (Nwajei et al., 2022). RRSI drives all stakeholders collaborate to achieve a "win–win" result, otherwise everyone will lose (Ross, 2003). There is no "win-lose" situation. Second, multi-agent RRSI contributes to closer team integration by bringing louder voices of downstream parties (i.e. main contractor and some key subcontractors) in the upstream design process. As the contractors' profits are commensurate with the project savings, they would be more willing to voice their opinions and solve the problems promptly. The active contractor involvement can stimulate creativity, improve the buildability, and smoothen the subsequent construction, thereby attaining value for money for the owner. Eventually, superior performances are observed in the IPD projects (Hanna, 2016; Ling et al., 2020). Reported by Cheng (2016), IPD enabled to save the owner's cost by about 20%. Moreover, the non-cost performances, such like project quality, schedule, safety and change management, were substantially improved by using IPD (El Asmar et al., 2013, 2016; Ibrahim et al., 2020).

The multi-agent RRSI used in IPD projects is different from the RRSI in any other procurement strategy. Some researchers argued that IPD imitated the RRSI from project alliancing (PA) (Lahdenperä, 2012), concluding that there are no actual differences between the RRSIs used in IPD and PA. Notwithstanding high level of similarities, a closer examination of these two collaborative project delivery methods revealed that the RRSI in IPD context is more ambitious for including the diverse expertise and bold in motivating creativity. A more significant number of parties are included in the IPD RRSI than PA RRSI. Specifically, the multi-agent RRSI in IPD extended the signatories from several key parties to dozens of risk/reward pool members (Cheng, 2016). Moreover, to encourage innovations, it is advocated for sharing the left contingency among IPD contracting parties (Liu et al., 2013).

Though widely recognized benefits of IPD, its uptake is still slow. The top IPD issue of concern is the use of RRSI (Ahmed et al., 2021). Moreover, few literatures have focused on how to design a suitable and satisfactory multi-agent incentivizing mechanism for IPD. To ease the reluctance of IPD participants to enter into the multi-agent RRIS, an optimum sharing model is proposed with a numerical example for detailed description.

To facilitate a better understanding of the incentivizing mechanism for IPD and aid in formulating a suitable RRSI for IPD practitioners, this chapter firstly reviews approaches used to design an acceptable multi-agent RRSI, based on which to further introduce an optimum sharing model. The second section reviews the RRSI, IPD definitions and principles, and then the multi-agent RRSI used in IPD. This is followed by an optimum sharing model on the theoretical basis of cooperative game theory and prospect theory. The practical implications and conclusion are presented in the final sections.

2   Overview of Risk/Reward Sharing Incentive

Risk/reward sharing incentive is also known as painshare/gainshare. The risk, or pain, refers to the monetary penalty caused by cost overrun, late completion or other underperformances. The reward, or gain, refers to the monetary reward caused by cost underrun, early completion, safety or other satisfactory results. The primary goal of risk/reward sharing incentive is to align the interests of project participants with the interest of project, thereby enhancing collaboration and ensure all the actions taken is for the "best of the project."

As for the constructs of RRSI, typically there are three limbs (Love et al., 2011; Ross, 2003) to form the RRSI (see Fig. 1). The first limb is the reimbursed cost, including the direct project costs and project overheads. It would be compensated to IPD partners no matter whether the project goals can be achieved or not. The second limb is called fee, also known as risk pool in IPD context. It consists of project profit and corporate overheads. The pre-established project objectives are the benchmark for the final payable fee, refunding all of which to the risk pool members if the project objectives are achieved otherwise a part or none at all. The third limb refers to the risk/reward sharing agreement, as shown in Fig. 1. It is also worth noting that the cost performance is not always linked with non-cost performances (as described in Component 3 in Fig. 1). In some cases, they are independent, wherein the owner would reserve an additional bonus for the non-cost performances.

When setting the RRSI, two issues concern the project participants who wanted to involve the RRSI. The first issue is whether they can set a reasonably challenging but also achievable target cost. When predicting the cost at the very early project stage, high level of uncertainty is characterised by the target cost. Moreover, it would seem to be beneficial for the owner to set the target cost as low as possible whereas beneficial for the contractor to have it as high as possible. The second issue is how to decide the sharing ratios of risk/reward among contracting parties. This issue can be analysed into two matters, i.e. how to set the sharing ratios between owner and non-owner participants, and the sharing ratios among non-owner participants. Considering the risk appetite (Hosseinian et al., 2020), pre-existing business relations (Melese et al., 2017) or some other factors (Han et al., 2019; Wang & Liu, 2015), some attempts have been made to optimize or propose reasonable sharing ratios of risk/reward. Besides, there are some generic frameworks that applied in the industry

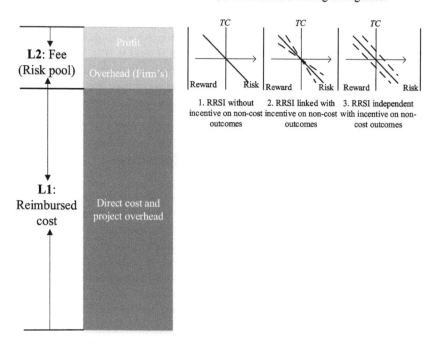

Fig. 1 Three components of risk/reward sharing incentive

for the risk/reward sharing (Department of Infrastructure & Regional development, 2015; Ross, 2003).

In the remainder of this section, the main focus is on the approaches to address these two concerns, i.e. (1) how to determine the target cost; (2) how to decide the sharing ratios of risk/reward among the contracting parties.

2.1 Setting Target Cost

Some approaches have been proposed to suggest how to set a target cost that's motivating, and also realistic. Underlying these approaches are two different ideas, i.e. (1) introduce competition for contractor selection to prevent a high target cost, (2) incentivize the contractor to propose a lower target cost.

One main criticism of collaborative delivery methods, e.g. IPD and project alliancing, is the lack of competition among service providers (mainly refer to the contractors). This is because almost all the key service providers are selected to involve the project at the outset of the project via interviews without price competition (Cheng, 2016). Moreover, the owner have shared the right/power with contractors to decide target cost. Mosey (2009) proposed a two-stage pricing approach to set the target cost by introducing competition into the early contractor involvement. The main contractor and key subcontractors are selected, based on competitive tendering, in the first of two stages to join the project design team, assisting in buildability, schedule and cost estimating (Mosey, 2009; Rahman, 2012). When the design is substantially complete and approved by the client, the contractor will offer a target cost with pain/gain incentivisation (or guaranteed maximum price) in the second of two stages to deliver the design. If the pricing offered is not satisfactory, an alternative contractor will be selected. The contractor automatically bears the risk of losing business when the pricing that he offered is unreasonably high. In this circumstance, a rational contractor will offer a reasonable target cost.

However, some researchers argued that introducing competition is against the spirit of collaboration advocated in an integrated team (Ross, 2003). Moreover, under the conventional contract agreements the tender price is only the starting point, and any extra cost will eventually be borne by the owner (Darrington & Lichtig, 2010). To incentivize the contractor to propose a reasonable target cost, Lahdenperä (2010, 2016a, 2016b) proposed a framework to formulate the two-stage target-cost arrangement. Before setting the target cost, a reference cost can be the initial target cost set by the owner or the pricing established by partial price competition (Lahdenperä, 2010), at the time of contractor involvement. In the first of two stages, the contractor involves in the project design and earns a reward (i.e. sharing of the difference between the reference cost and target cost, denoted as S_D) by suggesting a lower target cost than the reference cost after substantial design completion. In the second of two stages, the contractor shares the pain/gain (denoted the sharing ratio by S_R) according to the actual cost performance. To circumvent opportunistic promise, the second stage sharing ratio (S_R) for the contractor must be higher than the first stage (S_D) (Lahdenperä, 2016a).

As mentioned by Ross (2003), setting a reasonable target cost and making it agreed by all the contracting parties is the prerequisite to enter into the risk/reward sharing agreement. The next step for the contracting parties is to negotiate its key legal aspects.

2.2 Sharing Risk/Reward Among Contracting Parties

Assuming that the target cost is fairly established and the owner is very willing to proceed with the project that used RRSI, three key parts still needs to be decided to develop a risk/reward sharing agreement. They are (1) sharing ratios of risk/reward among the contracting parties, (2) incentivizing for non-cost performances, and (3) caps of risk/reward.

Sharing ratios: in practice the notional sharing of risk/reward is to share 50:50 between the owner and non-owner parties, while the sharing ratios among non-owner parties are decided according to the cost structure in Component 1 (see Fig. 1) (Ross, 1999), or the construct of fees that put into Component 2 (Department of Infrastructure & Regional development, 2015; Ross, 2003), and/or their influence on the project's outcome (Hall & Bonanomi, 2021; Love et al., 2011). Specifically, in some projects, the designers' risk/reward sharing ratios are increased given their impacts on project outcomes (Davis & Love, 2011).

When determining the sharing ratios according to the individual's cost structure in Component 1, the contractor would share most of the risk/reward. If the sharing ratio is determined to be proportional to the percentage of fees that kept in the risk pool, the designers' sharing ratios are expected to be improved, as their overheads and profit margins (15–55% according to Ross, 2003) are generally higher than the contractors' (around 10% according to Ross, 2003). In certain cases, the designers' sharing ratios of risk/reward are enlarged to recognize their high level of influence on the project outcomes (Love et al., 2011). Moreover, Ross (2003) stated that most of the designers are risk averse and reluctant to accept higher shares of risk/reward. Hence, to ease the reluctance and motivate the designers, it is suggested to increase the reward shared percentages and remain the risk shared percentages unchanged for the designers. When approaching the RRSI in such a way, however, the clients and contractors may consider that it violates fairness. This is because the risk sharing ratios should be in line with reward ratios, which is important to maintain the perceptions of equity (Love et al., 2011).

Non-cost performance: one of the main tasks during the agreement development period is to identify the non-cost key result areas (KRAs) for the project, and develop a benchmark to measure KRAs. It is encouraged to use the overall performance score (OPS) to measure the performances located in the KRAs (Love et al., 2011; Ross, 2003). Specifically, each KRA is assigned a weight (%) with the sum of weights being 100%. Each KRA is measured across a performance spectrum, whereby 50 refers to basic performance, and 0 and 100 refers to bottom end of failure and top end of outstanding respectively.

Concerning the approaches of sharing risk and reward among the contracting members, they can be categorized into linear and non-linear shared ratios. The linear shared ratios refer to the shared ratios of cost over or underruns. It keeps unchanged regardless of the non-cost performance. The non-linear one represents the shared ratios that are varied according to the key non-cost performance (Ross, 1999). In other words, the linear shared ratios mean that the incentivization for cost and non-cost performances are independent, whereas the non-linear shared ratios represent that non-cost performances linked with the cost performance (see Fig. 1). Nonlinear shared ratios are mainly used to motivate the non-owner parties to improve the non-cost project performance (Victorian Department of Treasury and Finance 2006), as the non-owner parties will not sacrifice key non-cost performance to achieve better cost outcomes if the cost and non-cost performances are interlinked (Ross, 1999; Thomsen et al., 2009). Moreover, Ross (2003) suggested to link the non-cost performance with the reward sharing.

Caps: to limit the loss of non-owner parties and safeguard the owner from the excessively high target cost, the downside risk and upside reward are usually capped in the RRSI (Love et al., 2011). In most of the cases, the cap of risk is the direct project cost, i.e. Component One described in Fig. 1. The upside cap for the non-owners, or the cap of reward should be negotiated and mutually agreed between the owner and non-owner parties. Normally, the maximum upside cap is OPS% × Target Cost (Ross, 2003). Besides, to stimulate innovations and real collaboration, some practitioners recommended that the cap on upside reward should be removed in the RRSI (Department of Infrastructure & Regional development, 2015; Department of Treasury & Finance, 2010; Ross, 2003). The tangible incentive would encourage the non-owner parties to endeavour for better performances.

3 Revisiting Multi-agent Risk/Reward Sharing Incentive for IPD

3.1 Definition, Origin and Characteristics of IPD

In general terms, IPD can be defined as (American Institute of Architects (AIA) California Council 2007) *a project delivery approach that integrates people, systems, business structures and practices into a process that collaboratively harnesses the talents and insights of all participants to reduce waste and optimize efficiency through all phases of design, fabrication and construction.*

IPD is considered to be initially evolved from an oil exploration project in the North Sea (Lahdenperä, 2012). At that time, the proposed oil exploration project was facing highly cost and technical uncertainty. In response to these uncertainty, the concept of PA was proposed as a collaborative project delivery method to cooperatively and flexibly deliver the project (Ashcraft, 2011). After its success, the capability of PA to deliver highly risky and complex projects has been recognized. Nowadays it has been extensively used in the Australian public infrastructure sector (Department of Infrastructure & Regional development, 2015; Walker et al., 2015). Moreover, the RRSI adopted in PA projects is the most acknowledged practice to achieve greater collaboration and closer integration (Ross, 2013).

Emigrating from Australia (Lahdenperä, 2012), the first term of IPD emerged in the US in 2003, when an enterprise, Westbrook Air Conditioning and Plumbing of Orlando, and its team members applied a delivery approach that seek to bond them together via a contractual agreement (Forbes & Ahmed, 2010; Matthews & Howell, 2005). Actually, the project delivery method used in their projects was design-build (DB), not "true IPD" (Lahdenperä, 2012; Matthews & Howell, 2005). In 2004 Sutter Health planned to test out a new procurement model to combat inefficiency of the construction industry (Cohen, 2010; Lichtig, 2005a; Mauck et al., 2009). The special counsel of Sutter Health, Mr. Lichtig, developed an integrated form of agreement

(IFOA) for Sutter Health (Kagioglou & Tzortzopoulos, 2010). IFOA initially incorporated the elements of lean project delivery and multi-party contract. It was primarily designed to align the interests of the project participants with the project goals, largely contributing to the project success (Lostuvali et al., 2014; Mauck et al., 2009). Meanwhile, using a multi-party contract, the "true IPD" can be considered to be adopted by Sutter Health.

Following the success of IFOA, a joint effort of 22 organizations, i.e. the ConsensusDocs 300, became the first standard contractual form for IPD in 2007 (ConsensusDocs LLC 2007). It is also a multi-party agreement that should be signed among owner, main contractor and designer and other key participants (ConsensusDocs LLC 2007; ConsensusDocs LLC 2012). At the same time, AIA published a series of supporting documents for IPD, including two of the most widely used multi-party contracts, AIA C191 and AIA C195. Moreover, Hanson Bridgett LLP also issued a tri-party IPD contract (i.e. Hanson Bridgett standard IPD agreement) among owner, architect and contractor. With the development and extensive use of IPD multi-party contract by practitioners, it seems like the euphoria around IPD came across the construction industry (Cheng, 2016; Korb et al., 2016; Ling et al., 2020; Walker & Rowlinson, 2019).

Except the definition of IPD stated above, some researchers have attempted to define IPD according to its most unique element, i.e. multi-party contract. For example, Cohen (2010) refers IPD to *a method of project delivery distinguished by a contractual arrangement among a minimum of owner, constructor and design professional that aligns business interests of all parties.* Moreover, they use multi-party contract to differentiate "true IPD" from "IPD-ish" projects. Specifically, the multi-party contract embodies a multi-agent RRSI among a minimum of owner, main contractor and architect, and equips the contracting parties with reduced liability and fiscal transparency (Cheng, 2016). If the multi-party contract is applied in an IPD project, it can be considered to be "true IPD", otherwise "IPD-ish" project (National Association of State Facilities Administrators (NASFA) et al. 2010).

Even though diverse definitions of IPD have been given by researchers, the similar IPD elements have been found in IPD relevant researches (Kent & Becerik-Gerber, 2010). There are three most well-recognized IPD elements, including early involvement of key participants, multi-agent RRSI, and multi-party contract (NASFA et al. 2010; Kent & Becerik-Gerber, 2010; Azhar et al., 2015; Lahdenperä, 2012; AIA California Council 2009).

Early involvement of key parties: designer, main contractor, key subcontractors and suppliers should be engaged and involved from the early design stage (AIA National & AIA California Council, 2007). One of the fundamental changes of IPD from the traditional procurement methods is to build integrated organization, in which the knowledge of all experts including the downstream contractors and suppliers can be brought together (Laryea and Watermeyer, 2016). The early involvement of contractors can contribute to more reliable cost estimation, better construction methods and sequences, while the suppliers can help with the equipment and material selection (Cheng, 2016; Cheng et al., 2012; Cohen, 2010). In particular, the contractors can help increase the cost predictability by providing cost estimation of alternative designs, updating cost estimating; improve the schedule reliability by creating and updating the construction milestone schedule (Hall & Scott, 2016). The contractor is also the main contributor to conduct constructability review and help the architects incorporate the relevant findings into the designs (Aashto Subcommittee on Construction, 2000).

In addition, continuous owner involvement is of the top criticality to IPD success (Ma et al., 2022a). The main difference between IPD and DB is that the IPD owner has been actively and continuously involved, especially during the intensified design process, identifying the value of design items and selecting the most suitable design alternative. Some practitioners even claimed that IPD or PA should be recommended as the preferred project delivery approach only if the owner is smart and capable (Department of Infrastructure & Regional development, 2015; Walker & Rowlinson, 2019).

Multi-party contract: the key parties within the RRSI structure are bound together through a multiparty agreement including a minimum of the owner, main contractor and architect. The agreement should also include key consultants and trade contractors. They can be brought into the IPD agreement through subagreements with the contractor and architect, or can be included as signatories by "joining agreement" amendments (AIA, 2007; NAFSA, 2010). In this contract, all elements are clearly stated and normally include incentives and risk sharing, payment method, dispute resolutions, and the responsibilities of all involved parties (O'Connor, 2009). Moreover, a single multiparty contract can facilitate the communication more efficiently and smoothly between the non-client participants, including architect, contractor, architect's consultant, subcontractors and suppliers (AIA California Council, 2007).

IFOA, the ConsensusDocs 300, AIA C191 and AIA C195, and Hanson Bridgett IPD standard agreement are the most commonly used contracts for IPD projects in the present day (Kent & Becerik-Gerber, 2010; Liu et al., 2013). IFOA, ConsensusDocs 300 and Hanson Bridgett IPD contract are multi-party contract wherein three or more parties can directly enter into the agreement. AIA C191 is a tripartite contractual agreement that would be signed by owner, main contractor and architect at the inception of the IPD project, and in AIA C195 a single purpose entity would be established, with which the key participants should sign the contracts respectively (AIA, 2008, 2009; Dal Gallo et al., 2009).

Multi-agent RRSI: As stated above, the multi-agent RRSI for IPD is operationalized by constructing the risk pool (i.e. Component 2 in Fig. 1) through retaining all or part of the profits from multiple non-owner parties who involved in the RRSI (Zhang & Li, 2014). If there are some cost savings, some portions will be added to risk pool. If there is cost overrun, risk pool will be used to pay for the project cost until its exhaustion (Ashcraft, 2011). The owner bears the risk of cost overrunning the amount of risk pool. The profile of the multi-agent RRSI, i.e. the risk pool members, risk/reward sharing ratios, incentive and disincentive for non-cost performance, and caps for risk/reward, should be negotiated in the preconstruction stage (Cheng et al., 2012, Dal Gallo et al., 2009). Generally, the risk pool members include at least architect and main contractor (AIA National & AIA California Council, 2009). In certain cases, other parties, such as design consultants, trade contractors and suppliers can also attend the risk pool team with the permission of its primary members (Lichtig, 2005a, 2005b). In the following section, the key features of multi-agent RRSI for IPD is described in detail.

3.2 Key Features of Multi-agent Risk/Reward Sharing Incentive in IPD Context

Multi-agent RRSI is the core tenant of IPD, and also the key trait that distinguishes IPD from other procurement strategies. Though the RRSI is also adopted in PA projects, a widely accepted definition of PA that explicitly includes multi-party contract or multi-agent RRSI is lacked. Moreover, some of the early PA adopters (like in 1990s) only involved two parties (i.e. the owner and the main contractor) in their RRSI (Ross, 2003; Walker et al., 2002). In contrast, analyzing the IPD cases reported by Cheng (2016), all projects used some form of IPD agreement with a RRSI that included more than three parties, some even included 24 parties (Akron Childen's Hospital, Kay Jewelers Pavilion, see Table 1). That being said, IPD is the most ambitious procurement strategy to harness the talents and insights of all participants. Another noteworthy matter is that in addition to the cost savings and non-cost performance incentive, the contingency funds that remained after the project completion are encouraged to be shared among the team members (Liu et al., 2013; Thomsen et al., 2009). Hence, from the incentivizing view, the incentive pool in IPD is greater in amount.

To summarize, two key features of multi-agent RRSI for IPD are (1) large size of risk/reward pool, and (2) great amount of incentive pool. It should be noted that the risk/reward pool and risk pool are completely different terms. The risk/reward pool is referred to a collection of parties that involve the multi-agent RRSI in IPD projects. Risk pool represents the 2rd component of RRSI (see Component 2 in Fig. 1).

First, to disclose the construct and size of the risk/reward pool in IPD projects, we analysed the IPD cases reported by Cheng (2016) (See Table 1). It was found that all projects used some form of IPD agreement with a RRSI that included more than three parties, some even 23 parties (Akron Childen's Hospital, Kay Jewelers Pavilion). As presented in Fig. 2, the multi-party contract ties together the parties in the signatory pool, which can be largened by bringing the trade contractors, engineers or even some suppliers (such like furniture manufacturer) through subagreements, i.e. extended risk/reward pool. A larger reward/risk pool team is advantageous to bringing different voices. For instance, the design firms claimed that the involvement of trade contractors in the design process empowered to model collaboration for architects, who can more easily obtain the updated projected budget and identify the value-added items.

Two commonly occurred issues concerned the risk/reward pool members when the pool team is larger than using the conventional procurement strategies. First, the coordination of a large risk/reward pool team required some extra time and

Table 1   Multi-agent RRSI for IPD projects

No	Project name (building type)	Form of contractual agreement	Parties involved in the multi-agent RRSI		
			Signatory pool	Extended risk/reward pool	Sum
1	Akron children's hospital, kay jewelers pavilion (healthcare)	Customized multi-party agreement with elements of ConsensusDocs 300 and AIA 195	Five parties: Owner; National architect, Local architect; National contractor, Local contractor	18 parties: Five engineers; 13 trade contractors	23 parties
2	Autodesk building innovation learning and design space (office)	Customized Hanson Bridget IPD agreement	Seven parties: Owner; Architect; Main contractor; Two engineers; Two trade contractors	None	Seven parties
3	Mosaic centre for conscious community and commerce (office)	Customized Hanson Bridget IPD agreement	Three parties: Owner; Architect; Main contractor	11 parties: Three engineers; Eight trade contractors	14 parties
4	Quail run behavioral health hospital (healthcare)	ConsensusDocs 300	Seven parties: Owner; Architect; Main contractor; Four trade contractors	None	Seven parties
5	Rocky mountain institute innovation center (office)	Customized Hanson Bridget IPD agreement	Three parties: Owner; National architect; Main contractor	Nine parties: Local architect; Two engineers; Five trade contractors; Three consultants	22 parties
6	St. Anthony hospital (healthcare)	Customized two-party agreement with an IPD owner, program manager (owner's representative), architect, and contractor	Four parties: Owner; Owner representative; architect; Main contractor	None	Four parties

(continued)

Table 1 (continued)

No	Project name (building type)	Form of contractual agreement	Parties involved in the multi-agent RRSI		
			Signatory pool	Extended risk/reward pool	Sum
7	Sutter medical office building: Los Gatos (office)	Customized IFOA	Three parties: Owner; National architect; Main contractor	Four parties: Four trade contractors	Seven parties
8	T. Rowe price owings mills campus building 1 (office)	Customized Hanson Bridget IPD agreement	Eight parties: Owner; Architect; Main contractor; One engineer; Four trade contractors	One party: trade contractor	Nine parties
9	Wekiva springs center expansion (healthcare)	ConsensusDocs 300	11 parties: Owner; Architect; Main contractor; One interior design; seven trade contractors	None	11 parties

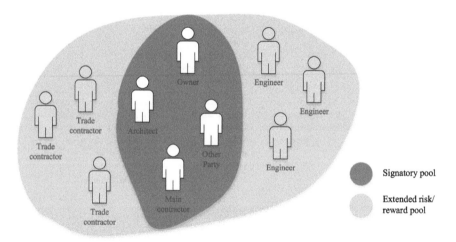

Fig. 2 Signatories and risk pool members in IPD projects (adapted from Cheng, 2016)

efforts. Moreover, the consensus decision making is encouraged to be applied in IPD context. It takes more time to arrive at the final solution than the authority decision making, sometimes leading to late timing. Hence, the big size of risk/reward pool made the decision making more inefficient and slower. At this time, it is important to have a clearly defined, strong and sustained leadership to organize the team of highly diversity (Ma et al., 2022a). Second, the involvement timing of risk/reward pool members is of importance. Because of the complex consensus decision making process or lack of integration experience, some key trade partners were involved late. It is believed that earlier timing would be more beneficial to the value creation.

Concerning the great amount of incentive pool, it can be proved that the remained contingency funds are encouraged to be shared among the risk/reward pool members (Thomsen et al., 2009). According to Liu et al. (2013), the contingency of IPD projects can serve two functions, i.e. covering uncertainty and incentivization, when it is not included in the target cost (see Option C, p7). Some IPD experts argued that freeing the contingency funds from the incentive pool (for example, separating the contingency from target cost or owning contingency it by owner) would seem like to encourage the non-owners build in additional contingency to monetize the project risk. In contrast, sharing the reserved contingency can make the stakeholders have more "skin in the game", thereby achieving the greatest extent of innovations.

3.3   Research Models for Multi-agent Risk/Reward Sharing Incentive

It has become a heated debate about how to set the RRSI, especially determining the sharing ratios of risk/reward among the contracting parties. Attempts have been made to provide the optimum RRSI from different research directions. RRSI is primarily used to solve the principal-agent problems. Thus, the principal-agent theory has been often used to identify the optimal sharing ratio (Chang, 2014; Hosseinian et al., 2020; Liu et al., 2016; Wang & Liu, 2015). The principle underlying their proposed principal-agent models is to maximize the client's utility considering the contractor is rational. The second school of research is to use the concepts of game theory, e.g. evolutionary game (Zheng et al., 2017), Nash equilibrium (Medda, 2007; Zhang & Li, 2014), cooperative game (Pishdad-Bozorgi & Srivastava, 2018), to produce a fair and stable risk allocation between the contracting parties. Moreover, some researchers aim to solve the risk sharing problems using real option valuation analysis (Boukendour & Hughes, 2014; Han et al., 2019; Liu et al., 2014). This analytical approach can provide a value of the options offered by certain parties and then functions as benchmark for risk allocation (Liu & Cheah, 2009).

However, considering the features of IPD projects, e.g. high level of integration and uncertainty, it may be inappropriate to apply the existing models in the multi-agent RRSI for IPD. First, the risks and uncertainties embedded in IPD projects were overlooked. Second, the collaborative nature of IPD, especially the continuous involvement of owner, was not fully considered. Third, some researches modelled the agents' economic behaviour based on expected utility, which were found to be inconsistent with the individual's real-world choices.

4 Optimum Model of Multi-agent Risk/Reward Sharing Incentive

4.1 Model Development

The RRSI for IPD is modelled as a cooperative game involving at least three risk/reward pool members. We aim to provide the optimal sharing solution that is acceptable, impartial and of the maximum utility for every risk/reward pool member. The theoretical bases are stochastic cooperative game theory (Suijs et al., 1999b, 1998) and prospect theory (Kahneman & Tversky, 1979). Stochastic cooperative game theory considers the uncertainty embedded in IPD projects and the extra value created by collaboration. Prospect theory illustrates the subjective utility of the uncertain outcomes.

To describe the agents' risk attitudes, a commonly used parametric form, i.e. the power utility function is applied, as shown below

$$v_i(x) = \begin{cases} x^{\alpha_i}, x \geq 0, \alpha_i > 0 \\ -\lambda_i(-x)^{\beta_i}, x < 0, \beta_i > 0, \lambda_i > 1 \end{cases} \tag{1}$$

where α_i is i's gain concavity coefficient of for the player,

λ_i represents the loss aversion coefficient,

and β_i describes the loss convexity coefficient.

According to Suijs et al. (1999a, 1999b), the sharing of stochastic payoff X^S to the team members can be represented by the pair of $(d_i, \hat{r}_i)$ for each agent $i \in S$. $\hat{r}_i$ represents the team member i's optimal sharing ratios of the maximum utility. d_i is the transfer payment that the team member i should transfer to others for the difference between the optimal and notional shares. Specifically, if the optimal share is different from the notional one, "transfer payment" is performed to make up the gap. The commonly used notional share (denoted as r_i'), as stated above, refers to sharing the risk/reward according to the fee structure (see Component 2 in Fig. 1). The notional share (Cheng, 2016; Ross, 2003) is seen as an impartial sharing as it is proportional to the cost structure of risk/reward pool members. When the optimal and notional shares are different, transfer payment can be introduced to ensure both optimality and impartiality of the RRSI.

Thus, to get the optimal RRSI, three steps should be taken, i.e. quantify the uncertain project outcome, obtain the optimal risk/reward sharing ratio $\hat{r}_i$, and calculate the transfer payment d_i in terms of r_i' and $\hat{r}_i$.

Quantification of uncertain outcome

According to prospect theory (Tversky & Kahneman, 1992), the indi-
vidual/organization usually sensitizes to the states of the outcomes in regard to the
reference point. The reference point refers to the target cost herein. Specifically, if
the project cost underruns the target cost, the project outcome is regarded as reward
or gain; if the project cost overruns the target cost, the project outcome is regarded
as risk or pain.

Let the prospect reward and risk perceived by the individual i be denoted by G_i
and L_i respectively. r_i represents the sharing ratio for the agent i. The prospect value
for i, therefore, can be presented as

$$v_i(r_i X^S) = \int_0^{TC} (r_i(TC - x_{AC}))^{\alpha_i} f(x_{AC}) dx_{AC}$$

$$- \lambda_i \int_{TC}^{+\infty} (-r_i(TC - x_{AC}))^{\alpha_i} f(x_{AC}) dx_{AC} \qquad (2)$$

where TC and x_{AC} refers to the target cost and actual cost of the IPD project
respectively,

$\alpha_i (0 \leq \alpha_i \leq 1)$ represents i's gain concavity coefficient,

$f(x_{AC})$ is the probability distribution function of the project actual cost,

λ_i represents i's loss aversion coefficient,

and $\beta_i (0 \leq \beta_i \leq 1)$ represents i's loss convexity coefficient.

Optimal RRSI

According to the stochastic cooperative game theory, the sharing solution is Pareto
optimal if and only if it maximizes the sum of certainty equivalent for the contracting
parties. The Pareto optimality can be achieved if X^S is the stochastic payoff for the
team S and it is impossible to improve the prospect value of one member without the
loss of another one (Suijs, 2000; Zhang & Li, 2014). Because d_i is a deterministic
payment, the certainty equivalent of d_i(i.e. $m_i(d_i)$) is d_i by itself. Moreover, as the
payments are transferred within the team S, $\sum_{i \in S} d_i = 0$. Thus, $\max \sum_{i \in S} m_i(d_i + r_i X^S)$
can be further deduced as $\max \sum_{i \in S} m_i(r_i X^S)$. The function of the certainty equivalent
is therefore presented as:

$$\max \sum_{i \in S} m_i(r_i X^S) \qquad (3)$$

s.t. $\sum_{i \in N_n \cup N_o} r_i = 1$;

$r_i > 0$, for all $i \in S$.

Transfer payments

When the optimal and notional shares are different, the best sharing solution can be achieved by sharing in the optimal ratios. The commonly used notional share (Cheng, 2016; Ross, 2003), however, is seen as an impartial sharing as it is proportional to the cost structure of risk/reward pool members. To ensure impartiality and bridge the difference between the optimal sharing ratio and the notional one, the transfer payment is introduced here. The transfer payment d_i for i can be understood as the side payment for the exchange of shares of risk/reward between team members to achieve the optimal sharing. Specifically, the concept of risk sharing zone (Melese et al., 2017) is applied.

Assume the transfer payments are performed among multiple agents, the zone of transfer payment between the agent i and j can be presented as,

$$v_j^{-1}(v_j((\hat{r}_j + \Delta r)x)) - v_j^{-1}(v_j(\hat{r}_j x)) \leq -d_i$$
$$= d_j \leq v_i^{-1}(v_i((r_i' + \Delta r)x)) - v_i^{-1}(v_i(r_i' x)) \tag{4}$$

where i and j refers to the agent who buys and sells the share of risk/reward respectively, d_i and d_j is the transfer payment for the agent i and j respectively, Δr is the share of risk/reward for exchange between two agents.

4.2 Numerical Example

An example of modelling results is provided for illustration purposes. Specifically, this illustrative example is used to demonstrate the value that the optimal risk/reward sharing solution provides for the contracting parties with divergent risk propensity. A stylized multi-agent RRSI is used for demonstration. IFOA contract was used for this IPD project and the contracting parties, i.e. owner, architect and major contractor involved in the RRSI. The risk/reward was shared according to the notional sharing approach.

The IPD project outcome and the risk aversion of the agents are shown as below:

- The target cost of this IPD project is set to be 2730 (million HKD), the actual cost is evaluated to be uniformly distributed on (2630, 300); the actual cost is evaluated to be uniformly distributed on (2730, 300) if DBB is used;
- The fee (including profit and firm's overheads) for architect and contractor is 1:9; Proportional to the fee structure, the notional risk/reward sharing for the owner, architect and main contractor is [0.5, 0.05, 0.45];
- Parameters describing owner's risk preference, gain coefficient $\alpha_o = 0.91$, loss coefficient $\beta_o = 0.92$, loss aversion index $\lambda_o = 2.27$;
- Parameters describing contractor's risk preference, gain coefficient $\alpha_c = 0.89$, loss coefficient $\beta_c = 0.93$, loss aversion index $\lambda_c = 1.86$;

Table 2 RRSI and the certainty equivalent of prospect values

Sharing approaches	IPD (notional RRSI)	IPD (optimal RRSI)
Sharing ratios for [owner, contractor, architect]	[0.5, 0.45, 0.05]	[0.316, 0.681, 0.003]
Transfer payments for [owner, contractor, architect]	Nil	[0.504, -0.568, 0.064]
Certainty equivalent of (Owner, MC, Architect)	(1.583, 3.914, -0.215)	(1.667, 4.126, 0.084)

- Parameters describing architect's risk preference, gain coefficient $\alpha_a = 0.88$, loss coefficient $\beta_a = 0.92$, loss aversion index $\lambda_a = 2.3$.

According to Eqs. (3) and (4), the optimal sharing vector equals $R^* = [0.316, 0.681, 0.003]$, and the transfer payment for the owner, contractor and architect is 0.504, -0.568, and 0.064 million respectively. This is to say, the optimal RRSI is to allocate 31.6%, 68.1% and 0.3% of the final project outcome for the owner, contractor and architect respectively; meanwhile, the contractor's compensation would be decreased (0.568 million), and the architect's compensation would be increased (0.064 million).

Table 2 presents the certainty equivalent of agents' prospect values when the notional and optimal sharing approaches are used. Apparently, the certainty equivalent is higher for every agent when adopting the optimal RRSI. Therefore, the optimal RRSI can provide a multi-win solution for the contracting parties.

5 Practical Implications

5.1 Design of Incentive Mechanism for IPD Projects

To formulate a suitable incentive mechanism for IPD projects, three aspects are of vital importance, i.e. select the right risk/reward pool members, set an optimal sharing ratios of risk/reward, and have a clear-cut benchmarking of non-cost performance for incentive and disincentive. It would not be that inefficient to make the consensus decisions if the right risk/reward pool members are selected. The risk/reward pool members can be incentivized should the sharing ratios be set optimal. Moreover, an easily understood measurement metrics of key non-cost performance aids in achieving the value for money for the owner.

Regarding the selection of risk/reward pool members, the firms with flatter organization charts, good understanding and experience of IPD, experience of multi-agent RRSI (or multi-party agreement), integrated design, lean practices and Building Information Modelling (BIM), sometimes with the previous partnership are preferred (Cheng, 2016). It was found that the industry practitioners' (with VS without IPD experience) understanding of the use of IPD is different (Ma et al., 2022b). The experienced IPD participants are more optimistic about IPD adoption, and more acknowledged the advantages that IPD affords. Moreover, having integration experience, the practitioners enable to understand the key factors in and real challenges to IPD success, smoothening the project delivery. Understanding the importance of integration experience, the owner would like to select the experienced partners, and limit the number of novice IPD users in his team. However, with some legal and industrial cultural barriers, the IPD uptake is slow and not all the construction firms are equipped with integration experience, especially the firms in the developing countries, e.g. China and Malaysia (Durdyev et al., 2019; Ma et al., 2022b; Walker & Rowlinson, 2019). Under this circumstance, Rowlinson (2017) contended that for the early IPD adopters, the willingness to use IPD, or adopt the integrated practices is considered to be an important criterion to select IPD partners. In addition, some foreign IPD consultants can be invited to assist the owner ensuring that the team is adhered to IPD principles. Another interesting discovery to mention is that despite that having previous partnership is believed to expedite cultivating a collaborative team, no evidence was found to support this view (Ma et al., 2022a). Instead, the future collaboration opportunities between IPD partners were proved to have positive impacts on risk management, especially for integration specific risks, e.g. project participants' resistance to take extra/new responsibilities in the integrated project.

As for the sharing ratios of risk/reward, this chapter proposed an optimal sharing model to increase the participants' willingness to join the multi-agent RRSI. Currently, the most commonly used sharing approach is the notional sharing method whereby the owner and non-owner share 50:50 risk/reward, and the sharing ratios among non-owner parties are proportional to their fee structure (see Component 2 in Fig. 1). The notional sharing approach without consideration of participants' risk attitudes may scare away the parties who are interested in IPD but may be risk averse. Indeed, some design firms are reluctant to join the RRSI. The researchers argued for higher reward sharing portions with the same risk portions for designers, considering their risk attitudes and influence on project outcomes (Love et al., 2011; Ross, 2003, 2013). However, this is against the interests of owner and contractors and also the violates the "equity" principle (Zhu & Cheung, 2021). The proposed sharing model set the notional approach as a fair stating point, and further use the "Pareto Optimality" to achieve the optimal share of risk/reward given the risk attitudes of risk/reward pool members. However, the optimal sharing ratios may be different from the notional one which is regarded as fair from the view of "equity". To bridge the difference between the optimal and notional sharing, the concept of transfer payment is introduced. Thus, the most satisfactory solution is to share the risk/reward according to the optimal sharing ratios, with payments transferred between the risk/reward pool members. In such a way, the fairness can be incorporated in an optimal RRSI.

Lastly, it is critical to have a well-defined benchmark metrics to boost the non-cost performances. The owner is suggested to develop clear, ambitious and also achievable project goals tied to the metrics, during the process of establishing a risk pool. For example, in the project of Akron Children's Hospital, Kay Jewelers Pavilion, the incentive is tied to the metrics of safety, quality and user-group satisfaction. It is believed that the straightforward and specific success metrics developed is the key to effectuate the incentivizing mechanism (Cheng, 2016, p 46).

5.2   Coupling BIM and Blockchain with IPD Incentivising Mechanism

In recent years, some attempts have been devoted to integrating BIM and blockchain with the use of IPD (Elghaish et al., 2019, Hunhevicz et al., 2020). BIM aims to attain greater collaboration for IPD (Ma & Ma, 2017), and the blockchain technology can be used to achieve the fiscal transparency and ease mistrust among participant for IPD (Hunhevicz & Hall, 2019, Hunhevicz et al., 2020).

BIM is poised to provide an integrated platform for IPD because of its capability to radically improve collaboration among the wide ranging and expertise (Kent et al., 2010). The implementation of BIM in IPD projects includes interdisciplinary communication and coordination, cost estimation and control, scheduling, virtual fabrication and construction (Goulding et al., 2014; Monteiro et al., 2014). Greater use of BIM was found to not only result in enhanced IPD performances, but contribute to a more positive perception of BIM empowering IPD (Ahmad et al., 2019; Azhar et al., 2015). Moreover, BIM has been applied in reducing the cash flow risk for IPD projects (Elghaish et al., 2019, 2023). As the profits of all risk/reward pool members should be put in the risk pool until the project completion, the cash flow risk concerns the parties involved the multi-agent RRSI (Ma et al., 2022a). Linking cost and schedule data to BIM elements, the model proposed by Elghaish et al. (2019) can provide risk/reward pool members with the estimated cash inflow according to the updated information of project execution.

Blockchain represents an evolving technology for distributed and secure recording and sharing of information. As blockchain enables to generate immutable and trusted storage of data, it is suggested to build the IPD incentive system with smart contract (Hunhevicz et al., 2020). Moreover, Elghaish et al. (2020) used the Blockchain technology to automatically perform the financial transactions and reveal the effect of project outcomes on the three components of IPD RRSI (see Fig. 1). It seems like the main application of blockchain suggested is to monitor the IPD cost performance. Observations of the commons between blockchain and IPD suggest that more attempts can be made in the future to explore the potential of blockchain, such like the distributed governance in IPD context (Hunhevicz et al. 2020, Hunhevicz & Hall, 2019).

6 Summary

IPD as a novel project delivery approach aims to optimize problem solving and decision making by harnessing diverse talents, knowledge and skills from the outset of project. Unlike traditional procurement methods, IPD align stakeholder success with achievement of the pre-established project objectives via multi-agent RRSI (Darrington & Lichtig, 2010; El-adaway et al., 2017). Tying the stakeholder interests, multi-agent RRSI directly addresses the problem of adversaries, leads to physiological safety and mutual trust, facilitates real collaboration and effective teamwork, and eventually brings superior project performances (Ibrahim et al., 2020; Ling et al., 2020). After more than two decades' development, multi-agent RRSI has been recognized as an effective incentivizing mechanism to boost IPD success (Hughes et al., 2012; Meng & Gallagher, 2012; Su et al., 2021; Whang et al., 2019).

By overviewing the elements of RRSI and multi-agent RRSI, as well as the characteristics of IPD, this chapter identifies two features of multi-agent RRSI for IPD, i.e. large size of risk/reward pool and great amount of imcentive pool. This reveals that IPD is the most ambitious procurement strategy to provide a diversity of voices, talents and insights. However, the notional sharing approach of RRSI is not as appealing as expected. To increase the willingness of IPD participants to join the multi-agent RRSI, we proposed an optimal sharing model based on stochastic cooperative game theory and prospect theory. Using the concept of "Perato Optimality" and introducing "transfer payment", the fairness and optimality can be incorporated in a "multi-win" sharing solution.

This chapter is useful for the practitioners to design a suitable multi-agent RRSI for their IPD projects. A novel risk/reward sharing method is also proposed on the basis of behavioral analysis and from perspective of cooperative game. Setting an optimal multi-agent RRSI, with the aid of BIM and blockchain technology, the IPD adoption is expected to be enhanced.

References

Aashto Subcommittee on Construction. (2000). Constructability review best practices guides. *Construction* (August).

Ahmad, I., Azhar, N., & Chowdhury, A. (2019). Enhancement of IPD characteristics as impelled by information and communication technology. *Journal of Management in Engineering, 35*(1), 4018055.

Ahmed, M. O., Abdul Nabi, M., El-adaway, I. H., Caranci, D., Eberle, J., Hawkins, Z., & Sparrow, R. (2021). Contractual guidelines for promoting integrated project delivery. *Journal of Construction Engineering and Management, 147*(11), 5021008.

AIA National and AIA California Council. (2009). *Experiences in collaboration—On the path to IPD*. Washington, DC, pp. 1–17.

AIA National and AIA California Council. (2007). *Integrated project delivery: A guide*. AIA National and AIA California Council.

American Institute of Architects. (2008). *AIA Document C195—2008, standard form single purpose entity agreement for integrated project delivery, exhibit D (Work Plan). AIA*. Washington, DC, Sacramento, CA.

American Institute of Architects (AIA). (2009). *AIA Document C191^{TM}—2009, Standard form multi-party agreement for integrated project delivery. AIA*. Washington, DC.

American Institute of Architects. (AIA). California Council (2007) *Integrated project delivery: A working definition*. Sacramento, CA.

Ashcraft, H.W. (2011) "Negotiating an Integrated Project Delivery Agreement". in *Hanson Bridgett LLP*. Hanson Bridgett LLP, 1–34

Asmar, M. E., Asce, A. M., Hanna, A. S., Ph, D., Asce, F., Loh, W., Ph, D., & Asce, A. M. (2016). Evaluating integrated project delivery using the project quarterback rating 3. *Journal of Construction Engineering and Management, 142*(1), 1–13.

Azhar, N., Kang, Y., & Ahmad, I. (2015). Critical look into the relationship between information and communication technology and integrated project delivery in public sector construction. *Journal of Management in Engineering, 31*(5).

Boukendour, S., & Hughes, W. (2014). Collaborative incentive contracts: Stimulating competitive behaviour without competition. *Construction Management and Economics, 32*(3), 279–289.

Chang, C. -Y. (2014). Principal-agent model of risk allocation in construction contracts and its critique. *Journal of Construction Engineering and Management, 140*(1), 04013032. http://www.scopus.com/inward/record.url?eid=2-s2.0-84891417456&partnerID=tZOtx3y1

Cheng, R. (2016) *MOTIVATION AND MEANS: How and why IPD and lean lead to success*. <Research report>

Cheng, R., Dale, K., Aspenson, A., & Salmela, K. (2012). *IPD case studies*. AIA, AIA Minnesota, School of Architecture—University of Minnesota.

Cohen, J. (2010). Integrated project delivery: Case studies. In *AIA National, AIA California Council, AGC California and McGraw-Hill*.

ConsensusDocs LLC. (2007). *ConsensusDOCS 300 standard form of tri-party agreement for collaborative project delivery*. https://www.leanconstruction.org/media/docs/deliveryGuide/Appendix3.pdf

ConsensusDocs LLC (2012). *ConsensusDocs Guidebook*. (April).

Dal Gallo, L., O'Leary, S. T., & Louridas, L .J. (2009). *Comparison of integrated project delivery agreements*.

Darrington, J. W., & Lichtig, W. A. (2010). Rethinking the 'G ' in GMP: Why estimated maximum price contracts make sense on collaborative projects. *The Construction Lawyer, 30*(2), 1–12.

Davis, P., & Love, P. (2011). Alliance contracting: adding value through relationship development. *Engineering, Construction and Architectural Management, 18*(5), 444–461. http://www.scopus.com/inward/record.url?eid=2-s2.0-80054951213&partnerID=40&md5=d3d29f1cb23ec1ea7bacc90211d65df8

Department of Infrastructure and Regional development. (2015). *National alliance contracting guidelines: Guide to alliance contracting*. Canberra, Australia. www.infrastructure.gov.au

Department of Treasury and Finance (2010) *The Practitioners ' Guide to Alliance Contracting*. (October). Victoria

Durdyev, S., Hosseini, M. R., Martek, I., Ismail, S., & Arashpour, M. (2019). Barriers to the use of integrated project delivery (IPD): A quantified model for Malaysia. *Engineering, Construction and Architectural Management, 27*(1), 186–204.

Egan, J. (1998). Rethinking construction the report of the construction task force. *Construction, 38*. http://scholar.google.com/scholar?hl=en&btnG=Search&q=intitle:RETHINKING+CONSTRUCTION#0

El-adaway, I., Abotaleb, I., & Eteifa, S. (2017). Framework for multiparty relational contracting. *Journal of Legal Affairs and Dispute Resolution in Engineering and Construction, 9*(3), 4517018.

El Asmar, M., Hanna, A. S., & Loh, W. (2013). Quantifying performance for the integrated project delivery system as compared to established delivery systems. *Journal of Construction Engineering and Management, 139*(11), 1–14.

Elghaish, F., Abrishami, S., Abu Samra, S., Gaterell, M., Hosseini, M. R., & Wise, R. (2019). Cash flow system development framework within integrated project delivery (IPD) using BIM tools. *International Journal of Construction Management, 0*(0), 1–16. https://doi.org/10.1080/15623599.2019.1573477

Elghaish, F., Abrishami, S., & Hosseini, M. R. (2020). (2020) Integrated project delivery with blockchain: An automated financial system. *Automation in Construction, 114*, 103182.

Elghaish, F., Pour Rahimian, F., Brooks, T., Dawood, N., & Abrishami, S. (2023). Web-based management system to share risk/reward for IPD projects. In *Blockchain of things and deep learning applications in construction*. Springer, pp. 83–97.

Forbes, L. H., & Ahmed, S. M. (2010). *Modern construction: Lean project delivery and integrated practices*. CRC Press.

Goulding, J. S., Rahimian, F. P., & Wang, X. (2014). Virtual reality-based cloud BIM platform for integrated AEC projects. *Journal of Information Technology in Construction, 19*, 308–325. http://www.scopus.com/inward/record.url?eid=2-s2.0-84927510857&partnerID=40&md5=620670f300ac7530513e48af57982823

Hall, D., & Scott, W. R. (2016). *Early stages in the institutionalization of integrated project delivery*.

Hall, D. M., & Bonanomi, M. M. (2021). Governing collaborative project delivery as a common-pool resource scenario. *Project Management Journal, 52*(3), 250–263.

Han, J., Rapoport, A., & Fong, P. S. W. (2019). Incentive structures in multi-partner project teams. *Engineering, Construction and Architectural Management, 27*(1), 49–65.

Hanna, A. S. (2016). Benchmark performance metrics for integrated project delivery. *Journal of Construction Engineering and Management, 142*(9), 04016040.

Hosseinian, M., Farahpour, E., & Carmichael, D. G. (2020). Optimum outcome-sharing construction contracts with multiagent and multioutcome arrangements. *Journal of Construction Engineering and Management, 146*(7), 4020067.

Hughes, D., Williams, T., & Ren, Z. (2012). Is incentivisation significant in ensuring successful partnered projects? *Engineering, Construction and Architectural Management, 19*(3), 306–319.

Hunhevicz, J. J., Brasey, P. -A., Bonanomi, M. M. M., & Hall, D. (2020). Blockchain and smart contracts for integrated project delivery: Inspiration from the commons. *EPOC 2020 Working Paper Proceedings*.

Hunhevicz, J. J., & Hall, D. M. (2019). Managing mistrust in construction using DLT: A review of use-case categories for technical design decisions. In *Proceedings of the European Conference for Computing in Construction*. held 2019, pp. 100–109.

Ibrahim, M. W., Hanna, A., & Kievet, D. (2020). Quantitative comparison of project performance between project delivery systems. *Journal of Management in Engineering, 36*(6), 4020082.

Kagioglou, M., & Tzortzopoulos, P. (2010). *Improving healthcare through built environment infrastructure*. Wiley.

Kahneman, D., & Tversky, A. (1979). Prospect theory: An analysis of decision under risk. *Econometrica, 47*(2), 263–292.

Kent, D. C., Asce, S. M., Becerik-gerber, B., & Asce, A. M. (2010). Understanding construction industry experience and attitudes toward integrated project delivery. *Journal of Construction Engineering and Management* (August), 815–825.

Kent, D. C., & Becerik-Gerber, B. (2010). Understanding construction industry experience and attitudes toward integrated project delivery. *Journal of Construction Engineering and Management, 136*(8), 815–825. http://www.scopus.com/inward/record.url?eid=2-s2.0-77955332881&partnerID=40&md5=b69eaa86114f50e80e9460d2aea038f8

Korb, S., Haronian, E., Sacks, R., Judez, P., & Shaked, O. (2016). Overcoming 'but We're Different': An Ipd Implementation in the Middle East". In *IGLC 2016—24th Annual Conference of the International Group for Lean Construction*, held 2016, pp. 3–12. https://www.scopus.com/inward/record.uri?eid=2-s2.0-84995976560&partnerID=40&md5=73209777e10ea6e08cb5260d4b9d2fca

Lahdenperä, P. (2016a). Preparing a framework for two-stage target-cost arrangement formulation. *International Journal of Managing Projects in Business, 9*(1), 123–146. http://www.emeraldin sight.com/doi/abs/https://doi.org/10.1108/IJMPB-07-2015-0049#.VyZIthluGKY.mendeley

Lahdenperä, P. (2016b). Formularising two-stage target-cost arrangements for use in practice. *International Journal of Managing Projects in Business, 9*(1), 147–170. http://www.emeraldinsight.com/doi/abs/https://doi.org/10.1108/IJMPB-07-2015-0050#.VyZDgbmANnA.mendeley

Lahdenperä, P. (2012). Making sense of the multi-party contractual arrangements of project partnering, project alliancing and integrated project delivery. *Construction Management and Economics, 30*(1), 57–79.

Lahdenperä, P. (2010). Conceptualizing a two-stage target-cost arrangement for competitive cooperation. *Construction Management and Economics, 28*(7), 783–796.

Laryea, S., & Watermeyer, R. (2016). Early contractor involvement in framework contracts. In *Proceedings of the 24th Annual Conference of the International Group for Lean Construction*, held 2016 at Boston, MA, USA, pp. 13–22.

Latham, M. (1994). *Constructing the Team*. HM Stationery Office London.

Lichtig, W. (2005a). Sutter health: Developing a contracting model to support lean project delivery. *Lean Construction Journal, 2*(1), 105–112.

Lichtig, W. (2005b). *Integrated Form of Agreement for Integrated Lean Project Delivery Among Owner, Architect & CM/GC.*

Ling, F. Y. Y., Teo, P. X., Li, S., Zhang, Z., & Ma, Q. (2020). Adoption of integrated project delivery practices for superior project performance. *Journal of Legal Affairs and Dispute Resolution in Engineering and Construction, 12*(4), 5020014.

Liu, J., & Cheah, C. Y. J. (2009). Real option application in PPP/PFI project negotiation. *Construction Management and Economics, 27*(4), 331–342.

Liu, J., Gao, R., Cheah, C. Y. J., & Luo, J. (2016). Incentive mechanism for inhibiting investors' opportunistic behavior in PPP projects. *International Journal of Project Management, 34*(7), 1102–1111.

Liu, J., Yu, X., & Cheah, C. Y. J. (2014). Evaluation of restrictive competition in PPP projects using real option approach. *International Journal of Project Management, 32*(3), 473–481.

Liu, M., Griffis, F. H., & Bates, A. J. (2013). Compensation structure and contingency allocation in integrated project delivery. In *ASEE Annual Conference and Exposition, Conference Proceedings* [online] held 2013 at Atlanta, Georgia, USA, p 6695. http://www.scopus.com/inward/record.url?eid=2-s2.0-84884316423&partnerID=40&md5=e5e8988cc7c9c48023cfe338d79a8124

Lostuvali, B., da Alves, T., & C.L., and Modrich, R.-U. (2014). Learning from the Cathedral hill hospital project during the design and preconstruction phases. *International Journal of Construction Education and Research, 10*(3), 160–180.

Love, P. E. D., Davis, P. R., Chevis, R., & Edwards, D. J. (2011). Risk/reward compensation model for civil engineering infrastructure alliance projects. *Journal of Construction Engineering and Management, 137*(2), 127–136.

Ma, Q., Li, S., & Cheung, S. O. (2022a). Unveiling embedded risks in integrated project delivery. *Journal of Construction Engineering and Management, 148*(1), 4021180.

Ma, Q., Shan, L., Teo, P. X., & Ling, F. Y. Y. (2022b). Barriers to adopting integrated project delivery practices. *Engineering Construction and Architectural Management* ahead-of-p (ahead-of-print).

Ma, Z., & Ma, J. (2017). Formulating the application functional requirements of a BIM-based collaboration platform to support IPD projects. *KSCE Journal of Civil Engineering, 21*(6), 2011–2026.

Matthews, O., & Howell, G. A. (2005). Integrated project delivery an example of relational contracting. *Lean Construction Journal, 2*(1), 46–61.

Mauck, R., Lichtig, W. A., Christian, D. R., & Darrington, J. (2009). Integrated project delivery: Different outcomes, different rules. In *The 48th Annual Meeting of Invited Attorneys*. held 2009. Victor O. Schinnerer & Company, Inc., pp 105–128.

Medda, F. (2007). A game theory approach for the allocation of risks in transport public private partnerships. *International Journal of Project Management, 25*(2007), 213–218.

Melese, Y., Lumbreras, S., Ramos, A., Stikkelman, R., & Herder, P. (2017). Cooperation under uncertainty: Assessing the value of risk sharing and determining the optimal risk-sharing rule for agents with pre-existing business and diverging risk attitudes. *International Journal of Project Management, 35*(3), 530–540.

Meng, X., & Gallagher, B. (2012). The impact of incentive mechanisms on project performance. *International Journal of Project Management, 30*(3), 352–362. http://dx.doi.org/https://doi.org/10.1016/j.ijproman.2011.08.006

Mesa, H. A., Molenaar, K. R., & Alarcón, L. F. (2016). Exploring performance of the integrated project delivery process on complex building projects. *International Journal of Project Management, 34*(7), 1089–1101. https://www.scopus.com/inward/record.uri?eid=2-s2.0-849 76869568&partnerID=40&md5=c13e45c72eb84fd101ac1d3563daffa7

Monteiro, A., Mêda, P., & Poças Martins, J. (2014). Framework for the coordinated application of two different integrated project delivery platforms. *Automation in Construction, 38*, 87–99. http://www.scopus.com/inward/record.url?eid=2-s2.0-84890513766&partne rID=40&md5=76c98db115fdec4fda7a77bea8511f86

Mosey, D. (2009). Chapter four: Early contractor involvement in design , pricing and risk management. In *Early Contractor Involvement in Building Procurement: Contracts, Partnering and Project Management*, pp 57–94.

National Association of State Facilities Administrators. (NASFA). Construction Owners Association of America (COAA); APPA: The Association of Higher Education Facilities Officers; Associated General Contractors of America (AGC); and American Institute of A. (2010). *Integrated Project Delivery For Public and Private Owners*. http://www.agc.org/galleries/projectd/IPD%2520for%2520Public%2520and%2520Private%2520Owners.pdf

Nwajei, U. O. K., Bølviken, T., & Hellström, M. M. (2022). Overcoming the principal-agent problem: The need for alignment of tools and methods in collaborative project delivery. *International Journal of Project Management, 40*(7), 750–762.

O'Connor, P. J. (2009). *Integrated project delivery: collaboration through new contract forms.* Minneapolis.

Pishdad-Bozorgi, P. (2017). Case studies on the role of integrated project delivery (IPD) approach on the establishment and promotion of trust. *International Journal of Construction Education and Research, 13*(2), 1–23.

Pishdad-Bozorgi, P., & Srivastava, D. (2018). Assessment of integrated project delivery (IPD) risk and reward sharing strategies from the standpoint of collaboration: A game theory approach. In *Construction Research Congress 2018*. held 2018, pp 196–206.

Rahman, M. (2012). A contractor's perception on early contractor involvement. *Built Environment Project and Asset Management, 2*(2), 217–233. http://www.emeraldinsight.com/https://doi.org/10.1108/20441241211280855

Ross, J. (2013). Gainshare/Painshare regime—Guidance paper with sample model/drafting. *Www. Pci-Aus. Com/Downloads/PCI_Gainshare-GuideModel_A_07Feb2011. Pdf (Accessed January 11, 2013).*

Ross, J. (2003). Introduction to project alliancing. *Project Control International Pty Limited.*

Ross, J. (1999). Project alliancing in Australia. In *Industry Summit on Relationship Contracting in Construction*. held 1999 at Sydney, Australia, pp 1–24.

Rowlinson, S. (2017). Building information modelling, integrated project delivery and all that. *Construction Innovation, 17*(1), 45–49.

Su, G., Hastak, M., Deng, X., & Khallaf, R. (2021). Risk sharing strategies for IPD projects: Interactional analysis of participants' decision-making. *Journal of Management in Engineering, 37*(1), 4020101.

Suijs, J., & Borm, P. (1999a). Stochastic cooperative games: Superadditivity, convexity, and certainty equivalents. *Games and Economic Behavior, 27*(2), 331–345.

Suijs, J., Borm, P., De Waegenaere, A., & Tijs, S. (1999b). Cooperative games with stochastic payoffs. *European Journal of Operational Research, 113*(1), 193–205.

Suijs, J., De Waegenaere, A., & Borm, P. (1998). Stochastic cooperative games in insurance. *Insurance: Mathematics and Economics, 22*(3), 209–228.

Thomsen, C., Darrington, J., Dunne, D., & Lichtig, W. (2009). Managing integrated project delivery. In *Construction Management Association of America (CMAA), McLean, VA*.

Tversky, A., & Kahneman, D. (1992). Advances in prospect theory cumulative representation of uncertainty. *Journal of Risk and Insurance, 5*(4), 297–323.

Victorian Department of Treasury and Finance. (2006). *Project Alliancing Practitioners Guide*.

Walker, D., & Rowlinson, S. (2019). *Routledge handbook of integrated project delivery*. Routledge.

Walker, D. H. T., Hampson, K., & Peters, R. (2002). Project alliancing versus project partnering: A case study of the Australian national museum project. *Supply Chain Management: An International Journal, 7*(2), 83–91.

Walker, D. H. T., Harley, J., & Mills, A. (2015). Performance of project alliancing in Australasia: A digest of infrastructure development from 2008 to 2013. *Construction Economics and Building, 15*(1), 1–18. http://www.scopus.com/inward/record.url?eid=2-s2.0-84924982046&partnerID=40&md5=99960e53a55feadc637fa9a2858e466e

Wang, Y., & Liu, J. (2015). Evaluation of the excess revenue sharing ratio in PPP projects using principal-agent models. *International Journal of Project Management, 33*(6), 1317–1324. http://dx.doi.org/https://doi.org/10.1016/j.ijproman.2015.03.002

Whang, S.-W., Park, K. S., & Kim, S. (2019). Critical success factors for implementing integrated construction project delivery. *Engineering, Construction and Architectural Management, 26*(10), 2432–3244.

Zhang, L. and Li, F. (2014). Risk/reward compensation model for integrated project delivery. *Engineering Economics, 25*(5), 558–567. http://www.scopus.com/inward/record.url?eid=2-s2.0-84920942889&partnerID=40&md5=3c9b9064c09b186892c9f693ddc7a64c

Zheng, L., Lu, W., Chen, K., Chau, K. W., & Niu, Y. (2017). Benefit sharing for BIM implementation: Tackling the moral hazard dilemma in inter-firm cooperation. *International Journal of Project Management, 35*(3), 393–405.

Zhu, L., & Cheung, S. O. (2021). Equity gap in construction contracting: Identification and ramifications. *Engineering, Construction and Architectural Management*.

Qiuwen Ma (Lizzy) Ma is a postdoctoral fellow at the Construction Dispute Resolution Research Unit, Department of Architecture and Civil Engineering, City University of Hong Kong. Dr. Ma conducted her PhD study with the Construction Dispute Resolution Research Unit, Department of Architecture and Civil Engineering of the City University of Hong Kong and graduated in 2020. Her research focuses on integrated project delivery (IPD), risk/reward sharing incentive and risk management. She has published journal and conference papers in IPD.

Chapter 7
Would Raising Psychological Well-Being Incentivize Construction Workers?

Keyao Li

Abstract Psychological well-being problems have raised concerns in the construction industry with reported high levels of mental health illness and suicide rate. Worse yet, the global COVID-19 pandemic has deteriorated the situation and caused more anxiety and depression cases. When basic psychological needs are not met, workers tend to experience less autonomous engagement at work. Thus, it is vital that management in the construction industry develop procedures, mechanisms, and interventions to improve worker experience. In this chapter, construction workers' experiences at work are examined by conceptualising the construct of psychological well-being in the context of construction community. Three types of well-being outcomes and their antecedents are discussed: Hedonic (i.e. job satisfaction, life satisfaction), Eudaimonic (i.e. work-life balance, job engagement) and Negative (i.e. Stress, burnout, psychological symptoms). The association between construction worker well-being experience and motivation at work is highlighted, emphasizing the importance of managerial commitment for a motivated and engaged workforce. More practically, hands-on prevention-focused leadership practices are suggested to support resilience and mitigate risks to health and well-being in times of disturbance. Management implications are recommended for decision makers to improve worker well-being and engagement in the construction community.

Keywords Psychological well-being · Construction industry · Motivation · Worker experience · Leadership

1 Introduction

Psychological well-being is a key component of individual overall health and well-being (WHO, 2021). Psychological well-being problem is a pervasive public health issue impacting over 2 in 5 Australians during their lifetime (ABS, 2022). The nature

K. Li (✉)
Future of Work Institute, Curtin University, Perth, WA, Australia
e-mail: keyao.li@curtin.edu.au

© The Author(s), under exclusive license to Springer Nature Switzerland AG 2023
S. O. Cheung and L. Zhu (eds.), *Construction Incentivization*, Digital Innovations in Architecture, Engineering and Construction,
https://doi.org/10.1007/978-3-031-28959-0_7

of construction work, high demands and rigid work practices had worsened the situation in the construction industry, resulting in pervasiveness of psychological illness across construction sectors. In Australia alone, more than two construction workers die as a result of suicide every working day in the past 10 years (Jenkin & Atkinson, 2021). And it has been reported that the construction workforce suffers from higher rates of suicide and mental illness issues, when compared to other industries (Kotera et al., 2020; PwC, 2014). Notorious psychological well-being problems in the construction industry might have deteriorated its low employee retention and made it a less attractive career choice for young professionals (Park et al., 2021).

Unfortunately, the COVID-19 pandemic has worsened the well-being crisis in the construction industry. Social restrictions during the pandemic, such as physical distancing, social isolation and remote working policies have triggered more loneliness, anxiety, and depression situations (OECD, 2021). According to Li and Griffin (2022a), experience of the pandemic could link to lower level of job satisfaction, when workers experienced increased psychological uncertainty and perceived less managerial support to their safety. With lower perceived leadership commitment, the uncertainties and pandemic-induced changes in the workplace might cause less role clarity and higher workload, thus resulting in poor well-being outcomes (Li & Griffin, 2022b). Moreover, when the global pandemic hit the labour market, workers perceived heightened job insecurity, causing a greater cognitive load that led to psychological exhaustion, reduced fulfilment, and higher levels of work stress (DeGhetto et al., 2017; Godinic et al., 2020). In times of uncertainty, construction workers have this increasing demand for more resources and corresponding mechanisms to support them to cope with psychological problems and well-being issues. It is vital that construction organizations focus on the challenge, provide well-being services and interventions to help employees thrive at work, and make the industry vibrant again.

2 Conceptualisation of Psychological Well-Being in the Construction Industry

The concept of psychological well-being was first highlighted by Aristotle when he argued that the ultimate life fulfilment was achieved by realizing one's true potential (Miller & Marjorie, 1986; Stones & Kozma, 1989). This provided insights for the following scholars to unpack and deepen the understanding of psychological well-being in different situations. Ryff and Singer (2008) suggested that psychological well-being should be studied based on individual's perceptions towards life, more specifically from the six aspects: self-acceptance, positive relations with others, personal growth, purpose in life, environmental mastery, and autonomy. Taking the perspective of individual's relationship with the external world, Fisher (2010) classified psychological well-being indicators into four categories: personal, such as meaning in life; communal, such as trust between individuals; environmental, such as connection with nature; and transcendental, such as peace with God. Despite the different streams of well-being literature, there is a common understanding that well-being represents not only the status of illness-free, but also positive experiences, such as engagement and satisfaction (Robertson & Flint-Taylor, 2008; Rousseau et al., 2008). Two types of approach were widely used to study positive psychological well-being, and these are hedonic approach and eudaimonic approach (Robertson & Flint-Taylor, 2008; Ryan & Deci, 2001). Hedonic approach has a focus on positive perspectives, feelings, experiences, and overall satisfaction. In contrast, eudaimonic approach emphasizes the fulfillment of living a life that is full of meaningfulness and value. With different research foci, these two approaches were complementary in the formation of psychological well-being foundations: the combination of positive experiences and sense of purpose (Robertson & Cooper, 2010).

Although psychological well-being has been well-studied as a multidimensional concept in the organizational psychology literature, there remains a lack of frameworks in the construction literature to capture the indicators of construction worker well-being and the respective influencing factors. Li et al. (2022) filled this gap and proposed a categorisation of psychological well-being for the construction workforce to inform their experiences at work. They first conducted bibliometric analysis to map and visualize the chronological patterns, journal sources, fundamental theories, and methodologies of the reviewed articles. Then, thematic analysis was further applied to identify their theoretical connections and networks. In their study, five themes of influencing factors for the construction community psychological well-being were identified: motivational, relational, working environment, personal attributes, and social cognitive. Theoretically, their study introduced more clarity to well-being theories in the construction literature. More practically, their findings offered managerial insight into decision making in the construction industry to proactively develop measures, mechanisms, and interventions to improve health and safety.

3   Psychological Well-Being and Motivation at Work

An important theory of human motivation and their application in work organization is self-determination theory (SDT) (Deci & Ryan, 1985, 2000; Ryan & Deci, 2001). SDT differentiated two types of motivation: (1) autonomous motivation, including both intrinsic motivation and fully internalized extrinsic motivation; and (2) controlled motivation, such as externally and internally controlled extrinsic motivation (Deci et al., 2017). Autonomous motivation and controlled motivation are both intentional, and they are starkly juxtaposed with amotivation, which represents a paucity of intention and self-determination (Gagne & Deci, 2005). More related to the workplace, autonomous motivation happens when workers are willingly participating in tasks and have a comprehensive understanding of its worth and meaningfulness. While when motivation is controlled, the extrinsic nature might result in short-term gains on projected achievements yet have negative impact on employee long-term engagement (Deci et al., 2017). Deci et al. (2017) further postulated that all employees have three basic psychological needs: the need for competence (White, 1959), the need for relatedness (Leary & Baumeister, 2000), and the need for autonomy (DeCharms, 1972). The satisfaction of these psychological needs could stimulate autonomous motivation.

Deci et al. (2017) expanded the STD model in the workplace context, and they argued that the influence from workplace context variables and individual differences to workplace health, well-being and behaviours were mediated by the psychological needs and different types of motivations. Moreover, the fulfillment of the three basic psychological needs could increase autonomous motivation. The following research further supported that satisfaction of the three needs led to less exhaustion and great enjoyment at work (Van den Andreassen et al., 2010; Broeck et al., 2008). DeCooman et al. (2013) examined basic need satisfaction and autonomous motivation at the same time, and they found that employees who felt greater need satisfaction at work also had a higher level of autonomous motivation. Adding more evidence to this, with a meta-analysis, Van den Broeck (2016) indicated that each of the three basic needs satisfaction explained independent variance in intrinsic motivation and well-being.

When applying the same in the construction industry, it is reasonable to draw the link between the fulfilment of psychological needs and construction workers' autonomous motivation. Based on Li et al. (2022), when efforts are made to redesign the work, optimize the job characteristics, and provide more job resources, construction workers tend to have greater satisfaction of their psychological needs. Put it in another way, the efforts to boost worker psychological well-being experience will contribute to the enhancement of worker autonomous motivation and engagement. This has been demonstrated in other industries with studies showing the positive correlations between worker well-being experience and motivation at work (Björklund et al., 2013; Ratanawongsa et al., 2008).

4   What Impacts Worker Psychological Well-Being in the Construction Industry?

In Li et al. (2022), construction worker psychological well-being outcomes are sorted into three groups: Hedonic (i.e. job satisfaction, life satisfaction), Eudaimonic (i.e. work-life balance, job engagement) and Negative (i.e. Stress, burnout, psychological symptoms). Five themes of antecedents were identified in their study, which could influence worker experience at work and their well-being. The dimensions, constructs and antecedents of construction worker psychological well-being are presented in Fig. 1. In this section, we will discuss each theme of these influencing factors and provide practical examples of how these are demonstrated in the construction workplace. The underpinning theories and antecedents under each theme were summarized and presented in Table 1.

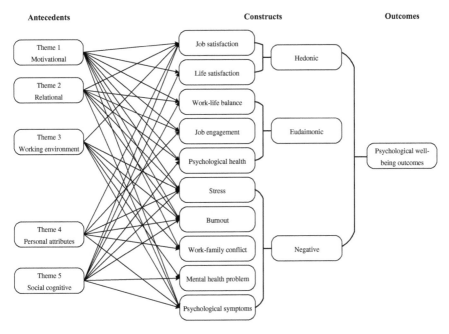

Fig. 1 Psychological well-being in the construction literature (Li et al., 2022)

Motivational theme describes how work conditions and characteristics could impact construction workers' motivational experience at work. The job demands-resources (JDR) theory is the key theoretical foundation that helps explain the rationales. The factors in this theme were categorized into job demands and job resources. Job resources refer to conditions, support and control at work and could enhance worker satisfaction and engagement (Hsu & Liao, 2016; Park & Jang, 2017; Zaniboni et al., 2016). Having adequate job resources is a prerequisite for construction workers to thrive at work. Examples of construction worker job resources include role clarity at work, feedback received at work, job support, and so on. Specifically in the construction community, decision-making autonomy, job security, compensation and rewards, and career development opportunities were identified as important job resources that could be beneficial for construction workers (Li et al., 2022). Job demands describe broadly all the efforts, physical, cognitive, and emotional, that are necessary for workers to do their tasks (Demerouti et al., 2001). High levels of demands are usually linked to negative psychological symptoms and health impairment (Arnold et al., 2007; Leiter, 1993; Steiner, 2018). Workload, as the principal job demand experienced by construction workers, has been found associated with intensified distress, lower satisfaction, exhaustion, and many other mental illness results (Li et al., 2022).

Relational theme describes workplace relationship, behaviours, and leadership practice. Having a good workgroup relationship and a social support network formed by co-workers and managers might improve worker well-being in the construction

Table 1 Antecedents of psychological well-being in the construction literature (Li et al., 2022)

Theme	Underpinning theory	Example
Theme 1: Motivational	Job demands-resources theory (Bakker & Demerouti, 2014); Job characteristics (Hackman & Lawler, 1971; Hackman & Oldham, 1976); Job demand-control model (Karasek, 1979)	Autonomy at work (Ling & Loo, 2015); Feedback (Hsu and Liao 2016); Workload (Idrees et al., 2017)
Theme 2: Relational	Leader-member-exchange model (Dienesch & Liden, 1986); Social exchange theory (Cook et al., 2013)	Workgroup relationships (Leung & Chan, 2012); Leadership support (Kerdngern & Thanitbenjasith, 2017)
Theme 3: Working environment	Organizational climate theory (James & Jones, 1974); Working condition and mental health (Ariza-Montes et al., 2019)	Culture and climate (Malone & Issa, 2013; Sutherland & Davidson, 1993; Toor & Ofori, 2009); Organizational Values (Panahi et al., 2016); Physical job demands (Janssen et al., 2001)
Theme 4: Personal attributes	Personality and well-being (DeNeve & Cooper, 1998); Family interference with work (Greenhaus & Beutell, 1985)	Tension in relationship with spouse/partner (Ligard and Francis 2007); Seniority (Lian & Ling, 2018); Physical health (Zaniboni et al., 2016)
Theme 5: Social cognitive	Social cognitive theory (Bandura, 1997); Social cognitive career theory (Lent, 2004; Lent & Brown, 2008); Social norms (Venkatesh & Davis, 2000; Ventakesh, 1999)	Career fit (Chew et al., 2020); Conflict difference between personal and organizational values (Panahi et al., 2016); Social influence (Fung et al., 2016); Psychological contract breach (Chih et al., 2016)

domain (Chan et al., 2016; Leung & Chan, 2012). Especially that the nature of construction work might require workers relocating to remote project sites, thus having supportive social connections plays an instrumental role in protecting on site workers' mental health. Due to the extreme gender stratification and macho nature of construction work, female construction workers and professionals were often found with limited job opportunities in the construction industry. Previous studies argued that gender-based harassment, discrimination, and bullying could link to not only physical symptoms (i.e. insomnia, stomach disorders, and headaches), but also severe mental illness (Greed, 2000; Bowen et al., 2013; Chew et al. 2020). The above interpersonal relationship factors in this theme highlighted the need for having more supportive leadership practice in the construction industry to improve workplace support and networks, in turn safety, health and well-being of construction workers.

Working environment theme represents another category of factors that could shape construction worker experience. Working environment involves not only physical working conditions but also workplace climate and culture, which could be determined by organization characteristics. It is not uncommon that construction projects are conducted in remote areas, with poor on site conditions such as extreme temperature, potential hazards, noise and poor light. These undesirable situations were found associated with worker physical health impairment and mental illness (Adhikary et al., 2018; Leung & Chan, 2012). Beyond physical working environment, organization attributes, culture and dynamics could all shape workplace climate. Studies have found that culture and value of an organization could affect how workers perceive the managerial support, and ultimately their engagement and satisfaction in the organization (Shan et al., 2017; Toor & Ofori, 2009). More recently, safety climate at workplace was found causing lower employee job satisfaction and less engagement in safety behaviours in times of turbulence, such as global pandemic (Li & Griffin, 2022a).

Of course, construction worker psychological well-being perceptions at work are affected by their **personal attributes and characteristics**. Type A personalities were found helpful in supporting construction workers to cope with challenges, difficulties, and stress at work (Çelik & Oral, 2021; Kamardeen & Sunindijo, 2017). When work under stress, construction workers with a marital status of separated, divorced, or widowed were relatively more vulnerable in developing mental illness symptoms (Kamardeen & Sunindijo, 2017). Relationship with partner, impact and support from family have also been found affecting worker experience and performance at work (Lingard & Francis, 2007; Pidd et al., 2017). Besides, worker smoking habits (Sutherland & Davidson, 1993), bad sleep status (Dong, 2018), and bad physical health (Holden & Sunindijo, 2018) might further worsen mental health situations. Notably, mindfulness attention training was reported helpful to support construction workers in reducing the negative influence from the construction site, such as loud noises (Boschman et al., 2013). Therefore, mindfulness techniques were suggested as effective strategies for workers to deal with stress (Carmody & Baer, 2008).

Comparing to other themes, **social cognition** theme is less studied yet underlies worker perception towards their jobs, their colleagues and organization environment. Career fit captures how well a worker considers himself or herself suitable for the job position. Career fit is an important part of workers' experiences and has the potential to affect their attitudes and observations at work, and ultimately their well-being status, especially for female construction workers in this male-dominated industry (Chew et al., 2020). How construction workers see their colleagues could also influence their work experience; for example, their stress level were increased when they perceived their colleagues having inadequate professional skills to complete the tasks (Leung & Chan, 2012). Workers' perceptions towards organization capture their emotional relations and connections with the organization. A poor organizational relationship, where there is a lack of trust and less organizational commitment could reduce worker job satisfaction and increase their turnover intentions (Leung et al. 2008; Idrees et al., 2017). In addition, workers' perceptions and feelings could be affected by their understanding of community expectations (Venkatesh & Davis,

2000; Ventakesh, 1999). For example, construction workers' attitudes and satisfaction at work can be influenced by their organizations' attitudes towards employee health and safety (Fung et al., 2016). Kotera et al. (2020) found that construction workers, who are in organizations with a shame-based attitude towards mental health issues, are the ones more likely to struggle with mental illness.

5 How to Improve Construction Worker Well-Being and Motivate Engagement at Work?

As discussed in the previous sections, worker psychological well-being experience and their motivation at work are closely related. Therefore, it is important that management in the construction industry develop procedures, mechanisms, and interventions to enhance worker experience, thus their motivated engagement at work. Job satisfaction is a well-studied indicator of worker psychological well-being, and it measures a pleasurable emotional state, describing workers' positive perception of their jobs, tasks, and work environment (Ali et al., 2014; Wang & Jing, 2018; Weiss et al., 1999). Using social exchange theory and organizational support theory, Michael et al (2005) argued that the increase of management commitment to safety could enhance workers' positive experience, because people formed their beliefs and attitudes based on their observations of whether their organization is valuing them. Similarly, Ayim Gyekye (2005) found that when workers noticed that their organization was promoting workplace safety, rewarding safe work, providing safe equipment, and responding to safety concerns, they would tend to experience higher levels of job satisfaction. Therefore, the positive association between managerial safety commitment and worker job satisfaction was supported and verified.

The influence of management safety commitment highlighted the need to deepen the understanding of leadership impact on worker experience and engagement. Especially in a time of uncertainty when the construction industry is navigating new paths for growth in a post-pandemic world. When the global pandemic unprecedentedly changed business and social landscape, the "new normal" would likely look a lot different comparing to the previous work routines in the construction industry. Safety leadership is important in times of turbulence when such leadership practices could contribute to workplace health, safety and performance (Griffin & Neal, 2000; Griffin & Talati, 2014). Li and Griffin (2022b) specifically proposed two types of leadership strategy to support resilience and adaptivity in times of crises, namely prevention-focused adapt strategy and defend strategy. The aim of prevention-focused adapt strategy is to understand the current crisis and use it as a stimulus to encourage active learning and improvement, so that the organization will be better equipped to reduce loss and damage in the similar situations in the future. This could be achieved by having open discussions at different levels of organization, on the mistakes and errors that led to the current setbacks, and strategies to avoid future occurrence. Therefore, leaders with an adapt strategy encourage transparent communications and redirect organizational resources to build capability in facing new challenges in the future (De Smet et al., 2021). In contrast, the goal of defend strategy is to manage risks. With a defend strategy, leaders take initiatives to proactively identify and manage risks, through fostering a culture of vigilance, integrating frequent auditing, conducting repeated assessments, and maintaining constant preparedness for potential hazards. Therefore, defend strategy behaviours underscore the danger of disregarding safety procedures and legislated obligations. With empirical evidence, Li and Griffin (2022b) claimed that increases in both adapt and defend strategies could link to positive well-being outcomes, through the improvement of role clarity and employees' perceptions of leadership in the workplace. Therefore, safety leadership practices were vital to engage workforce and minimise risks to workplace health and well-being in times of turbulence. More practically, hands-on prevention-focused leadership practices were suggested by Li and Griffin (2022b) that might shape worker well-being. These are summarized and presented in Table 2.

As the global pandemic drastically brought changes to the workplace, these changes were found increasing psychological uncertainty for workers and negatively impacting their experiences and emotions at work (DiFonzo & Bordia, 1998; Rafferty & Griffin, 2006). Cullen et al. (2014) explained that the heightened psychological uncertainty was caused by inadequate knowledge about the impact of sudden changes, thus might reduce their sense of being in control and supported by the organization, in turn, impaired satisfaction at work. In addition, when workers are feeling

Table 2 Prevention-focused leadership practices for worker well-being

Leadership dimension	Leadership practice
Lead for vigilance	• Communicate safety standards and procedures regularly
	• Foster a vibrant safety culture
	• Build accountability
	• Conduct regular monitoring
	• Provide safety training
Lead for adaptability	• Promote adjustment
	• Highlight learning opportunities
	• Create formal and informal communication channels
	• Reflect on safety visions and policies
	• Cultivate learning mindset
Lead with compassion	• Be compassionate and understanding
	• Provide timely support
	• Show gratitude and appreciation
	• Foster belonging in the organization

insecure in their job, they might also feel like being deprived of a safe working environment, resulting in frustration of psychological needs (Khan & Ghufran, 2018; Nelson et al., 1995; Vander Elst et al., 2012). Not only mental status could be affected by increased psychological uncertainty, such as job insecurity (Ashford et al., 1989), low commitment (Hui & Lee, 2000), mistrust (Schweiger & DeNisi, 1991); but also workers' physical health conditions, such as systolic blood pressure (Pollard, 2001). Most recently, Li and Griffin (2022a) further supported that workers' experience during the pandemic decreased their job satisfaction by intensifying their perceived uncertainty. This added more evidence to the negative association between uncertainty and positive well-being outcomes, especially in times of crises.

To boost psychological well-being outcomes in general, Li et al. (2022) suggested a checklist for management and decision makers in the construction community. For instance, good work design practices, which serve to improve role clarity, responsibility, and transparency at work, could increase job commitment and engagement. Developing a healthy workplace culture and workgroup relationships could strengthen social connections, providing necessary resources to cope with stress brought by long working hours and heavy workload. Enhancing construction worker organizational commitment and psychology attachment would ultimately improve workers' sense of belonging and ownership mindset. These could be achieved by building a positive organization culture that priorities worker health and safety and acknowledges their contributions to the organization. In addition, it is important that organizations have policies in place to support worker work-life balance and provide relevant trainings to assist workers in developing healthy working habits. A list of managerial implications is presented in Table 3.

Table 3 Examples of managerial implications (Li et al., 2022)

Purpose	Managerial practice
Improve motivation at work	Optimise workplace design practices, examples include reducing ambiguity and improving feedback
Improve workplace relationship	Build workplace social support networks
Improve organization culture	Foster a compassionate culture; Prioritise worker health and safety
Improve personal support	Provide relevant training, support, and care
Improve worker organizational commitment	Introduce and communicate organizational value and vision
In general	Deploy a whole-of-organization approach to improve mental health and well-being

6 Summary

Worker psychological well-being issues have raised concerns in the construction industry with high prevalence of mental illness issues (Kotera et al. 2019; PwC 2014). Worse yet, the global COVID-19 pandemic has worsened the situation and further exacerbated anxiety and depression (OECD, 2021). It has been found that there is a positive correlation between worker well-being experience and motivation at work (Björklund et al., 2013; Ratanawongsa et al., 2008). When workers' basic psychological needs are satisfied, they tend to have more autonomous motivation (Deci et al., 2017; Gagne and Deci, 2005). Therefore, it is important that management in the construction industry develop procedures, mechanisms, and interventions to improve worker experience, thus their motivated engagement at work.

It was highlighted in this chapter the vital role of prevention-focused safety leadership in engaging workforce and minimising risks. Two types of prevention-focused leadership strategy in Li and Griffin (2022b) were introduced to improve resilience and encourage adaptivity in times of crises, namely adapt strategy and defend strategy. More practically, hands-on safety managerial practices were suggested (see Table 2) that might shape employee well-being. Furthermore, Li et al. (2022) conceptualized worker psychological well-being in the construction industry with three types of well-being outcome: Hedonic (i.e. job satisfaction, life satisfaction), Eudaimonic (i.e. work-life balance, job engagement) and Negative (i.e. Stress, burnout, psychological symptoms). Five themes of antecedents were identified that could impact worker well-being, and these are motivational, relational, working environment, personal attributes, and social cognitive. Based on these, managerial implications were suggested in this chapter (see Table 3) for management and decision makers in the construction community to promote worker experience and reduce negative outcomes.

References

ABS. (2022). *National Study of Mental Health and Wellbeing*, ABS, Accessed 27 July 2022.

Adhikary, P., Sheppard, Z. A., Keen, S., & van Teijlingen, E. (2018). Health and well-being of Nepalese migrant workers abroad. *International Journal Migration Health Social Care, 14*(1), 96–105. https://doi.org/10.1108/IJMHSC-12-2015-0052.

Ali, A. E. I., Kertahadi, M. C., & Nayati, H. (2014). Job satisfaction, organizational behavior, and training to improve employees performance a case: Public hospitals-Libya. *Journal of Business and Management, 16*(8), 75–82.

Andreassen, C. S., Hetland, J., & Pallesen, S. (2010). The relationship between 'workaholism', basic needs satisfaction at work and personality. *European Journal of Personality: Published for the European Association of Personality Psychology, 24*(1), 3–17.

Ariza-Montes, A., Hernández-Perlines, F., Han, H., & Law, R. (2019). Human dimension of the hospitality industry: Working conditions and psychological well-being among European servers. *Journal Hospitality Tourism Manage, 41*, 138–147. https://doi.org/10.1016/j.jhtm.2019.10.013.

Arnold, K. A., Turner, N., Barling, J., Kelloway, E. K., & McKee, M. C. (2007). Transformational leadership and psychological well-being: The mediating role of meaningful work. *Journal of Occupational Health Psychology, 12*(3), 193.

Ashford, S. J., Lee, C., & Bobko, P. (1989). Content, causes, and consequences of job insecurity: A theory-based measure and substantive test. *Academy of Management Journal, 32*, 803–829.

Ayim Gyekye, S. (2005). Workers' perceptions of workplace safety and job satisfaction. *International Journal of Occupational Safety and Ergonomics, 11*(3), 291–302.

Bakker, A. B., & Demerouti, E. (2014). Job demands–resources theory. In *Wellbeing: A complete reference guide, 1–28*. New York: Wiley. https://doi.org/10.1002/9781118539415.wbwell019

Bandura, A. (1997). *Self-efficacy: The exercise of control*. Freeman.

Björklund, C., Jensen, I., & Lohela-Karlsson, M. (2013). Is a change in work motivation related to a change in mental well-being? *Journal of Vocational Behavior, 83*(3), 571–580.

Boschman, J. S., van der Molen, H. F., Sluiter, J. K., & Frings- Dresen, M. H. W. (2013). Psychosocial work environment and mental health among construction workers. *Applied Ergonomics, 44*(5), 748–755. https://doi.org/10.1016/j.apergo.2013.01.004

Bowen, P., Edwards, P., & Lingard, H. (2013). Workplace stress among construction professionals in South Africa: The role of harassment and discrimination. Engineering, Construction and Architectural Management.

Carmody, J., & Baer, R. A. (2008). Relationships between mindfulness practice and levels of mindfulness, medical and psychological symptoms and well-being in a mindfulness-based stress reduction program. *Journal of Behavioral Medicine, 31*(1), 23–33. https://doi.org/10.1007/s10 865-007-9130-7

Çelik, G., & Oral, E. (2021). Mediating effect of job satisfaction on the organizational commitment of civil engineers and architects. *International Journal Construcation Manage, 21*(10), 969–986.

Chan, I. Y. S., Leung, M. Y., & Liu, A. M. M. (2016). Occupational health management system: A study of expatriate construction professionals. *Accident Analysis Prevention, 93*, 280–290.

Chew, Y. T. E., Atay, E., & Bayraktaroglu, S. (2020). Female engineers' happiness and productivity in organizations with paternalistic culture. *Journal of Construction Engineering and Management, 146*(6), 05020005.

Chih, Y.-Y., Kiazad, K., Zhou, L., Capezio, A., Li, M., & Restubog, S. L. D. (2016). Investigating employee turnover in the construction industry: A psychological contract perspective. *Journal Construction Engineering Manage, 142*(6), 04016006. https://doi.org/10.1061/(ASC E)CO.1943-7862.0001101

Cook, K. S., Cheshire, C., Rice, E. R., & Nakagawa, S. (2013). Social exchange theory. In *Handbook of social psychology*, 61–88. Dordrecht, Netherlands: Springer.

Cullen, K. L., Edwards, B. D., Casper, W. C., & Gue, K. R. (2014). Employees' adaptability and perceptions of change-related uncertainty: Implications for perceived organizational support, job satisfaction, and performance. *Journal of Business and Psychology, 29*(2), 269–280.

De Cooman, R., Stynen, D., Van den Broeck, A., Sels, L., & De Witte, H. (2013). How job characteristics relate to need satisfaction and autonomous motivation: Implications for work effort. *Journal of Applied Social Psychology, 43*(6), 1342–1352.

De Smet, A., Mysore, M., Reich, A., & Sternfels, B. (2021). Return as a Muscle: How Lessons from COVID-19 Can Shape a Robust Operating Model for Hybrid and beyond, McKinsey Paper.

DeCharms, R. (1972). Personal causation training in the schools 1. *Journal of Applied Social Psychology, 2*(2), 95–113.

Deci, E. L., & Ryan, R. M. (1985). The general causality orientations scale: Self-determination in personality. *Journal of Research in Personality, 19*(2), 109–134.

Deci, E. L., & Ryan, R. M. (2000). The" what" and" why" of goal pursuits: Human needs and the self-determination of behavior. *Psychological Inquiry, 11*(4), 227–268.

Deci, E. L., Olafsen, A. H., & Ryan, R. M. (2017). Self-determination theory in work organizations: The state of a science. *Annual Review of Organizational Psychology and Organizational Behavior, 4*, 19–43.

DeGhetto, K., Russell, Z. A., & Ferris, G. R. (2017). Organizational change, uncertainty, and employee stress: Sensemaking interpretations of work environments and the experience of politics and stress, Power, Politics, and Political Skill in Job Stress (Research in Occupational Stress and Well Being, Vol. 15), Emerald Publishing Limited, Bingley, pp. 105–135. https://doi.org/10.1108/S1479-355520170000015002.

Demerouti, E., Bakker, A. B., Nachreiner, F., & Schaufeli, W. B. (2001). The job demands-resources model of burnout. *Journal of Applied Psychology, 86*(3), 499.

DeNeve, K. M., & Cooper, H. (1998). The happy personality: A metaanalysis of 137 personality traits and subjective well-being. *Psychological Bulletin, 124*(2), 197–229. https://doi.org/10.1037/0033-2909.124.2.197

Dienesch, R. M., & Liden, R. C. (1986). Leader-member exchange model of leadership: A critique and further development. *Academy of Management Review, 11*(3), 618–634. https://doi.org/10.2307/258314

DiFonzo, N., & Bordia, R. (1998). A tale of two corporations: Managing uncertainty during organizational change. *Human Resource Management, 37*, 295–303. https://doi.org/10.1002/(SICI)1099-050X(199823/24

Dong, R. R. (2018). Study on mental health status and life quality of migrant workers in construction industry. *Journal of Environmental Protection and Ecology, 19*(4), 1864–1872.

Fisher, J. (2010). Development and application of a spiritual well-being questionnaire called SHALOM. *Religions, 1*(1), 105–121. https://doi.org/10.3390/rel1010105

Fung, I. W. H., Tam, V. W. Y., Sing, C. P., Tang, K. K. W., & Ogunlana, S. O. (2016). Psychological climate in occupational safety and health: The safety awareness of construction workers in South China. *International Journal Construction Manage, 16*(4), 315–325. https://doi.org/10.1080/15623599.2016.1146114

Gagné, M., & Deci, E. L. (2005). Self-determination theory and work motivation. *Journal of Organizational Behavior, 26*(4), 331–362.

Godinic, D., Obrenovic, B., & Khudaykulov, A. (2020). Effects of economic uncertainty on mental health in the COVID-19 pandemic context: Social identity disturbance, job uncertainty and psychological well-being model. *International Journal of Innovation and Economic Development, 6*(1), 61–74.

Greed, C. (2000). Women in the construction professions: Achieving critical mass. *Gender Work Organism, 7*(3), 181–196.

Greenhaus, J. H., & Beutell, N. J. (1985). Sources of conflict between work and family roles. *Academy of Management Review, 10*(1), 76–88. https://doi.org/10.2307/258214

Griffin, M. A., & Neal, A. (2000). Perceptions of safety at work: A framework for linking safety climate to safety performance, knowledge, and motivation. *Journal of Occupational Health Psychology, 5*(3), 347.

Griffin, M. A., & Talati, Z. (2014). Safety leadership. In D. V. Day (Ed.), *The Oxford Handbook of Leadership and Organizations* (pp. 638–656). Oxford University Press.

Hackman, J. R., & Lawler, E. E. (1971). Employee reactions to job characteristics. *Journal of Applied Psychology, 55*(3), 259–286. https://doi.org/10.1037/h0031152

Hackman, J. R., & Oldham, G. R. (1976). Motivation through the design of work: Test of a theory. *Organizational Behaviour Human Perform, 16*(2), 250–279. https://doi.org/10.1016/0030-5073(76)90016-7

Holden, S., & Sunindijo, R. Y. (2018). Technology, long work hours, and stress worsen work-life balance in the construction industry. *International Journal Integration Engineering, 10*(2), 13–18. https://doi.org/10.30880/ijie.2018.10.02.003.

Hsu, L. C., & Liao, P. W. (2016). From job characteristics to job satisfaction of foreign workers in Taiwan's construction industry: The mediating role of organizational commitment. *Human Factors and Ergonomics in Manufacturing & Service Industries, 26*(2), 243–255.

Hui, C., & Lee, C. (2000). Moderating effects of organization-based self-esteem on organizational uncertainty: Employee response relationships. *Journal of Management, 26*, 215–232.

Idrees, M., Hafeez, M., & Kim, J.-Y. (2017). Workers' age and the impact of psychological factors on the perception of safety at construction sites. *Sustainability, 9*(5), 745. https://doi.org/10.3390/su9050745

James, L. R., & Jones, A. P. (1974). Organizational climate: A review of theory and research. *Psychological Bulletin, 81*(12), 1096. https://doi.org/10.1037/h0037511

Janssen, P. P. M., Bakker, A. B., & de Jong, A. (2001). A test and refinement of the demand–control–support model in the construction industry. *International Journal Stress Manage, 8*(4), 315–332. https://doi.org/10.1023/A:1017517716727

Jenkin, G., & Atkinson, J. (2021). *Construction Industry Suicides: Numbers, characteristics and rates: Report prepared for MATES in Construction NZ.* University of Otago Wellington, Wellington.

Kamardeen, I., & Sunindijo, R. Y. (2017). Personal characteristics moderate work stress in construction professionals. *Journal Construction Engineering Management, 143*(10), 04017072. https://doi.org/10.1061/(ASCE)CO.1943-7862.0001386

Karasek, R. A., Jr. (1979). Job demands, job decision latitude, and mental strain: Implications for job redesign. *Administrative Science q, 24*(2), 285–308. https://doi.org/10.2307/2392498

Kerdngern, N., & Thanitbenjasith, P. (2017). Influence of contemporary leadership on job satisfaction, organizational commitment, and turnover intention: A case study of the construction industry in Thailand. *International Journal Engineering Business Management, 9,* 184797901772317. https://doi.org/10.1177/1847979017723173.

Khan, R. U., & Ghufran, H. (2018). The mediating role of perceived organizational support between qualitative job insecurity, organizational citizenship behavior and job performance. *Journal of Entrepreneurship and Organization Management, 7*(228), 2.

Kotera, Y., Green, P., & Sheffield, D. (2020). Work-life balance of UK construction workers: Relationship with mental health. *Construction Management and Economics, 38*(3), 291–303.

Leary, M. R., & Baumeister, R. F. (2000). The nature and function of self-esteem: Sociometer theory. In *Advances in experimental social psychology* (Vol. 32, pp. 1–62). Academic Press.

Leiter, M. P. (1993). Burnout as a developmental process: Consideration of models. In *Professional burnout: Recent developments in theory and research*, 237–250. Boca Raton, FL: CRC Press.

Lent, R. W. (2004). Toward a unifying theoretical and practical perspective on well-being and psychosocial adjustment. *Journal of Counseling Psychology, 51*(4), 482–509. https://doi.org/10.1037/0022-0167.51.4.482

Lent, R. W., & Brown, S. D. (2008). Social cognitive career theory and subjective well-being in the context of work. *Journal of Career Assessment, 16*(1), 6–21. https://doi.org/10.1177/1069072707305769

Leung, M. Y., & Chan, I. Y. S. (2012). Exploring stressors of Hong Kong expatriate construction professionals in Mainland China: Focus group study. *Journal of Construction Engineering and Management, 138*(1), 78–88.

Leung, M., Chan, Y., Chong, A., & Sham, J.F.-C. (2008). Developing structural integrated stressor–stress models for clients' and contractors' cost engineers. *Journal Construction Engineering Management, 134*(8), 635–643. https://doi.org/10.1061/(ASCE)0733-9364(2008)134:8(635)

Li, K., & Griffin, M. A. (2022a). Safety behaviours and job satisfaction during the pandemic: The mediating roles of uncertainty and managerial commitment. *Journal of Safety Research.*

Li, K., & Griffin, M. A. (2022b). Prevention-focused leadership and well-being during the pandemic: Mediation by role clarity and workload. *Leadership & Organization Development Journal*, (ahead-of-print).

Li, K., Wang, D., Sheng, Z., & Griffin, M. A. (2022). A deep dive into worker psychological well-being in the construction industry: A systematic review and conceptual framework. *Journal of Management in Engineering, 38*(5), 04022051.

Lian, J. K. M., & Ling, F. Y. Y. (2018). The influence of personal characteristics on quantity surveyors' job satisfaction. *Built Environment Project Assistance Management, 8*(2), 183–193. https://doi.org/10.1108/BEPAM-12-2017-0117

Ling, F. Y. Y., & Loo, C. M. C. (2015). Characteristics of jobs and jobholders that affect job satisfaction and work performance of project managers. *Journal of Management Engineering, 31*(3), 04014039. https://doi.org/10.1061/(ASCE)ME.1943-5479.0000247

Lingard, H., & Francis, V. (2007). 'Negative interference' between Australian construction professionals' work and family roles: Evidence of an asymmetrical relationship. *Engineering Construction Architecture Management, 14*(1), 79–93. https://doi.org/10.1108/096999807107 16990

Malone, E. K., & Issa, R. R. A. (2013). Work-life balance and organizational commitment of women in the U.S. construction industry. *Journal of Professional Issues Engineering Education Practice, 139*(2), 87–98. https://doi.org/10.1061/(ASCE)EI.1943-5541.0000140

Michael, J. H., Evans, D. D., Jansen, K. J., & Haight, J. M. (2005). Management commitment to safety as organizational support: Relationships with non-safety outcomes in wood manufacturing employees. *Journal of Safety Research, 36*(2), 171–179.

Miller, J. F., & Marjorie, J. (1986). Development of an instrument to measure hope. *Nursing Research, 37*(1), 6–10.

Nelson, A., Cooper, C. L., & Jackson, P. R. (1995). Uncertainty amidst change: The impact of privatization on employee job satisfaction and well-being. *Journal of Occupational and Organizational Psychology, 68*(1), 57–71.

OECD (Organisation for Economic Co-operation and Development). (2021). *Tackling the mental health impact of the COVID-19 crisis: An integrated, whole-of-society response.* OECD Publishing.

Panahi, B., Preece, C. N., & Wan Zakaria, W. N. (2016). Personalorganisational value conflicts and job satisfaction of internal construction stakeholders. *Construction Economics and Buildings, 16*(1), 1–17. https://doi.org/10.5130/AJCEB.v16i1.4811

Park, C. J., Kim, S. Y., & Nguyen, M. V. (2021). Fuzzy TOPSIS application to rank determinants of employee retention in construction companies: South Korean Case. *Sustainability, 13*(11), 5787.

Park, R., & Jang, S. J. (2017). Mediating role of perceived supervisor support in the relationship between job autonomy and mental health: Moderating role of value–means fit. *The International Journal of Human Resource Management, 28*(5), 703–723.

Pidd, K., Duraisingam, V., Roche, A., & Trifonoff, A. (2017). Young construction workers: Substance use, mental health, and workplace psychosocial factors. Advances in Dual Diagnosis.

Pollard, T. M. (2001). Changes in mental well-being, blood pressure and total cholesterol levels during workplace reorganization: The impact of uncertainty. *Work and Stress, 15*, 14–28.

PwC [Beyondblue (Organisation) PricewaterhouseCoopers Australia]. (2014). Creating a mentally healthy workplace: return on investment analysis. Adelaide, Australia: PwC.

Rafferty, A., & Griffin, M. (2006). Perceptions of organizational change: A stress and coping perspective. *Journal of Applied Psychology, 91*, 1154–1162.

Ratanawongsa, N., Roter, D., Beach, M. C., Laird, S. L., Larson, S. M., Carson, K. A., & Cooper, L. A. (2008). Physician burnout and patient-physician communication during primary care encounters. *Journal of General Internal Medicine, 23*(10), 1581–1588.

Robertson, I. T., & Cooper, C. L. (2010). Full engagement: The integration of employee engagement and psychological well-being. *Leadership and Organizational Development Journal, 31*(4), 324–336. https://doi.org/10.1108/01437731011043348

Robertson, I. T., & J. Flint-Taylor. (2008). "Leadership, psychological wellbeing and organisational outcomes." In Oxford handbook on organisational well-being, edited by S. Cartwright and C. L. Cooper. Oxford, UK: Oxford University Press.

Rousseau, V., Aubé, C., Chiocchio, F., Boudrias, J.-S., & Morin, E. M. (2008). Social interactions at work and psychological health: The role of leader–member exchange and work group integration. *Journal of Application Social Psychology, 38*(7), 1755–1777. https://doi.org/10.1111/j.1559-1816.2008.00368.x

Ryan, R. M., & Deci, E. L. (2001). On happiness and human potentials: A review of research on hedonic and eudaimonic well-being. *Annual Review of Psychology, 52*(1), 141–166. https://doi.org/10.1146/annurev.psych.52.1.141

Ryff, C. D., & Singer, B. H. (2008). Know thyself and become what youare: A eudaimonic approach to psychological well-being. *Journal of Happiness Studies, 9*(1), 13–39. https://doi.org/10.1007/s10902-006-9019-0

Schweiger, D., & DeNisi, A. (1991). Communication with employees following a merger: A longitudinal field experiment. *Academy of Management Journal, 34*, 110–135.

Shan, Y., Imran, H., Lewis, P., & Zhai, D. (2017). Investigating the latent factors of quality of work-life affecting construction craft worker job satisfaction. *Journal of Construction Engineering Management, 143*(5), 04016134. https://doi.org/10.1061/(ASCE)CO.19437862.0001281

Steiner, S. (2018). Burnout culture shift: Strategies and techniques for preventing and addressing library worker fatigue and demotivation. *International Information Library Review, 50*(4), 319–327. https://doi.org/10.1080/10572317.2018.1526832

Stones, M. J., & Kozma, A. (1989). Happiness and activities in later life: A propensity formulation. *Canadian Psychology, 30*(3), 526–537. https://doi.org/10.1037/h0079827.

Sutherland, V., & Davidson, M. J. (1993). Using a stress audit: The construction site manager experience in the UK. *Work and Stress, 7*(3), 273–286. https://doi.org/10.1080/02678379308257067

Toor, S.-R., & Ofori, G. (2009). Ethical leadership: Examining the relationships with full range leadership model, employee outcomes, and organizational culture. *Journal of Business Ethics, 90*(4), 533–547. https://doi.org/10.1007/s10551-009-0059-3

Van den Broeck, A., Ferris, D. L., Chang, C. H., & Rosen, C. C. (2016). A review of self-determination theory's basic psychological needs at work. *Journal of Management, 42*(5), 1195–1229.

Van den Broeck, A., Vansteenkiste, M., De Witte, H., & Lens, W. (2008). Explaining the relationships between job characteristics, burnout, and engagement: The role of basic psychological need satisfaction. *Work & Stress, 22*(3), 277–294.

Vander Elst, T., Van den Broeck, A., De Witte, H., & De Cuyper, N. (2012). The mediating role of frustration of psychological needs in the relationship between job insecurity and work-related well being. *Work & Stress, 26*(3), 252–271.

Venkatesh, V., & Davis, F. D. (2000). A theoretical extension of the technology acceptance model: Four longitudinal field studies. *Management Science, 46*(2), 186–204. https://doi.org/10.1287/mnsc.46.2.186.11926

Ventakesh, V. (1999). Creation of favorable user perceptions: Exploring the role of intrinsic motivation. *MIS Quarterly, 23*(2), 239–260.

Wang, Z., & Jing, X. (2018). Job satisfaction among immigrant workers: A review of determinants. *Social Indicators Research, 139*(1), 381–401.

Weiss, H. M., Nicholas, J. P., & Daus, C. S. (1999). An examination of the joint effects of affective experiences and job beliefs on job satisfaction and variations in affective experiences over time. *Organizational Behavior and Human Decision Processes, 78*(1), 1–24.

White, R. W. (1959). Motivation reconsidered: The concept of competence. *Psychological Review, 66*(5), 297.

WHO (World Health Organization) (2021) *Mental health action plan 2013–2030*, WHO, Geneva, accessed 1 August 2022.

Zaniboni, S., Truxillo, D. M., Rineer, J. R., Bodner, T. E., Hammer, L. B., & Krainer, M. (2016). Relating age, decision authority, job satisfaction, and mental health: A study of construction workers. *Work, Aging and Retirement, 2*(4), 428–435.

Dr. Keyao (Eden) Li is a research fellow at the Future of Work Institute, Curtin University. Eden conducted her Ph.D. study on biases in dispute resolution with the Construction Dispute Resolution Research Unit, Department of Architecture and Civil Engineering of the City University of Hong Kong and graduated in 2019. Her research seeks to understand workplace psychology issues at different levels of organization, and how psychological barriers affect rational negotiation, worker well-being, safety challenges, and innovation decisions at workplace. Eden is also a research fellow at the Australian Research Council (ARC) Training Centre for Transforming Maintenance Through Data Science. Her research at the centre aims to translate advances in data science arising from the research projects into workplace improvement for the future mining industry but also applicable to other industries.

Chapter 8
Revamping Incrementalism to Incentivize the Land and Housing Policy Agendas in Hong Kong

Pui Ting Chow

Abstract This chapter explores the concepts of incrementalism and incentivization in the context of land and housing policy agendas. Given ongoing challenges in land and housing shortages and a rapidly changing environment, status quo orientation of government will lead to success or otherwise failure of new people-based and result-oriented strategies. On one hand, incremental land and housing policies seemingly fail to "muddle through" the status quo. On the other hand, public administrators are exposed to more uncertainties in increasingly complex policy mixes and a fragmented sociopolitical and economic context, without properly incentivized, they will eventually lose their job satisfaction. As such, there is a pressing need to develop a model to improve applicability of the theory of incrementalism as a commonplace accounting of recent effort in changing policymaking process. The chapter addresses three main questions: why are virtues of incrementalism remaining valuable, how can incremental policy changes and unfavourable policy outcomes be explained, and what can be done to reduce vices of incrementalism? First, the chapter argues that incrementalism, as a "branch method" of decision-making, offers a more realistic and effective approach to land and housing policymaking compared to classic bounded rationality model. This "branch method" describes power of small, marginal, momentous and accommodated steps to achieve policy goals. The virtues of incrementalism, such as its resourcefulness in overcoming cognitive limitations, diverging interests, and changing policy goals, make it a valuable tool in complex policy situations. Second, the chapter acknowledges that accumulative incrementalism recognizes the long periods of policymaking stasis without theorizing the co-existence of very seldom events of drastic policy changes. The empirically predominant form of accumulative incrementalism comes at certain cost in its explanatory power. This proposition guides this study to draw on Atkinson's intellectual inquiries of institutionalism and behavioural economics to analyze the dynamic of incremental policy changes and unfavourable policy outcomes and view punctuated equilibria as part of policy continuity. Third, incentivization is identified as one of the crucial factor in the effectiveness of incrementalism in a rapid changing environment. The chapter proposes a

P. T. Chow (✉)
HKU SPACE Po Leung Kuk Stanley Ho Community College, Hong Kong, China
e-mail: pt_rptchow@teacher.hkuspce-plk.hku.hk

© The Author(s), under exclusive license to Springer Nature Switzerland AG 2023
S. O. Cheung and L. Zhu (eds.), *Construction Incentivization*, Digital Innovations in Architecture, Engineering and Construction,
https://doi.org/10.1007/978-3-031-28959-0_8

169

framework that incorporates normative, affective and calculative incentives. Overall, the chapter presents a conceptual model that analyses the dynamic of incrementalism, intellectual inquiry and incentivization in the context of land and housing policy agendas.

Keywords Incrementalism · Status Quo · Stasis · Punctuated equilibria · Intellectual inquiry · Incentivization

1 Introduction

Public administration plays a pivotal role in shaping land and housing policies. The provision of high-quality land and housing relies heavily on effective public service in the Hong Kong Special Administrative Region (HKSAR). A capable public administration is crucial for fostering public trust in the government and creating a favourable business environment. In this context, incrementalism has emerged as an important approach to decision-making, whereby policy changes are made through small, marginal, momentous, and accommodated steps. Incremental approach was predominant throughout the twentieth century. However, Adam et al. (2021), Atkinson (2011) and Pal (2011) argued that incrementalism also has become one of the key factors in rarity of drastic policy changes. This approach, while it's accumulative effect continuously adds to comprehensiveness and complexity of policy mixes, can lead to policy changes that are insufficient, status quo-oriented, and unable to keep pace with the rapid changing socio-economic and political situations. In extreme cases, it results in an outgrowth of selective public policy implementation and ineffective debate on the substance of public policy (Howlett & Migone, 2011). To address these concerns, there is growing interest in revamping incrementalism by introducing new incentives that motivate public administrators to be proactive and respond to changes more promptly (Druskienė & Sarkiunaite, 2018). Studies have shown that public administrators are primarily driven mostly by altruistic and idealistic motives, which are not sufficient for even small changes in the current fragmented sociopolitical and economic context (Allan, 2019). Therefore, the HKSAR Government (HKSARG) has to explore new incentives to revamp incrementalism, or risk losing the job satisfaction and their effectiveness of civil servants, ultimately affecting the success of new people-based and result-oriented strategies (HKSARG, 2022a; Lee, 2011, 20122012).

Both the topics of incrementalism and incentivization have attracted significant attention in the field of public administration in recent years. According to Adam et al. (2021), the government has faced criticism for its insufficient nature and status quo orientation. This frustration has been particularly prominent in the context of incremental land and housing policies (Van Noorloos et al., 2020). Such a negative assessment aptly opposes Lindblom's seminal proposition that incrementalism is an important virtue of democracy and has the ability to "muddle through" and gradually achieve sweeping changes in public policy (Bendor, 1995; Lindblom, 1959). In this study, debates on revamping incrementalism and incentivization of the land and housing policy agendas have been revisited. By reviewing theoretical and empirical aspects of these concepts, the major obstacles to governmental decision-making in an environment of increasingly complex policy mixes are identified. The observations are made through an eventful, anecdotal and archival analysis of peer reviewed publications in top-tier outlet regarding incentivization of land and housing supply (Zhang & Pearlman, 2004). The discussion led to the development of a conceptual model of revamping incrementalism and incentivization in land and housing policy agendas.

2 Incrementalism

Herbert Simon's classic theory of bounded rationality has been found to be inadequate in predicting the decision-making process of public administrators involved in land and housing policies in Hong Kong (Baldwin et al. 2012; Ostrov, 2002; Wescott & Bowornwathana, 2008). In contrast, the incrementalism model, as it was termed by Lindblom (1959), has been regarded as a more practical model that takes roles of the policy environment more seriously and provides an alternative that centres the ideal of bounded rationality and satisficing activity among policymakers (Cairney, 2012). Incrementalism is described as a "branch method" of decision-making in which policy changes are made through small, marginal, momentous and accommo-dated steps (Jones et al., 1997). It has been alternatively named gradualism (Qizil-bash, 2010), articulated rationality (Hayes, 2002), seriality (Balla et al., 2015) and disjointed problem solving (Atkinson, 2011; Pal, 2011). Lindblom (1959) asserted that the solutions and outcomes of incrementalism were superior to those of the "root method", which involved redesigning public policy from scratch for every problem-solving attempt. The intellectual underpinning of incrementalism reveals its resource-fulness in overcoming cognitive limitations, diverging interests, and changing policy goals in policy-making situations. The conceptualization of these three constructs in the land and housing policy context is discussed hereinafter.

2.1 Cognitive Limitations (P1 in Fig. 1)

The concept of incrementalists acknowledge that mistakes are inevitable in every policy-making process. However, it emphasizes the importance of avoiding the possibility of large mistakes and limiting unpredictable consequences by under-going a satisficing process of trial-and-error learning (Migone & Howlett, 2015:85). Civil servants' doctrine and practices are, then, in large part designed to take these characteristics of bureaucracy into account (Sayre, 1954). During recruitment and training, civil service newcomers are exposed to similar policies' experiences to antic-ipate possible consequences and better understand the impact of policy changes due to

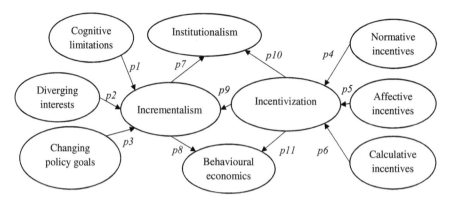

Fig. 1 The conceptual model

their incremental nature (Yun, 2020). Civil servants are thus trained to be "wise" and "intelligent" and are capable of taking necessary measures to mitigate risks resulting from cognitive limitations (Barratt, 2015; Fry, 2002). Incremental changes, to some extent, can lead the public to attribute undesirable outcomes of policy changes to societal changes instead of policy intervention (Priedolset al. 2022). This perception buys time for policy correction, and Lindblom (1959) highlighted this aspect of incrementalism as inertia of ideological conservatism and its advantage and leeway for policymakers' risk aversion.

In the context of land and housing policy agendas, incrementalism is often viewed as a laissez-faire approach. For instance, John Cowperthwaite, the fifth financial secretary of the colonial ruling, followed Arthur Clarke's lead and outlined his doctrines in his 1961 maiden budget speech, stating that "*In the long run the aggregate of decisions of individual businessmen, exercising individual judgement in a free economy, even if often mistaken, is less likely to do harm than the centralized decisions of a government; and certainly, the harm is likely to be counteracted faster*" (Wong, 2013: 59). In 1993, when Chris Patten first met John Cowperthwaite, he remarked "*So, you are the architect of all this?*" To which Cowperthwaite responded, "*I did very little.*" "*...all I did was to try to prevent some of the things that might undo it*" (Murphy, 2020). These statements aptly align with the underlying premises of incrementalism, wherein decision-makers are aware of their limitations and take necessary measures to mitigate risks with significant unintended consequences that are difficult to reverse (Adam et al., 2021).

2.2 Diverging Interests (P2 in Fig. 1)

Incrementalists recognize the importance of promoting the common good in a pluralist democracy of diverging interests (Hagan et al., 2001). In cases where achieving a consensus as an ideal is almost impossible, policymakers will aim to reach a compromise that is honourable, fair or Pareto-improving, which it is generally better than having no agreement at all (van Parijs, 2012). Due to the diversity of values and ideologies held by policymakers, it is particularly challenging to obtain common ground that everyone can agree on and gain a majority vote in supporting drastic reforms. Over time, the existence of a particular policy has been the resultant of a history of conflict and bargaining among policy actors. They and their successors are so familiar with the policy development and tend to reinforce existing patterns of distribution rather than encourage redistribution negotiations (Gutmann & Thompson, 2013). This helps to explain why incrementalism is not only described as "intelligent" but also "dominant" in the public administration regime (Lindblom, 1959).

Civil servants are responsible for considering not only the public interests, but also the balance of power within legislature and the reactions to policy changes by interest groups. They must persuade political parties, vested interests and the public that policy changes are appropriate and ensure that policies are implemented properly. However, reaching agreements over values, standards, principles and criteria of "good" land and housing policies on an abstract level is unattainable for the government and vested interests (including developers, contractors, construction professionals, property agents, investors, speculators and homeowners) (Poon, 2011). This is due to the narrower range of feasible options in diverging interests among policy actors and increasingly complex policy mixes the land and housing policymakers faced (Greener, 2002). Policy actors have adapted the current policies to their daily operations, which has resulted in a path dependence. They tend to protect the exiting agendas, follow past decisions, and encourage policy continuity (Pierson, 2000). When policy proposals are considered, most of them are rejected with some alleged justification to continue on the same path (Hansen, 2002). Hansen (2002) further suggested that these allegations could be made without reference to the structure of costs and incentives created by the original policy choices. Policy actors emphasize what the public could lose if the policy is implemented, and it is impractical for decision-makers to construct and propose a comprehensive grand design due to disagreements over core values and unpredictable consequences (Arnold, 2002). Consequently, small adjustments to the status quo are made premised on what is practical and what is possible (Atkinson, 2011). Without making trade-offs between governmental objectives and those favoured by other actors, they face a relatively high level of resistance from affected policy actors, i.e., the inertia of *veto points* as termed by Lindblom (1959). Failing to take advantage of an opportunity to change, they intend to miss it completely (Poon, 2011). As such, decision making processes are thus largely compromising rather than consensual in a time of crises.

2.3   Changing Policy Goals (P3 in Fig. 1)

In policymaking situations, prioritizing policy goals, associating them with policy agendas, and connecting them to tools to achieve those objectives can be challenging (Peter, 2018). Policymakers need to cater for changing policy goals during the decision-making process. However, assigning weights to changing criteria for making choices is often difficult (Rezaei et al., 2021). Lindblom's partisan mutual adjustment suggests that incremental change is more effective than any other effort to accomplish this task via central command by connecting resources and collective preferences (Atkinson, 2011; Lindblom, 1990).

Policymakers face difficulties in exploring options, prioritizing policy goals, and striking a balance among them, especially in the complex and conflict-prone area of land and housing policies. These difficulties are compounded by high goal conflict (e.g., quantity and quality), competing belief system (e.g., urban development and ecological conservation), high technical certainty (e.g., how to build and who to build), a homogeneity of policy actors from fewer levels of government bureaus, departments and the public (e.g., environmental protectionists and local community), and the existence of threat to the public (e.g., talent attraction and development of industries) (Arnott, 2018; Poon, 2011). These characteristics have disabled drastic policy changes (Wilsford, 1994). Incremental adjustment is a practical approach that can thus help policymakers focus on marginal alternatives to the existing policies. However, Fullan (2000) notes that policy agendas as intentional interventions may not necessary generate change. In other words, policy agendas can simply be rhetorical expressions that did not emerge from beliefs to change. Policymakers can term policy agendas as reforms but still focus on management principles not prominent governance values (Huque & Jongruck, 2020). As such, implementation of policy agendas was prominent more in rhetoric than substance and policymakers can secure status quo, which is masked by constant changing physical, social, economic and environmental conditions. Some alleged policy agendas, as placebos, give hope to the public who is looking for policymakers' effort to some desirable outcomes in which what is desirable itself continues to evolve under reconsideration (Adam et al., 2021).

3   Revamp

Lindblom (1990) lamented impaired quality of inquiry that characterizes public decision-making, but he also made it clear that incrementalism was not the source (Atkinson, 2011). Nevertheless, a failure of overly restrictive incrementalism is observed. Thus, Lindblom's heuristics should be tested against other forms of applied rationality. Both marginal and large scale policy changes come from interaction of multilevel institutions and behavioural decision-making. It shows a combination of policy movements that create patterns of stasis (stability), incrementalism (mobilization) and punctuated equilibria (large-scale departures) in contemporary public administration and management (Jones & Baumgartner, 2005; Lindblom, 1990; True et al., 2018). Atkinson (2011) further identified two lines of intellectual inquiry, institutionalist inquiry and behavioural economics inquiry, to resolve problems rooted in Lindblom's lament. The incorporation of institutionalist inquiry and behavioural inquiry in the model and an emphasis on the status quo as a focal point for policy change offer more empirical insights into how policy change occurs in practice, especially in the context of land and housing policy (Cheung, 2002; Cheung & Wong, 2019; Chiu et al., 2022). The conceptualization of these two constructs is discussed hereinafter.

3.1 Institutionalist Inquiry (P4 in Fig. 1)

Incrementalism legitimizes and reinforces a conservative approach that closely resembles empirical realities of decision-making (Lindblom, 1990). It allows policymakers to form articulated rationality and to gain unconstrained trust in uncertain situations by gradually making small adjustments to the existing policy framework (Hayes, 2002). This approach reinforce the status quo and marginalize disadvantaged groups. With narrowing effects of socialization and social inequalities, phenomena of (1) insidious effects of conformity, (2) enervating effects of docility, and (3) tactical use of incrementalism by elites over marginalized groups are obvious (Hayes, 2002). All of these have combined to produce responses to pressing problems that are seldom opportune (Atkinson, 2011). It is too much for complacent acceptance of imperfections in the current system (Adam et al., 2021). While Lindblom originally downplayed power dimension that is now in full view, incrementalism as a strategy has become congenial to those who seek nothing more than agreeable and manageable adjustments to the status quo (Atkinson, 2011; Lindblom, 1959). In other words, institutionalism, which imposes a set of formalized rules that may be enforced by calling upon a third party (Streeck & Telen, 2005) can further reinforce the conservative approach. Its formality moulds itself into a higher form of rational intelligence that policy proposals on grand issues are usually to confirm volition rather than to change volition (Coccia & Benati, 2018). Policymakers need to identify enduring constellations of incentives that both limit and facilitate changes, and be open to setting aside incrementalism when a bloder agenda is necessary (Atkinson, 2011).

With the facts about (1) a 10-year steady but low home ownership ratio of approximately 51% (CenstatD, 2022), (2) a median of just 16m^2 per capita living area (Lau, 2022), (3) an average of 6.1 years waiting time for subsidized public housing (HKHA, 2022), and (4) an average price of above HK\$162,000/m^2 for the world's least affordable small private flat (RVD, 2022), it is obvious that there is market failure and that shortages of land and housing will persist for years. It is difficult to imagine that an incremental solution will be adequate. Adopting familiar strategies can only meet procedural requirements or democratically define public engagement (Poon, 2011). However, entire belief systems are currently being challenged, problem definitions are being contested, and cracks in social consensus are being revealed (Fitzpatrick et al., 2014). The land and housing policy portfolios have been shifting to different bureaus under the purview of either the chief secretary for administration or the financial secretary in the rounds of government reorganizations over the years (AUD, 2007; Lee, 2001). Fairly frequent changes have also taken place in the division of functions among departments (Arnott, 2018). It is clear that assignment of policy tasks to different secretariat policy branches is problematic but simple. The social, economic and political aspects are intertwined, and policy formulation involves inputs from departments across bureaus (Ng et al., 2021). The current Hong Kong administrative system on land and housing does not run contrary to or substantially deviate from the one designed by the British colonial regime, which is centred on a bureaucrat-dominated approach (Cutber & Dimitrious, 1992; Lau & Kuan, 2002; Luk, 2018).

3.2 Behavioural Economics Inquiry (P5 in Fig. 1)

Lindblom (1959) argued that (1) incrementalism is ubiquitous and overwhelmingly prevalent and (2) nothing could be more rational than to reduce risk by engaging in trial and error. However, many land and housing policies were inherently nonincremental, e.g., MacLehose's ten-year housing policy in 1972, the Tenants Purchase Scheme in 1997, and the establishment of the Urban Renewal Authority in 2000 to replace the Land Development Corporation. These policies entailed a significant departure or even a paradigm shift from the long existing ones (Hall, 1993). As such, incrementalism places too much emphasis on the predominance of path dependence and gradual adjustment (Thelen, 2004) and ignores the radical changes resulting from crises and external shocks (Jones & Baumgartner, 2005). In fact, when policymakers are confronted with deficiencies in the current policy, they view negative outcomes caused by action as more problematic than negative outcomes caused by inaction (Atkinson, 2011). This is known as omission bias that is common in public administration. The public generally perceives that undesirable outcomes caused by omissions are natural and takes fault finding in a reduced degree of personal responsibility (Spranca et al., 1991). This is opposite to what Lindblom (1959) presumed regarding decision-makers' actions. Omission bias resulting in policy inactivity (McConeel & Hart, 2019) is the very nature of public policy in terms of "whatever governments choose to do or not to do" (Dye, 2014). Long decision-making processes could make no progress, and the consequences can be nearly nil as a result of diverging interests. While Lindblom (1959) explained how individuals performed in an environment characterized by significant levels of complexity, it did not directly address the requirement of accountability. Tetlock and Boettger (1994) added that those who are obliged to explain and justify incremental decisions, i.e., top politically appointed officials in this study, are more susceptible to omission bias and status quo effect owing to their accountability for failure of policy initiatives.

By examining the path of governance and public administration of the HKSARG over a fairly long period, researchers have found the land and housing agendas to be incremental (Cheung, 2002; Dimitrious & Cook, 2018). The observation points to the central role of the civil servant' workforce in the generation, adaptation and diffusion of technical and organizational changes (Alexander, 2006). Incrementalism helps to explain the predominant form of policy changes of the HKSARG, whereas the same institutional system and ruling of the government bureaus and departments produce a plethora of small accommodations (Adams, 1987; Cheung & Wong, 2019; Lai & Wang, 1999). The policymaking process seemingly fails to deal smoothly and seamlessly with new information and instead falls prey to sporadic and irrelevant debates and remorseful arguments (Poon, 2011).

The decision-making process of public policy emphasizes two important elements: issue definition and agenda setting (Balla et al., 2015; Jones & Baumgartner, 2005). For issue definition, the issues of land and housing agendas are defined in public discourse in different ways (Huang, 2021; Nissim, 2021). Incrementalism leads policymakers to think that "what could be more reasonable" (Bendor, 2015:196) rather than to (1) start by defining a policy objective, (2) identify all alternatives for reaching this objective, and (3) choose the best of these alternatives. As such, the policymaking process fails to pursue the strategy of rational, systematic comparisons among all options and to allow for a synoptic assessment of optimal policy solutions. This inevitably leads to policy failure at worst or to suboptimal policy solutions at best (Allan, 2019). Agenda setting takes place in two ways: 'from above', through top politically appointed officials, and 'from below', through the working groups of policy bureaus and departments formulating specific proposals. That is, the HKSARG has evolved into a set of policy subsystems that are important in making policy, but macrolevel policymaking forces are also at play (Wong, 2012).

4 Incentivization

Incentivization refers to the external force to direct, energize and maintain an action (Grat, 2008). An organization uses different tools and techniques to reward employees for their works performed at the microlevel, i.e., to meet their needs and expectations and ensure their job satisfaction (Perry et al., 2010). The ultimate goal of incentivization is to ensure that an organization can successfully implement its agendas at the macrolevel (Behn, 1995). In the current study, incentivization is an additional impetus of reinforcement, so policymakers can significantly motivate themselves to plan for departures from the incremental past (Atkinson, 2011). Drawing on the literature on organizational behaviour and human resource management, the studies of incentivization have gone beyond the traditional dichotomies, i.e., intrinsic/extrinsic (Bullock et al., 2015; French & Emerson, 2014; Groeneveld et al., 2009; Manolopoulos, 2008; Taylor, 2010), being responsible/being held responsible

(Cooper & Jayatilaka, 2006), moral/ monetary (Druskienė & Šarkiūnaitė, 2018), intangible/tangible (Thom & Ritz, 2004), sociopsychological/material (Žilinskas & Zakarienė, 2007), and nonfinancial/financial (Šavareikienė, 2008). As these dimensions and constructs have been continuously circumscribed, widened and reformulated, this study reassembles the concept of incentivization into normative incentives, affective incentives and calculative incentives to increase the applicability of the concept to current situations.

4.1 Normative Incentives (P6 in Fig. 1)

From the public management and governance perspective, public service is always confronted by a host of rational and moral issues. Research shows that civil servants are reliably moved by normative incentives (Yung, 2014). Once they are normatively incentivized, civil servants will deliberate and internalize their normative judgements about what is right or wrong, good or bad (Rosati, 2016). These normative judgements typically motivate civil servants, at least to some degree, to act in accordance with norms, rules and regulations (Wang et al., 2020). The normative incentives encourage civil servants to make normative judgements that something is good for society. Interest and enjoyment in being a civil servant and in the task itself drive his/her intrinsic motivation (Cooper & Jayatilaka, 2006). The incentive of making normative judgements is thus regarded as the key feature that marks them as normative, thereby distinguishing them from mathematical and empirical judgements (Svavarsdottir, 1999). Mathematical judgements provide a reason to act in a particular way and empirical judgements provide a specific and rational course of action. In fact, the motivating force of normative judgements, i.e., normative incentives, is the most prominent and distinguishing feature in the case of public services (Forte et al., 2022). Normative incentives have the most significant effect in the case of narrowly moral judgements in public administration because civil servants as agents are more susceptible to the opposition between self-interest and morality. However, current studies have not explained the puzzling failures of the normative incentive to motivate in case of mixed policies (Wang et al., 2020). What is the precise nature of the connection among normative incentive, motivation and judgement? The failures of normative motivation are witnessed not only among unsuccessful and confused governments, but also among those apparently sound and self-possessed ones (Polidano, 2001; Vandenabeele et al., 2004). Additionally, there are growing phenomena of rational and strong-willed governments that seemingly make normative judgements while remaining utterly indifferent and amoralistic (Rosati, 2016). It suggests that normative incentives can drive performance only if civil servants are satisfied with the learning and pleasure that they expect to receive. It helps explain why these governments might fail to assign tasks that satisfy civil servants' curiosity or that are personally rewarding or contributive (Guzman & Espejo, 2015). In examining these situations, the common phenomena of incentivizing land and housing policy agendas are given as scenarios to discuss the sharply differing views of normative incentives, and these views have important implications for foundational issues in this revamping exercise. More precisely, differing views of normative incentives involve commitment to particular propositions and the existence of the perceived moral properties that have been brought to bear on questions of semantics, ontology and epistemology (Rosati, 2016).

Scenario 1: Civil servants are sympathetic about the discourse of the universal moral *right to housing* for the homeless, but they are sceptical about enforceability of legal rights (Fitzpatrick et al., 2014; King, 2003; LegCo, 2022).

Civil servants express an ethical belief that is a proposition about objectively prescriptive properties, i.e., *the right to housing*, which is built-in "to-be-pursuedness" (Rosati, 2016). However, they eventually reject the moral properties, and due to this presupposition of failure, they are systematically in error in their moral judgements. They lose their motivation to overcome any opposing desires or inclinations. The understanding of this situation must confront the central question, that is, whether normative incentive motivates moral judgements on their own or by intermediation of desire or other cognitive states. It is clear that the situation falls within the latter answer. As moral non-cognitivism maintains and describes, these civil servants hold favourable pro-social attitudes and simply express a motivating state that their group already holds. If these civil servants are normatively incentivized, they will internalize (1) being responsible for the task (i.e., to fight for the right to housing) without constant supervision of top officials, (2) being aware of their situations as well as focused on delivering the task beyond expectation and resolving challenges, and (3) without being demotivated by such challenges (Cooper & Jayatilaka, 2006).

Scenario 2: The government is in a position to break the deadlock between to conserve and not to conserve (Chan & Chiu, 2020).

According to applied normative theory, when civil servants need to make a decision on whether or not to take action to conserve, they must justify whether (1) consequences are good in utilitarian terms and/or (2) action follows the rule in deontologist terms. However, the view that moral judgements, obligations and beliefs, and the sentences that express them, can be true or false provides the correct account of moral semantics and of what moral judgements mean (Cooper & Jayatilaka, 2006; Rosati, 2016). In other words, they seem to represent the society to a certain extent and to express a moral belief, attributing particular moral properties (i.e., the values of conserving or not conserving) or normative characteristics to the action or state of affairs. Taking the apparent representational form of moral judgements as their lead, their apprehension of moral properties becomes the motivation for their actions. They will do the right thing, unaided by any additional source of motivation; their motivational power depends on no individual desire or disposition. Additionally, the apprehension of the values not only motivates them but also provides overriding motivation. Once civil servants as agents apprehend moral properties, their motivating power overcomes any opposing desires or inclinations (Rosati, 2016).

4.2 Affective Incentives (P7 in Fig. 1)

Affective incentives are one of the most important motivation mechanisms that explain the prosocial behaviour of civil servants (Perry & Wise, 1990; Sun, 2021). The antecedents and consequences of affective incentives to policy change have attracted interest from researchers, especially in studies of organizational behaviour (Oreg et al., 2018; Wright et al., 2013). Affective incentives are grounded in human emotions and are manifested as a desire and willingness to help others and to be useful to society (Camilleri, 2007). They also describe civil servants' predisposition to respond to intrinsic incentives grounded primarily and uniquely in public institutions and organizations (Perry & Wise, 1990). Ertas (2014) found that affectively motivated civil servants are more likely to engage in volunteer work outside their scope of public services and thus are more capable of delivery good public services. They also value their work for the government, as they have seen government duties offer more public service opportunities (Luthans et al., 2015; Youssef & Luthans, 2007). Rainey et al. (2021) suggested that civil servants who value intrinsic rewards, e.g., the verbal appreciation from the public, over extrinsic ones, take more altruistic attitudes than their counterpart who work mainly under directives. Other studies have found when compared with employees in private sector civil servants are more self-determined, optimistic, resilient, hopeful, supportive to democratic values and committed to civic duty (Conway, 2000; Youssef & Luthans, 2007). Psychological capital derived from affective incentives forms a human asset of an organization (Fidelis et al., 2021). Affective motives are associated with greater effort, motivation and perseverance in better performance within organizations (Avey et al., 2010). The manifestation of affective motives has unique characteristics, such as being (a) a psychologically positive capacity, (b) a theory with validation in scientific research that is measurable and (c) a state-like construct, which is relatively stable over time and is open to change and development that has a positive impact on attitudes, behaviour, performance and well-being at work (Luthans & Youssef-Morgan, 2017; Youssef-Morgan & Luthans, 2015).

Scenario 3: Photos of suffocating scenes of overcrammed subdivided flats have popped up on international social media since 2013 and public officers frowned on the pathetic lives of "the deplorable" (Chen, 2021; Chiu & Siu, 2022: 240; Harper, 2017; Stacke, 2017).

The affective statements, which are expression of civil servants' "sympathy" in Scenario 1 and their "worry" in Scenario 3, reflect the way their affective incentives tie them to the society. Being a civil servant is seen as an "instrument of faith" and generally as a kind of calling; what is needed next is a balance of "practical action" and "creation of a spirit", i.e., an affective attachment (Sinclair, 2015). Affective attachment constitutes a commitment to a specific policy out of personal conviction (Perry & Wise, 1990). The affective incentives could have inspired civil servants to act in a way that is beneficial to people who are living in the "coffin cubicles" or "caged homes". However, Hong Kong is unlikely to be able to eliminate the problem of subdivided flats in the coming twenty or so years and no later than 2049 (FSO, 2021).

4.3 Calculative Incentives (P8 in Fig. 1)

Calculative incentives are the most traditional, primary and fundamental system in public administration and management (Bullock et al., 2015). They base on remuneration and reward and are perceived to be an instrumental link between civil servants' behaviour and their receipt of extrinsic rewards (Cooper & Jayatilaka; French & Emerson, 2014). They also refer to as monetary return or compensation for the time, energy, and effort that civil servants invested in the organization (Druskienė & Sarkiunaite, 2018). The current remuneration system for civil servant has been criticized for being static, conservative, restricted, limited to legal provisions, and difficult to change (Cheung, 2002; Fry, 2002; Lam, 2004). The media commented that civil servants become accustomed to pay increases and consider increases in wages to be the norms. The public has urged that pay adjustment mechanism for the civil service should be revamped. A merit system consisting of a fixed unit (e.g., official salary) and a variable unit (e.g., performance-based assessment) may be more appropriate and flexible (Brewer & Kellough, 2016; Perry et al., 2006). This two-unit merit system allows a better link between salary and performance; however, due to the nonprofit nature, specificity, complexity and multifunctionality of civil service activities, it is impossible to assess the contribution of a particular civil servant to a joint and collective task. Thus, application of the system will be complex, expansive and risk prone (Langbein & Knack, 2010; Walther, 2015).

Scenario 4: The government is developing key performance indicators (KPIs) to measure progress towards goals and to adopt a "result-oriented approach" to improve governance efficiency (HKSARG, 2022b). If more ambitious targets were set for the "Task Force on Public Housing Projects" and "Steering Committee on Land and Housing Supply", how would the assessment be made, and who would be held responsible and be rewarded or punished within the system?

In the previous round of civil service reforms worldwide, government bureaus and departments moved to a consumerist governance regime and towards serving the community, customer orientation, performance pledges and performance management in the 1990s (Burns & Li, 2015). Thus, KPIs are not new in the current system. If part of the pay of an individual civil servant depends on the quality and quantity of overall housing supply, measurement of achievement among responsible civil servants is ambiguous (Polidano, 2001; Wong, 2012). This is similar to the current situation, where objective KPIs have already been set, but discretionary extras and bonuses become automatic wages increase to civil servants without considering their performance and achievement. Factors such as abuse, selfishness, protectionism, and individual likes and dislikes cannot be avoided (Huque et al., 2000).

5   The Conceptual Model and Propositions

The increasing attention to pressing problems in land and housing cannot be met by incremental governmental strategies. Government has long extoled the status quo and long periods of policy stasis are accompanied by subterranean changes, with the government adjusting the roles of existing institutions, adding new ones, and reaching accommodations with nongovernment organizations and societal partners (HKSARG, 2022b; Lindblom, 1959; Thelen, 2003). In absence of strong veto players, the HKSARG should be able to initiate significant policy shifts now. Remaining veto players are able to accept short-term sacrifices of their policy interests in order to maintain long-term coalition success with the HKSARG (HKSARG, 2022a; Scharpf, 2000). The existing model of incrementalism relies heavily on the status quo and underestimates the possibility of policy punctuation. The literatures suggest that incentivization linked to institutional design yields compelling findings in policy punctuation (Jones & Baumgartner, 2005; Park & Sapotichne, 2019). However, much work needs to be done to understand the operationalization of incentiviza-tion in the institutional design and arrangements that make policy punctuation more or less likely. As such, the current study develops a conceptual model to revamp incrementalism and Fig. 1 shows the conceptual model of the study.

In the early stage of model development, cognitive limitations, diverging interests, and changing policy goals in policymaking situations are identified for their long-term relevance to the incrementalism of an organization (*p1*, *p2* and *p3* in Fig. 1), and incentivization comprises normative, affective and calculative incentives (*p4*, *p5* and *p6* in Fig. 1). As an advancement of previous studies, institutionalism and behavioural economics in Atkinson's inquiries are systematically incorporated into the conceptual model (*p7* and *p8* in Fig. 1). One of the most important findings of the study is that incrementalism explains slow and uneven progress in land and housing policy agendas in Hong Kong. By adding the incentivization dimension to the model, revamping becomes possible (*p9*, *p10* and *p11* in Fig. 1). Once the conceptual model is developed, each dimension should be transformed into operationalized and measurable constructs for empirical and practical testing. Using the typologies of the relevant studies, the lists of the constructs of each dimension are provided in Tables 1, 2 and 3.

6   Discussion

It is proposed that significant policy changes, i.e., punctuated equilibria, are more difficult to achieve if there are no new incentives in the government hierarchy. The question of what "significant" means is an issue, but the current situation of substandard living conditions for less privileged people has already activated both the public and top officials and called for a more revolutionary approach to deep-rooted problems. The HKSARG is now "bidding farewell" to subdivided flats and

Table 1 Operationalized and measurable constructs related to *p1* to *p3*

Proposed operationalized and measurable constructs	References
Incrementalism	Adam et al. (2021), Jones and Baumgartner (2005), Streeck and Thelen (2005), True et al. (2018)
• Retrospective measures of length and magnitude of policy changes	
• The process of the land and housing policy changes was □ abrupt —— □ incremental —— □ stable	
• The result of the land and housing policy changes was □ discontinuous ——————□ continuous	
• The outcome of the land and housing policy changes was □ reproduction by adaptation □ gradual transformation □ survival and return □ breakdown and replacement	
Cognitive limitations	Barratt (2015), Fry (2002), Migone ad Howlett (2015), Yun (2020)
• The government put strong emphases on mistake avoidance in the land and housing policy change	
• The government introduced a trial-and-error system to limit the unfavourable outcomes of the land and housing policy change	
• The civil servants have been trained to make no mistakes	
• The land and housing market is more effective	
• A "good" result is more likely to emerge from market forces rather than the change in land and housing policies	
Diverging interests	Gutmann and Thompson (2013), Hagan et al. (2001), Hansen (2002), van Parijs (2012)
• There are common ground for the government and vested interests (including developers, contractors, construction professionals, property agents, investors, speculators and homeowners) to reach agreements over the values, standards, principles and criteria of the land and housing policy changes	
• It is difficult to obtain a majority vote in support of drastic reforms	
• The policy actors are familiar with history of policy development	

(continued)

Table 1 (continued)

Proposed operationalized and measurable constructs	References
• The policy actors protect exiting agendas	
• The policy actors follow their past decisions	
• The government proposed the changes in land and housing policies based on: ☐ consensus decision ──────☐ compromising decision	
Changing policy goals	Adam et al. (2021), Atkinson (2011), Lindblom (1990), Fullan (2000), Huque and Jongruck (2020), Peter (2018), Rezaei et al. (2021), Wilsford (1999)
• The government has clear weightings on policy goals in land and housing	
• The government faces high level of conflict about goals of the land and housing policies	
• The government is certain about technicality of the land and housing policy	
• The government has engaged all the parties affected by the change of policies	
• The government has acknowledge the land and housing crises	

confidently envisions better land and housing policies in the long term. With the beauty of simplicity and parsimoniousness (pun intended) of incrementalism, the development of the proposed conceptual framework is also remarkably simple, with fewer veto players engaged in the decision-making process, and the results indicate that it is the right time to facilitate punctuated policy change (Tsebeli, 2000). The window of exceptional opportunity, i.e., critical conjuncture, is now opened (Capoccia & Kelemen, 2007; Wilsford, 1994). The model allows for the possibility of revamping incrementalism if incentives are properly implemented in response to the inquiry of institutionalism and behavioural economics. Incrementalism limits changes in land and housing policy agendas to rather deterministic and mechanical ones. In microlevel observations, it seems that decision-makers are firmly determined to make rational choices unencumbered by cognitive limitation, diverging interests and changing policy goals. In this study, the proposed conceptual model assumes incentives revamping incrementalism with a heavy emphasis putting on the punctuation of policy change. There is no alternative for adjustment beyond the government core on land and housing policy. Nevertheless, the institutional force remains central. On the one hand, it is the question of how the government initiates drastic and rapid policy changes; on the other hand, it is the observation that a pluralistic situation, with fewer policy actors and a variety of alternative veto points, is more congenial to coherent and timely policy choices. Thus, the systematic incorporation of institutionalism and behavioural economics in this study adds research value to the practice of incrementalism. The current study emphasizes impediments to decision-making and highlights the need for the design and arrangement of incentivization.

Table 2 Operationalized and measurable constructs related to *p4* to *p6*

Proposed operationalized and measurable constructs		References
Incentivization		
•	I have been motivated by the organization to complete the policy tasks	Behn (1995), Cooper and Jayatilaka (2006), Grat (2008), Perry et al. (2010)
Normative incentives		
•	My desire to fulfil the obligation to the public is promoted	Atkinson (2017), Hustedt and Salomonsen (2018), Bullock et al., (2015), French and Emerson (2014), Groeneveld et al. (2009), Manolopoulos (2008), Taylor (2010)
•	My desire to achieve my career goals is promoted	
•	My desire to ensure the welfare of impoverished members of the public is promoted	
•	I found the policy goal to be clear	
•	I am allowed to do the policy work in an expert manner	
•	I am allowed to do the policy work according to explicit, objective standards rather than to personal or party or other obligations and loyalties	
•	My devotion to work is promoted	
•	My commitment to the government is promoted	
•	My job satisfaction is promoted	
•	My desire to work autonomously is promoted	
Affective incentives		
•	My desire to participate in policymaking processes is satisfied	Forte et al., (2022), Žilinskas and Zakarienė (2007), Thom and Ritz (2004)
•	My desire to contribute to public affairs is promoted	
•	My desire to be altruistic to the public is promoted	
•	My desire to be respected, recognized and appreciated by the public is promoted	
•	My desire to develop social and psychological with my team is promoted	
•	My desire to develop my personality and work character is promoted	

(continued)

Table 2 (continued)

Proposed operationalized and measurable constructs	References
• My loyalty to the government is promoted	
Calculative incentives	
• My salary is adequate	French and Emerson (2014), Druskienė and Šarkiūnaitė (2018), Šavareikienė (2008)
• My salary is sufficient	
• My salary is fair	
• My job is secure	
• My desire to achieve my career goal is promoted	
• My supervisor has given clear and specified criteria for my performance evaluation	
• My supervisor has taken steps and procedures to ensure that the work process is fair, reasonable and proper	

Table 3 Operationalized and measurable constructs related to *p7* to *p11*

Proposed operationalized and measurable constructs	References
Institutionalism	
• The government has a clear hierarchy of authority in the land and housing policy related bureaus and departments	Adam et al. (2021), Atkinson (2011), Coccia and Benati (2018), Fitzpatrick et al. (2014), Hayes (2002), Streeck and Telen (2005)
• The entire belief systems of the land and housing policy are currently being challenged	
• The problem definitions in land and housing are being contested	
• The cracks in social consensus are being revealed	
Behavioural economics	
• The crises in land and housing are caused by the government	Atkinson (2011), Dye (2014), Hall (1993), Jones and Baumgartner [2005], Spranca et al. (1991), Tetlock and Boettger (1994), Thelen (2004)
☐ inactions——☐ neutral—☐ actions	
• Who is held accountable to the land and housing polices?	
☐ the government as a whole——☐ top appointed officials —☐ the policy bureaus and/or departments	
• The setting of the policy agendas in land and housing is:	
☐ a top-down approach ——☐ neutral—☐ bottom-up approach	

This study aims at developing a conceptual model to analyse two more causal dimensions of incrementalism in public administration, i.e., the intellectual inquiries and incentivization. Moreover, this study starts with the process of generalizing and operationalizing several constructs of each dimension as far as possible. A more holistic and comprehensive approach to describing the current situation of land and housing policy changes in Hong Kong is possible. To support more fruitful management implications, it is urged that a continuous time meta-analysis of empirical studies is needed to refine idea of when drastic and rapid policy changes are incentivized and how they take place. In addition, quantitative studies of the length and magnitude of the punctuated equilibria are also recommended for future studies.

7 Summary

This study gives the answers to the three main questions: why are the virtues of incrementalism remaining valuable? How can incremental policy changes and unfavourable policy outcomes be explained? What can the vices of incrementalism be reduced? To do so, the theoretical and empirical aspects of incrementalism and incentivization have been reviewed and the intellectual inquires as the major obstacle to governmental decision-making in an environment of increasingly complex policy mixes has been identified. An eventful, anecdotal and archival analysis of peer reviewed publications in top-tier outlets regarding incentivization of land and housing supply has been conducted. The discussion led to the development of a conceptual model of revamping incrementalism and incentivization in land and housing policy agendas. The study appreciates incrementalism of its core power to overcome cognitive shortcomings, diverging interests and changing policy goals in policymaking. However, empirical evidences show that policymakers are often susceptible to omission and status quo biases. They might simply reject change (policy inaction). Even the term of reforms is used, the policy agendas of these reforms can simply be rhetorical, or at most, be a by-product of a local search for small policy changes and they are not emerged from the belief to change. This proposition imposes the importance of incorporation of Atkinson's intellectual inquiries and views punctuated equilibria as part of the policy continuity. The conceptual model suggests that decision-makers will rarely launch new alternatives, which are being challenged today. They will launch punctuated policy change when they are properly incentivized. Thus, another contribution of this study is its operationalization of the constructs of dimensions, i.e., incrementalism, intellectual inquiries and incentivization.

The model helps to explain the contemporary phenomena that requires top appointed officials to reconsider the design and arrangement of incentivization. As the new government emphasizes through its people-based and result-oriented strategies, land and housing policies will be subject to prospects for change. It is important to build in institutionalism and behavioural economics as variables into the model of incrementalism. As there are fewer veto points in the current bureaucracy, the government generates opportunities to initiate drastic policy change, incremental or otherwise. It is clear that incentivization is the key to revamping incrementalism, to keep decision-makers pursuing consistent goals, and to stand in the way of comprehensive problem solving. Taking a more macrolevel view to see punctuated changes as a continuity of incremental changes will bring decision-makers to a path breaking height. This study connects both micro- and macro-views of incrementalism in a broad and empirical portrayal of the policy-making process.

References

Adam, C., Hurka, S., Knill, C., & Steinebach, Y. (2021). On democratic intelligence and failure: The vice and virtue of incrementalism under political fragmentation and policy accumulation. *Governance, 35*(2), 525–543.

Adams, D. (1987). The land development corporation in Hong Kong. *Planning Practice & Research, 2*(3), 13–16.

Alexander, E. R. (2006). *Evaluation in planning: Evolution and prospects*. Ashgate.

Allan, J. I. (2019). Dangerous incrementalism of the Paris agreement. *Global Environmental Politics, 19*(1), 4–11.

Arnold, R. D. (2002). *Fixing social security: The politics of reform in a polarized age*. Princeton University Press.

Arnott, C. J. (2018). Chapter 11 Institutional planning framework and effective land-use/transport planning. In Dimitriou, H. T., & Cook, A. H. S. (eds.), *Land-use/transport planning in Hong Kong: The end of an era: A review of principles and practices* (Routledge revivals), New York: Routledge, 297–322.

Atkinson, C. L. (2017). Competence in bureaucracy. In Farazmand, A. (eds.) *Global Encyclopedia of Public Administration, Public Policy, and Governance*. Cham: Springer.

Atkinson, M. M. (2011). Lindblom's lament: Incrementalism and the persistent pull of the status quo. *Policy and Society., 30*(1), 9–18.

Audit Commission Hong Kong (AUD). (2007). Chapter 12 - Lands Department - Temporary use of vacant government sites, https://www.aud.gov.hk/pdf_e/e49ch12.pdf.

Baldwin, R., Cave, M., & Lodge, M. (2012). *Understanding regulation: Theory, strategy, and practice* (2nd ed.). Oxford University Press.

Balla, S. J., Lodge, M., & Page, E. (2015). *The oxford handbook of classics in public policy and administration*. University Press.

Behn, R. (1995). The bog questions of public management. *Public Administration Review, 55*(4), 313–324.

Barratt, E. (2015). Liberal conservatism, 'boardization' and the government of civil servants. *Organization, 22*(1), 40–57.

Bendor, J. (1995). A model of muddling through. *American Political Science Review, 89*, 819–840.

Bendor, J. (2015). Incrementalism: Dead yet flourishing. *Public Administration Review, 75*(2), 194–205.

Brewer, G. A., & Kellough, J. E. (2016). Administrative values and public personnel management: Reflections on civil service reform. *Public Personnel Management, 45*(2), 171–189.

Bullock, J. B., Stritch, J. M., & Rainey, H. G. (2015). International comparison of public and private employees' work motives, attitudes, and perceived rewards. *Public Administration Review, 75*(3), 479–489.

Burns, J., & Li, W. (2015). The impact of external change on civil service values in post-colonial Hong Kong. *China Quarterly, 222*.

Cairney, P. (2012). Rationality and incrementalism. In P. Cairney (Ed.), *Understanding Public Policy: Theories and Issues* (pp. 94–110). Palgrave Macmillan.

Camilleri, E. (2007). Antecedents affecting public service motivation. *Personnel Review, 36*(3), 356–377.

Capoccia, G., & Kelemen, D. (2007). The study of critical junctures: Theory, narrative and counterfactuals in historical institutionalism. *World Politics, 59*(3), 341–369.

Census and Statistics Department (CenstatD). (2022). 2022. Table 5: Statistics on Domestic Households, https://www.censtatd.gov.hk/en/web_table.html?id=5.

Chan, H. M., & Chui, C. K. (2020). Land development in Hong Kong: To conserve or not to conserve? That's not the question. In Yung, B., & Yu, K-p., *Land and Housing Controversies in Hong Kong: Perspectives of Justice and Social Values*, 99–124.

Chen, J. (2021). Can SAR eliminate subdivided flats within a decade?. *China Daily*. https://www.chinadailyhk.com/article/251445#Can-SAR-eliminate-subdivided-flats-within-a-decade.

Cheung, A. B. L. (2002). The changing political system: Executive-led government or "disabled governance"? In Lau, S. K. (ed.), *The First Tung Chee-hwa Administration – The First Five Years of the Hong Kong Special Administrative Region*. Hong Kong: Chinese University Press.

Cheung, W. K. S., & Wong, K. S. K. (2019). Understanding governance of public land sales: An experiment from Hong Kong. *Regional Studies, Regional Science, 6*(1), 607–622.

Chiu, S. W. K., & Siu, K. Y. K. (2022). *Hong kong society: High-definition beyond the spectacle of east-meets-west (Hong Kong studies reader series)*. Palgrave Macmillan.

Conway, M. M. (2000). *Political participation in the united states* (3rd ed.). CQ Press.

Coccia, M., & Benati, I. (2018). Rewards in public administration: A proposed classification. *Journal of Social and Administrative Sciences, 5*(2), 68–80.

Cooper, R. B., & Jayatilaka, B. (2006). Group creativity: The effects of extrinsic, intrinsic, and obligation motivations. *Creativity Research Journal, 18*(2), 153–172.

Cuthbert, A. R., & Dimitrious, H. T. (1992). Redeveloping the fifth quarter: A case study of redevelopment in Hong Kong. *Cities, 9*(2), 196–204.

Dimitriou, H. T., & Cook, A. H. S. (2018). *Land-use/transport planning in Hong Kong: The end of an Era: A review of principles and practices*. Routledge.

Druskienė, A., & Sarkiunaite, I. (2018). Motivational incentives of civil servants in Lithuanian municipalities. *Public Policy and Administration, 17*(3), 344–370.

Dye, T. R. (2014). *Understanding public policy* (14th ed.). Pearson Prentice Hall.

Ertas, N. (2014). Public service motivation theory and voluntary organizations: Do government employees volunteer more? *Nonprofit and Voluntary Sector Quarterly, 43*(2), 254–271.

Fidelis, A. C. F., Fernandes, A. J., Rech, J., & Larentis, F. (2021). Relationship between psychological capital and motivation: Study in health organizations of Southern Brazil. *International Journal for Innovation Education and Research, 9*(3), 186–201.

Fry, G. K. (2002). The conservatives and the civil service: 'One step forward, two steps back'? *Public Administration, 75*(4), 695–710.

Financial Secretary's Office (FSO). (2021). Working together to get rid of subdivided flats, My Blog, https://www.fso.gov.hk/eng/blog/blog20210725.htm.

Fitzpatrick, S., Bengtsson, B., & Watts, B. E. (2014). Rights to housing: Reviewing the terrain and exploring a way forward. *Housing, Theory and Society, 31*(4), 447–463.

Forte, T., Santinha, G., Oliveira, M., & Patrão, M. (2022). The high note of meaning: A case study of public service motivation of local government officials. *Social Sciences, 11*(9), 1–18.

French, P. E., & Emerson, M. C. (2014). Assessing the variations in reward preference for local government employees in terms of position, public service motivation, and public sector motivation. *Public Performance & Management Review, 37*(4), 552–576.

Fullan, M. (2000). The three stories of education reform. Phi Delta Kappan, 581–584.

Grant, A. M. (2008). Employees without a cause: The motivational effects of prosocial impact in public service. *International Public Management Journal, 11*(1), 48–66.

Greener, I. (2002). Understanding NHS reform: The policy-transfer, social learning and path-dependency perspectives. *Governance, 15*(2), 161–183.

Groeneveld, S., Steijn, B., & van der Parre, P. (2009). Joining the Dutch civil service: Influencing motives in a changing economic context. *Public Management Review, 11*(2), 173–189.

Gutmann, A., & Thompson, D. F. (2013). Valuing compromise for the common good. *Dædalus.,* *142*(2), 185–198.

Guzman, F., & Espejo, A. (2015). Dispositional and situational differences in motives to engage in citizenship behaviour. *Journal of Business Research, 68*(2), 208–215.

Hagan, J. D., Everts, P. P., Fukui, H., & Stempel, J. D. (2001). Foreign policy by coalition: Deadlock, compromise, and anarchy. *International Studies Review, 3*(2), 169–216.

Hall, P. (1993). Policy paradigms, social learning, and the state: The case of economic policymaking in Britain. *Comparative Politics, 25*(3), 275–296.

Hansen, R. (2002). Globalization, embedded realism and path dependence: The other immigrants to Europe. *Comparative Political Studies, 35*(3), 259–283.

Harper. P. (2017). Boxed in - Inside Hong Kong's miserable coffin homes where thousands live in spaces barely bigger than their bed. The Sun. https://www.thesun.co.uk/news/3743131/hong-kong-coffin-homes-pictures-size/

Hayes, M. T. (2002). *The limits of policy change: incrementalism, worldview, and the rule of law.* Georgetown University Press.

HKSARG. (2022a). New team strives to serve citizens, HKSARG NEWS, https://www.news.gov.hk/eng/2022/06/20220619/20220619_174929_861.html.

HKSARG (2022b). *The Chief Executive's 2022b Policy Address.* Hong Kong: HKSARG. https://www.policyaddress.gov.hk/2022b/public/pdf/policy/policy-full_en.pdf.

Hong Kong Housing Authority (HKHA). (2022). Number of Applications and Average Waiting Time for Public Rental Housing, https://www.housingauthority.gov.hk/en/about-us/publications-and-statistics/prh-applications-average-waiting-time/index.html.

Howlett, M., & Migone, A. (2011). Charles Lindblom is alive and well and living in punctuated equilibrium land. *Policy and Society, 30*(1), 53–62.

Huang, Y. (2021). *Land and housing policy research report: Decisive moment: Can Hong Kong save itself from the land and housing supply crisis?* Our Hong Kong Foundation.

Huque, A. S., & Lee, G. O. M. (2000). *Managing public services: Crises and lessons from Hong Kong.* Ashgate.

Hustedt, T., & Salomonsen, H. H. (2018). From neutral competence to competent neutrality? Revisiting neutral competence as the core normative foundation of Western bureaucracy, *Bureaucracy and Society in Transition* (Comparative Social Research, Vol. 33), Bingley: Emerald Publishing. 69–88.

Jones, B. D., & Baumgartner, F. R. (2005). A model of choice for public policy. *Journal of Public Administration Research and Theory: J-PART, 15*(3), 325–351.

Jones, B. D., True, J. L., & Baumgartner, F. R. (1997). Does incrementalism stem from political consensus or from institutional gridlock? *American Journal of Political Science, 41*(4), 1319–1339.

True, J. L., Jones, B. D., & Baumgartner, F. R. (2018). Punctuated-equilibrium theory: Explaining stability and change in public policymaking. In C. Weible & P. A. Sabatier (Eds.), *Theories of the Policy Process* (pp. 155–188). Westview Press.

King, P. (2003). Housing as a freedom right. *Housing Studies, 18*(5), 661–672.

Lai, R. N., & Wang, K. (1999). Land-supply restrictions, developer strategies and housing policies: The case in Hong Kong. *International Real Estate Review, 2*(1), 143–159.

Lam, J. T. M. (2004). Ministerial system in Hong Kong: A strengthening of the executive leadership. *Asian Perspective, 28*(1), 183–216.

Langbein, L., & Knack, S. (2010). The worldwide governance indicators: Six, one, or none? *Journal of Development Studies, 46*(2), 350–370.

Lau, G. (2022). Regulation on minimum home size in London, Research Office, Information Services Division, Legislative Council Secretariat, https://www.legco.gov.hk/research-publications/english/essentials-2022ise09-regulation-on-minimum-home-size-in-london.htm.

Lau, S. K., & Kuan, H. C. (2002). Hong Kong's stunted political party system. *The China Quarterly, 172*, 1010–1028.

Lee, J. C. Y. (2001). The changing context of public sector reform and its implications in Hong Kong. In Cheung, A. B. L., & Lee, J. C. Y. (eds.), *Public Sector Reform in Hong Kong*. Hong Kong: Chinese University Press.

Lee, J. K. C. (2011). Housing policy: A neoliberal agenda? In Lam, N. M. K. & Scot, I. (Eds.), *Gaming, Governance and Public Policy in Macao*. Hong Kong: The University of Hong Kong Press.

Lee, J. K. C. (2012). Housing policy at a crossroad? Re-examining the role of Hong Kong government in the context of a volatile housing market. In Chiu, S. W.-K., & Wong, S.-l. (Eds.), *Repositioning the Hong Kong Government: Social Foundations and Political Challenges*. Hong Kong: Hong Kong University Press, 165–186.

Legislative Council (LegCo) (2022, August 9). LegCo Subcommittee visits families at sub-divided units and transitional housing projects (with photos) https://www.info.gov.hk/gia/general/202 208/09/P2022080900818.htm?fontSize=1

Lindblom, C. E. (1959). The science of "muddling through." *Public Administration Review, 19*(2), 79–88.

Lindblom, C. E. (1977). *Politics and markets: The world's political economic systems*. Basic Books.

Lindblom, C. E. (1990). *Inquiry and change: The troubled attempt to understand and shape society*. Yale University Press.

Luk, G. C. H. (2018). Straddling the handover: Colonialism and decolonization in British and PRC Hong Kong. In R. K. M. Yep (Ed.), *From a British to a Chinese Colony? Hong Kong before and after the 1997 Handover, Berkeley: Institute of East Asian Studies* (pp. 1–49). University of California.

Luthans, F., & Youssef-Morgan, C. M. (2017). Psychological capital: An evidence-based positive approach. *Annual Review of Organizational Psychology and Organizational Behavior, 4*, 339–366.

Luthans, F., Youssef-Morgan, C. M., & Avolio, B. J. (2015). *Psychological capital and beyond*. Oxford University Press.

Manolopoulos, D. (2008). Work motivation in the Hellenic extended public sector: An empirical investigation. *International Journal of Human Resource Management, 19*(9), 1738–1762.

McConnell, A., & Hart, P't. (2019). Inaction and public policy: Understanding why policymakers 'do nothing'. *Policy Sciences, 52*, 645–661.

Migone, A., & Howlett, M. (2015). Charles E. Lindblom, "The science of muddling through". In Balla, S. J., Lodge, M., & Page, E. C. (Eds.), *The Oxford Handbook of Classics in Public Policy and Administration*. Oxford: Oxford University Press.

Murphy, P. (2020). *The political economy of prosperity: Successful societies and productive cultures*. Routledge.

Ng, M. K., Lau, Y. T., Chen, H.-W., He, S. W. (2021). Dual land regime, income inequalities and multifaceted socio-economic and spatial segregation in Hong Kong. In: van Ham, M., Tammaru, T., Ubarevičienė, R., Janssen, H. (eds). *Urban Socio-Economic Segregation and Income Inequality*. The Urban Book Series. Cham: Springer International Publishing AG.

Nissim, R. (2021). *Land administration and practice in Hong Kong* (5th ed.). Hong Kong University Press.

Oreg, S., Bartunek, J. M., Lee, G., & Do, B. (2018). An affect-based model of recipients' responses to organizational change events. *Academic Management Review., 43*, 65–86.

Ostrov, B. C. (2002). Hong Kong's CyberPort—do government and high tech mix? *The Independent Review, 2*, 221–236.

Pal, L. A. (2011). Assessing incrementalism: Formative assumptions, contemporary realities. *Policy and Society., 30*(1), 29–39.

Park, A. Y. S., & Sapotichne, J. (2019). Punctuated equilibrium and bureaucratic autonomy in American city governments. *Policy Studies Journal, 48*(4), 896–925.

Perry, J. L., & Wise, L. R. (1990). The motivational bases of public service. *Public Administration Review., 50*, 367–373.

Perry, J. L., Hondeghem, A., & Wise, L. R. (2010). Revisiting the motivational bases of public service: Twenty years of research and an agenda for the future. *Public Administration Review, 70*(5), 681–690.

Perry, J. L., Mesch, D., & Paarlberg, L. (2006). Motivating employees in a new governance era: The performance paradigm revisited. *Public Administration Review, 66*(4), 505–514.

Peters, B. G. (2018). Chapter 1: The logic of policy design. In Peters, B. G. (ed.), *Policy Problems and Policy Design*. Cheltenham, UK: Edward Elgar.

Pierson, P. (2000). Increasing returns, path dependence and the study of politics. *American Political Science Review, 94*(2), 251–267.

Polidano, C. (2001). Why civil service reforms fail. *Public Management Review, 3*(3), 345–361.

Poon, A. (2011). *Land and the ruling class in Hong Kong* (2nd ed.). Enrich Professional Pub.

Priedols, M., Dimdins, G., Gaina, V., Leja, V., & Austers, I. (2022). Political trust and the ultimate attribution error in explaining successful and failed policy initiatives. *SAGE Open, 12*(2).

Qizilbash, M. (2010). Development ethics at work. Explorations, 1960 -2002. By Denis Goulet. *Economica, 77*(306), 410–411.

Rainey, H. G., Fernandez, S., & Malatesta, D. (2021). *Understanding and managing public organizations* (6th ed.). John Wiley & Sons.

Rating and Valuation Department (RVD). (2022). Private Domestic—Average Prices by Class (from 1982), https://www.rvd.gov.hk/en/publications/property_market_statistics.html.

Rezaei, J., Arab, A., & Mehregan, M. (2021). Equalizing bias in eliciting attribute weights in multiattribute decision-making: Experimental research. *Journal of Behavioral Decision Making, 35*(2), e2262.

Rosati, C. S. (2016) Moral motivation. In Zalta, E. N. (ed.) *Stanford Encyclopedia of Philosophy*. Stanford, C.A.: Metaphysics Research Lab.

Šavareikienė, D. (2008). *Motivation in the Management Process (Original: Motyvacija Vadybos Procese)*. Šiauliai University Publishing House.

Sayre, W. S. (1954). The recruitment and training of bureaucrats in the United States. *Annals of the American Academy of Political and Social Science, 292*, 39–44.

Scharpf, F. W. (2000). Institutions in comparative policy research. *Comparative Political Studies, 33*(6/7), 762–790.

Sinclair, G. F. (2015). The international civil servant in theory and practice: Law, morality, and expertise. *European Journal of International Law, 26*(3), 747–766.

Spranca, M., Minsk, E., & Baron, J. (1991). Omission and commission in judgment and choice. *Journal of Experimental Social Psychology, 27*, 76–105.

Stacke, S. (2017). Life inside Hong Kong's "coffin cubicles". *National Geographic*. https://www.nationalgeographic.com/photography/article/hong-kong-living-trapped-lam-photos

Streeck, W., & Thelen, K. (2005). Introduction: Institutional change in advanced political economies. In W. Streeck & K. Thelen (Eds.), *Beyond Continuity: Institutional Change in Advanced Political Economies* (pp. 1–39). Oxford University Press.

Sun, S. (2021). The relationship between public service motivation and affective commitment in the public sector change: A moderated mediation model. *Frontiers in Psychology., 10*, 3389.

Svavarsdottir, S. (1999). Moral cognitivism and motivation. *Philosophical Review, 108*, 161–219.

Taylor, J. (2010). Graduate recruitment in the Australian public sector: The importance of line managers. *Public Management Review, 12*(6), 789–809.

Tetlock, P. E., & Boettger, R. (1994). Accountability amplifies the status quo effect when change creates victims. *Journal of Behavioral Decision Making., 7*, 1–23.

Thelen, K. (2003). How institutions evolve: Insights from comparative historical analysis. In Mahoney, J., & Rueschemeyer, D. (eds.), *Comparative Historical Analysis in the Social Sciences*. Cambridge: Cambridge University Press.

Thelen, K. (2004). *How Institutions Evolve: The Political Economy of Skills in Germany, Britain, the United States, and Japan*. Cambridge University Press.

Thom, N., & Ritz, A. (2004). Public Management: An Innovative Outline of Public Sector Management (Original: Viešoji vadyba: inovaciniai viešojo sektoriaus valdymo metmenys). Vilnius: Lithuanian Law University Publishing Centre

Tsebelis, G. (2000). Veto players and institutional analysis. *Governance: An International Journal of Policy and Administration, 4*, 441–474.

Van Noorloos, F., Cirolia, L. R., Friendly, A., Jukur, S., Schramm, S., Steel, G., & Valenzuela, L. (2020). Incremental housing as a node for intersecting flows of city-making: Rethinking the housing shortage in the global South. *Environment and Urbanization, 32*(1), 37–54.

van Parijs, P. (2012). What makes a good compromise? *Government and Opposition, 47*(3), 466–480.

Vandenabeele, W., Depré, R., Hondeghem, A., & Yan, S.-f. (2004). The motivational patterns of civil servants. *Public Policy and Administration, 13*(1), 52–63.

Walther, F. (2015). New public management: The right way to modernize and improve public services? *International Journal of Business and Public Administration, 12*(2), 132–143.

Wang, T., van Witteloostuijn, A., & Heine, F. (2020). A moral theory of public service motivation. *Frontiers in Psychology., 11*, 517763.

Wescott, C. G., & Bowornwathana, P. (2008). *Comparative Governance Reform in Asia: Democracy, Corruption, and Government Trust* (1st ed.), Research in public policy analysis and management, vol. 17. Bingley: Emerald.

Wilsford, D. (1994). Path dependency, or why history makes it difficult but not impossible to reform health care systems in a big way. *Journal of Public Policy, 14*(3), 251–283.

Wong, R.-Y.-c. (2013). *Diversity and occasional anarchy* (1st ed.). Hong Kong University Press.

Wong, W. W. H. (2012). The civil service. In Lam, W. M., Lui, P. L-t., & Holliday, I. (eds.). *Contemporary Hong Kong Politics: Governance in the Post-1997 Era*. Hong Kong: Hong Kong University Press (2nd Ed.). 87–110.

Wright, B. E., Christensen, R. K., & Isett, K. R. (2013). Motivated to adapt? The role of public service motivation as employees face organizational change. *Public Administration Review., 73*, 738–747.

Youssef, C. M., & Luthans, F. (2007). Positive organizational behavior in the workplace: The impact of hope, optimism, and resilience. *Journal of Management, 33*(5), 774–800.

Youssef-Morgan, C. M., & Luthans, F. (2015). Psychological capital and well-being. *Stress and Health, 31*(3), 180–188.

Yun, C. (2020). Politics of Active Representation: The Trade-off Between Organizational Role and Active Representation. *Review of Public Personnel Administration, 40*(1), 132–154.

Yung, B. (2014). Differential public service motivation among Hong Kong public officials: A qualitative study. *Public Personnel Management, 43*(4), 415–441.

Zhang, S-m., & Pearlman, K. (2004) China's land use reforms: A review of journal literature. *Journal of Planning Literature*, CPL Bibliography 373, 19(1), 16–61.

Žilinskas, V. J., & Zakarienė, J. (2007). Darbuotojų skatinimas – aktuali mokslo ir praktikos problema (Original: Employee promotion - a relevant problem in science and practice). *Tiltai, 3*(1), 25–34.

Dr. Pui Ting Chow is a lecturer at the HKU School of Professional and Continuing Education Po Leung Kuk Stanley Ho Community College. Her Ph.D. studied withdrawal in construction dispute negotiation and was completed with the Construction Dispute Resolution Research Unit of the City University of Hong Kong. Her Ph.D. was awarded the 2011 Best Dissertation (Ph.D. Category) by the Hong Kong Institute of Surveyors. Dr. Chow's research centres on decision-making in construction dispute negotiation, intra- and inter- organizational behaviours, and self-regulation in construction management. Her research sheds light on the external motivation of result-oriented behaviours, promotes end users' well-being, and guides organizational design of effective dispute resolution system. Dr. Chow's research has been published in leading construction engineering and management journals.

Part III
Specific Uses

Chapter 9
Means to Incentivize Safety Compliance at Work

Tak Wing Yiu

Abstract Construction is one of the most dangerous sectors to work in; governments from various countries enact health and safety regulations to cultivate good health habits and impose stakeholders' duties to ensure the work environment is safe. However, these regulations always impose penalties on the stakeholders of construction organizations in attaining their objectives. This chapter gives an overview of these regulations in the United Kingdom, Australia, New Zealand and Singapore and discusses the effectiveness of these penalties that Deterrence Theory underpins. In addition, alternative means to incentivize safety compliance at work from literature are discussed. Recommendations are then given for further research on incentivizing safety compliance in the construction sector. These include developing incentive and penalty provisions in construction contracts, revisiting the applicability of Deterrence Theory and reinforcing the link between safety incentives and compliance at construction sites.

Keywords Safety compliance · Construction health and safety · Incentives · Penalty

1 Introduction

Construction is one of the most dangerous sectors to work in; accident fatalities are three to four times more likely in construction than in other sectors (International Labour Organization, 2015). Despite the endeavours of researchers to improve health and safety in the construction sector, tragic events often occur as a consequence of people not following set rules. In the past few decades, construction organizations and academics have devised a large number of new practices to maintain health and safety standards and to achieve effective safety performance. It has been found that effective safety performance can only be achieved through effective implementation

T. W. Yiu (✉)
School of Built Environment, University of New South Wales, Sydney, NSW, Australia
e-mail: kenneth.yiu@unsw.edu.au

© The Author(s), under exclusive license to Springer Nature Switzerland AG 2023
S. O. Cheung and L. Zhu (eds.), *Construction Incentivization*, Digital Innovations in Architecture, Engineering and Construction,
https://doi.org/10.1007/978-3-031-28959-0_9

of safety regulations, leadership, safety planning, safety compliance, performance measurement, risk assessment, safety inspection, and safety culture (Khalid et al., 2021). These factors are interrelated and cannot be implemented in isolation, but it is important to prioritize these factors in order to significantly improve safety performance on construction projects. Safety performance could be divided into safety compliance and safety participation by different knowledge, skills and motivation (Griffin & Neal, 2000).

Senior management can establish unique leadership behaviours to encourage safety participation, but how to encourage more employees to adhere to safety regulations needs to be further explored (Clarke, 2013; Griffin & Hu, 2013). Governments from various countries enact health and safety regulations with the aim of cultivating good safety habits and creating strong safety culture in construction workplaces. These regulations often define and impose duties of stakeholders such as employers, employees and manufacturers to make sure the work environment is safe, provide adequate safety measures, and develop systems/plans for dealing with emergency situations. Penalties are often imposed on those who breach these regulations, with offenders being liable to a fine or imprisonment. Consequently, it is a priority for any construction company to have a comprehensive culture of safety compliance, especially in fostering a behaviour of following safety regulations established by legislative and regulatory agencies which have a goal of protecting workers from various work-related hazards (Mishra, 2022). Theoretical work to exploit the concept of compliance was conducted a few decades ago. As suggested by Meier and Johnson (1977), there are two sources of compliance; namely, *'compliance produced by influences other than a legal threat'* and *'compliance produced by legal threats—deterrence'*. Imposing penalties and prosecutions are obviously a form of legal threat that become the deterring factors that the legal literature assumes will bring about compliance with the law. Tyler (1990) further developed these sources. He suggests that the two sources of compliance can be referred as normative and instrumental. A normative perspective states *'compliance is produced through personal morality and feelings of law being just'*, while instrumental perspective *'underlies deterrence literature'*. This view is also supported by (Brown, 1997). Typically, a government or authority often adopts an instrumental perspective to ensure safety compliance, with penalty and/or prosecution the means used to punish those who fail or refuse to comply with the law. However, there is little evidence to demonstrate the effectiveness of these penalties, herewith called negative incentives', in attaining the objectives of the law/regulations. Options for adopting instrumental approach to ensure compliance are subjects of research in the construction sector. This chapter is designed to explore how government drive safety compliance through legislation, and to discuss the means and their effectiveness to incentivize safety compliance at work. Recommendations are then given to further research on incentivizing safety compliance in construction sector.

2 Health and Safety Acts

United Kingdom

United Kingdom governments have enacted a number of health and safety bills to protect the health and safety of construction workers on construction sites, such as "Health and Safety at Work etc. Act 1974"; "The Construction (Health, Safety and Welfare) Regulations 1996" and "Building Safety Act 2022". These Acts provide detailed information on the safety risk management, development of safety procedures associated with construction work and the safety duties of the parties involved in the construction industry. The Health and Safety at Work Act 1974 is the primary legislation for securing workers' health and safety in the United Kingdom. Schedule 3A of this act summaries the mode of trial and maximum penalties applicable to a list of offences under Section 33. Breaches of this Act can incur fines of up to £20,000 or/and imprisonment for a term not exceeding two years.

Australia

Business in Australia must ensure workers' safety. Failure to comply with the Work Health and Safety Act 2011 can lead to penalties and/or prosecutions (Table 1).

New Zealand

New Zealand (NZ) Government is determined to reduce work related illness, injuries and deaths. In 2016, a new Health and Safety at Work Act (HSW Act) came into effect and has increased the liability of duty bearers. For instance, it introduced a new term called PCBU (Person Conducting a Business or Undertaking) who has

Table 1 Maximum penalties for offences under the Work and Health and Safety Act 2011 in Australia (*Source* Work Safe, Australia)

Categories of offence	Maximum penalties
1—Reckless Conduct (*Section 31*)	A fine of AUS$300,000 and/or imprisonment for 5 years for an individual, A fine of AUS$600,000 and/or imprisonment for 5 years for an individual as a person in control of a business or undertaking or as an officer, or A fine of AUS$3,000,000 for a body corporate
2—Failure to comply with Health and Safety Duty (*Section 32*)	A fine of AUS$150,000 for an individual, A fine of AUS$300,000 for an individual as a person in control of a business or undertaking or as an officer, or A fine of AUS$1,500,000 for a body corporate
3—Failure to comply with Health and Safety Duty (*Section 33*)	A fine of AUS$50,000 for an individual, A fine of AUS$100,000 for an individual as a person in control of a business or undertaking or as an officer, or A fine of AUS$500,000 for a body corporate

the primary duty of ensuring the safety of workers and anyone affected by work of the PCBU. The PCBU is not necessarily one person; it can also refer to a business entity, such as an organization or company. It '*must ensure, so far as is reasonably practicable, that the health and safety of other persons is not put at risk from work carried out as part of the conduct of the business or undertaking*' (Section 36 of HSW Act, New Zealand). The maximum penalties for offences under the Act are shown in Table 2.

Singapore

The Workplace Safety and Health (WSH) Act in Singapore focusses not only on compliance but also on workplace safety and health systems and outcomes. High penalties for non-compliance and risky behaviors are imposed (Table 3).

Table 2 Maximum penalties for health and safety duty offences (extracted from Worksafe, 2019 and HSW Act of New Zealand)

Sections of HSW Act (New Zealand)	Individual who is not a PCBU or Officer (e.g., a worker or other person at a workplace)	Officer of a PCBU or an Individual who is a PCBU (e.g., self-employed)	Anyone Else (e.g., an organization that is a PCBU)
Section 47 *Offence of reckless conduct in respect of duty*	A term of imprisonment not exceeding 5 years or a fine not exceeding NZ$300,000, or both	A term of imprisonment not exceeding 5 years or a fine not exceeding NZ$600,000, or both	A fine not exceeding NZ$3,000,000
Section 48 *Offence of failing to comply with duty that exposes individual to risk of death or serious injury or serious illness*	A fine not exceeding NZ$150,000	A fine not exceeding NZ$300,000	A fine not exceeding NZ$1,500,000
Section 49 *Offence of failing to comply with duty*	A fine not exceeding NZ$50,000	A fine not exceeding NZ$100,000	A fine not exceeding NZ$500,000

Table 3 Maximum (General) penalties for failing to comply with WSH Act (Extracted from the Ministry of Manpower, Singapore)

Type of Offender	Maximum Fine (SGD)	Maximum Imprisonment
Individual	First conviction: $200,000 Repeat offender: $400,000	2 years
Corporate body	First conviction: $500,000 Repeat offender: $1,000,000	N/A

3 Penalty and Prosecution

Penalties, a common means to standardize workers' safety behaviours in the construction sector (Teo & Ling, 2009; Wong et al., 2020), and prosecutions are routinely imposed following non-compliance with the law. As fines increase, construction workers will be increasingly aware of their safety conduct (Wu et al., 2022). The research of Man et al. (2017) shows that the number of financial safety penalties may be equivalent to workers' half- or whole-time wages. Therefore, workers are likely to refrain from taking risks at work to avoid receiving such safety penalties. Discipline is frequently used in addition to financial penalties to regulate workers' safe behaviours (Lipscomb et al., 2013; Mishra, 2022), with the severity of disciplinary action ranging from minor sanctions (such as verbal warnings and retraining) to major sanctions (such as suspension of work or dismissal) (Mishra, 2022). Furthermore, a safety-offence points system can keep track of all worker infractions and each safety violation can be given a score. Workers must undertake additional safety training whenever their safety-offence points reach a specific level, or they may face other disciplinary actions (Wong et al., 2020). Moreover, it has been shown that unsafe behaviour on construction sites could be contagious, which means that if a worker's behaviour does not meet safety compliance, such unsafe behaviour can easily be copied by other worker. However, penalties are an effective way of preventing this contagious behaviour (Jiang et al., 2018). Although penalties have been regarded as a basic safety management method, this negative reinforcement can only have effects in the short term and do not improve the overall worker safety compliance (Teo & Ling, 2009). Increasing fines have been criticized as having limited effectiveness in incentivizing construction workers to operate safely (Teo & Ling, 2009; Wu et al., 2022). For instance, there is a negative correlation between the amount of the violation fine and construction workers' safety compliance; that is, the higher the fine, the worse the effect, and eventually the construction workers are converted from safe operations to violations of safety rules (Wu et al., 2022). In addition, excessive financial fines can dampen workers' enthusiasm and lower morale (Teo & Ling, 2009).

The severe penalties in various Health and Safety Acts to date primarily originated from the concept of deterrence. In this concept, the potential offender considers the penalty for the crime before committing it and weighs the consequences (Biddle, 1969). It often refers to the *Deterrence Theory* that encourages people to obey the law on the assumption that *'people are rational actors who weigh the costs and benefits when deciding whether to offend sanction threats and imposed punishments are presumed to inhibit initial criminal activity and deter its subsequent recurrence by increasing the costs of crime'* (Piquero et al., 2011). Deterrence Theory is underpinned by three major predicators: (1) severity, (2) certainty and (3) celerity (Abramovaite et al., 2022; Bentham, 1789). Severity refers to the harshness of the penalty, while certainty refers to the probability of being caught. The last element, celerity, which is always overlooked, refers to the swiftness of punishment (Abramovaite et al., 2022; Pratt & Turanovic, 2018). The severity of a penalty decreases the level of illegal behavior, while a low or high probability of being caught is critical in a decision to act illegally (Herath & Rao, 2009). From occupational health and safety (OHS) perspectives, Dorrian and Purse (2011) quantitatively and qualitatively studied the deterrent effect of OHS enforcement on employer behaviour. This study revealed that traditional Deterrence Theory is not fully applicable to OHS. It further advocated that '*there are many gaps in the understanding of the role played by enforcement in promoting compliance with OHS obligations and in reducing work-related injury*'. As such Deterrence theory may need to be revisited and re-conceptualized regarding its application to OHS. Another interesting study, which was conducted by McCallum et al. (2012), examined the deterrent impact of occupational health and safety prosecutions in Australia. One of the key findings was that publicity given to OHS cases would improve the deterrence effects of OSH prosecutions. Sanctions such as publicity orders would be a better way to penalize offenders and inform the public about the outcome of an OHS case, as well as enhancing educative functions to the community.

Research on Deterrence Theory in relation to the enforcement of construction health and safety Acts remains limited. Its applicability should be re-visited. For example, construction practitioners may not have perfect knowledge of their 'risk assessment (e.g., the risks of being caught)' when a rational choice between the costs and benefits is made. Up to now, the most relevant research on studying law compliance measures for the construction sector was conducted by Brown (1997). This study discussed measures to ensure compliance based on theories from the law, psychology and sociology. The Health and Safety legislation within the United Kingdom construction industry adopts the instrumental perspective to impose legal threats for non-compliance; however, Brown (1997) recommended that the integration of deterrence theory and normative control such as legitimacy, morals, justice and fairness should be included to ensure maximum compliance levels. A possible tactic that offers a balanced approach between normative and instrumental approaches could be the introduction of safety incentive and penalty (I/P) provisions in construction contracts (Hasan & Jha, 2013).

4 Means to Incentivize Safety Compliance

Most of the Health and Safety Acts impose liabilities, in the form of penalties or prosecutions, on construction organizations. These penalties, often regarded as legal threats, can be regarded as negative incentives (or an instrumental perspective) that aim to stop construction practitioners from violating the provisions of these Acts. However, there is little evidence to justify the belief that these negative incentives help in improving safety. It is argued that the construction sector should encourage positive behaviors that fundamentally improve the safety culture of the sector. Rewards and incentives motivate employees to voluntarily comply with safety regulations (Health & Safety Executive, 2012) and improve safety performance (Zulkefli et al.,

2014). These can include recognition, time off, stock ownership, special assignments, advancement, increased autonomy, training and education, social gatherings, prizes and money (Zulkefli et al., 2014). Managers or safety practitioners have implemented a variety of interventions, including safety training, safety communication, safety management commitment, safety policy, safety incentives, etc., to encourage workers' adherence to safety compliance (Di Tecco et al., 2017; Wang et al., 2017; Kim et al., 2019; Ji et al., 2021; Zhang et al., 2022). Ghasemi et al. (2015) proposed a new incentive system, called 'surprise incentive system', to the construction sector. This system offers financial awards to employees who comply with predetermined performance measures such as 'proper use of PPEs', 'record and report near misses', 'record and report unsafe condition' and 'propose appropriate technical and managerial suggestions to correct unsafe conditions and behaviors'. Interestingly, it was found that this system improved safety performance in short term, but a declining trend in safety performance was recorded after 3 or 6 months. Similar results were obtained by the study of Ahmed and Faheem (2021) in that incentive and penalty programmes did not have a long-lasting positive effect on worker safety compliance. These results demonstrate that incentives should be regularly evaluated and modified to prevent their value declining over time (Ghasemi et al., 2015). The role model of managers and the method of distribution of incentives are the key factors in the success of incentive programme (Zulkefli et al., 2014). This study concluded that managers should encourage self-motivation to work in safe manner (e.g., by hosting events and organizing award granting committees), and that non-financial incentives (e.g., receiving awards and recognition for working safely) are more effective in comparison to financial incentives. In sum, giving workers financial or non-financial rewards is a crucial component of achieving safety compliance (Kim, 2018). These incentive measures should be frequently used by management (Guo et al., 2018), and the sector should incorporate means or schemes to incentivize safety compliance at work. A summary of these means or schemes is given in Tables 4 and 5.

4.1 Financial Incentives

Safety incentive programs that provide financial incentives can take many forms (Lipscomb et al., 2013); for instance, providing cash or bonuses to individual workers or work groups if no injuries occur within a certain working time (Lipscomb et al., 2013; Ji et al., 2021; Liu et al., 2022; Zulkefli et al., 2014). In addition, some studies (Idoro, 2008; Ji et al., 2021; Teo & Ling, 2009; Zulkefli. et al., 2014) have shown that dedicated financial incentives (cash or bonuses) to individual workers who have outstanding safety performance not only stimulate continued safety compliance by award-winning workers but also incentivize improvements in safety behaviors of non-awarded workers. Rather than the direct financial incentives, there are other non-cash means to encourage expected safety behaviours among workers, such as monthly gift cards (Ghasemi et al., 2015; Health & Safety Executive, 2012), free parking passes or petrol station gift vouchers (Sparer et al., 2016), free overseas trips or dining &

Table 4 A summary of financial incentives on safety compliance

Financial incentives	References
Financial Rewards to ALL workers if nobody get injured during the completion of certain work hours	Hu et al. (2012), Idoro (2008), Ji et al. (2021), Kim and Kim (2019), Lipscomb et al. (2013), Liu et al. (2022), Zulkefli et al. (2014)
Financial Rewards to individual worker due to his/her prominent safety behaviors (does not get injured in a particular period of time)	Idoro (2008), Teo and Ling (2009), Zulkefli et al. (2014)
Salary progression or bonus program: Field supervisors might be rewarded for working ahead of schedule, general contractors for contract management and overall profits, and so on. This keeps bonuses relevant and achievable for workers at any level	Kim (2018), Welles (2022)
Gift cards, a catered lunch, a raffle, or a gas station gift certificate, trip overseas, dinning & shopping vouchers, paid vacation	Ghasemi et al. (2015), Idoro (2008), Kim (2018), Kim and Kim (2019), Sparer et al. (2016), Teo and Ling (2009), Zulkefli et al. (2014)
Safety Inspection Score: With the goal of changing safety culture, the mechanics of employee safety incentive programs use a given safety performance threshold to reward workers when a certain performance criterion is achieved. Financial reward can be received if the workers exceed this predetermined threshold level of safety at the end of a reward period (i.e., one month or one quarter)	Liu et al. (2022)
Income-sharing mechanism: A safety incentive mechanism (e.g., income-sharing contract) combining reward and punishment with income sharing can be implemented	McDermott et al. (2017), Liu et al. (2022), Welles (2022)

shopping vouchers (Kim, 2018; Teo & Ling, 2009) and safety raffles (Aksorn & Hadikusumo, 2008; Health & Safety Executive, 2012). However, workers are more likely to anticipate raising wages or bonuses than non-cash incentives (Kim, 2018); introducing income-sharing contracts is one of the examples that could incentivize workers' safety compliance. This approach is mainly distributed to workers in the form of year-end bonuses. A specified portion of the total income of project can be collected after the construction workers meet or surpass the predetermined profit or safety objective (Liu et al., 2022; Welles, 2022).

Table 5 A summary of non-financial incentives on safety compliance

Non-financial Incentives	References
Internal promotion: Senior management rewards employees with internal promotion once they meet specific safety targets or have outstanding safety performance. This incentive method can effectively enhance the independent enthusiasm of employees ' safety compliance	Idoro (2008), Teo and Ling (2009), Lu and Yang (2010), Zulkefli et al. (2014), Welles (2022)
Oral Praise or Positive Evaluation: • To retain and encourage employees, it is crucial to verbally recognize them, both in private and in front of their peers • Cite specific examples of their hard work and how it positively impacted a project or the team • Diverse ways of praising such as venue praise, commendation meetings, newspapering, team meetings, etc	Lu and Yang (2010) Health and Safety Executive (2012), Zulkefli et al. (2014), Ghasemi et al. (2015), Ji et al. (2021), Welles (2022), Zhang et al. (2022)
Enhance safety training and safety education: • Proper labour training to avoid accidents (video/audio/posters display, mock drills, etc.) • Comprehensive and continuing safety training improve workers' risk perception, work abilities, and safety awareness, which ultimately encourages them to adhere to safety compliance	Lu and Yang (2010), Hasan and Jha (2013), Ji et al. (2021), Wu et al. (2022), Zhang et al. (2022)
Safety Communication with feedback: Senior leaders should regularly communicate with workers about safety behaviours and promptly correct workers ' unsafe behaviours. Leaders need to encourage workers to give feedback to improve communication	Teo and Ling (2009), Hu et al. (2012), Sparer et al. (2016), Ji et al. (2021), Zhang et al. (2022)
Moral incentive: Any form of reward should pay attention to respect for workers ' values, traditional culture, and religious beliefs. This moral incentive could improve workers ' sense of identity so that workers are more willing to comply with safety regulations issued by leaders	Teo and Ling (2009), Health and Safety Executive (2012)
Closer Supervision of Works The leadership should regularly visit the site for safety monitoring or inspection to ensure the safety compliance of workers in the construction site. Safety management without supervision cannot function properly	Griffin and Hu (2013), Man et al. (2017), Wang et al., (2017), Welles (2022), Wong et al. (2020)

(continued)

Table 5 (continued)

Non-financial Incentives	References
Safety policy: This policy refers to a clear code of conduct rather than a penalty. Workers with low levels of education may not be aware of the safety behaviour standards. They need the direction of these safety policies to enhance safety compliance	Mearns et al. (2003), Lu and Yang (2010), Di Tecco et al. (2017), Wong et al. (2020), Ji et al. (2021), Zhang et al. (2022)

4.2 Non-financial Incentives

4.2.1 Promotion and Praise

When workers realize that maintaining safety compliance might lead to promotion, they are more inclined to work more safely (Welles, 2022). Moreover, effective praise (whether in front of peers or in private) is essential to improving the safety motivation of construction workers (Lu & Yang, 2010; Welles, 2022). Leaders should frequently praise their workers' risk identification or safety behaviours instead of habitually criticizing workers' mistakes on construction sites (Zhang et al., 2022). There are numerous forms of praise, including through newspapers, posters, site briefings, team meetings or newsletters (Ghasemi et al., 2015; Health & Safety Executive, 2012). It has been shown that the incentives for recognition proposed on special occasions or toolbox meetings by the safety director or manager may be more efficient in improving workers' adherence to safety compliance (Zulkefli. et al., 2014). These positive reinforcements are important for promoting workers' safety compliance (Zhang et al., 2022).

4.2.2 Enhance Safety Training and Communication

According to Wu et al. (2022), construction workers are more willing to operate safely when they are aware of the potentially severe consequences of unsafe behaviour. Hence, correcting construction workers' fluke and optimism bias by enhancing safety education can facilitate their safety compliance. The method of strengthening workers' safety education is generally via safety training and safety communication (Ji et al., 2021). Safety training can make workers aware that compliance with safety policies is in their own interest, and it can be more effective if the training is conducted in a worker's mother language (Hasan & Jha, 2013). According to Zhang et al. (2022), the most significant aspects influencing safety compliance are safety training and safety communication with feedback, which directly and positively affect workers' safety compliance. Leaders could provide effective safety training and safety communication to fulfil workers' safety demands, which could incentivize their safety compliance. The relationship between safety compliance and frequency of safety training is also positive and significant (Lu & Yang, 2010).

4.2.3   Promote Moral Incentive

A UK government agency, the Health and Safety Executive, suggested one sort of incentive could a moral incentive (Health & Safety Executive, 2012). They recommend that incentive methods should consider the values, cultural traditions, and religious beliefs of employees. For instance, any food reward incentive should not conflict with workers' cultural and religious beliefs. In addition, special prayer rooms or subsidized pilgrimage trips could be offered to workers with religious beliefs.

4.2.4   Closer Supervision of Works

On construction sites, workers with poor safety awareness tend to violate safety requirements such as operating in a non-compliant manner, or the non-compliance use of equipment (Wu et al., 2022; Zahoor et al., 2016). Therefore, frequent (face-to-face) monitoring at construction sites is gradually being adopted to increase safety compliance (e.g., the proper use of personal protection equipment (Ghasemi et al., 2015; Wong et al., 2020). While some researchers have stated that passive monitoring may result in dissatisfaction among workers (Zhang et al., 2022), other researchers hold the opposite view (Griffin & Hu, 2013; Wang et al., 2017; Welles, 2022). For instance, to achieve proactive monitoring, supervisors should physically be present at the construction site and to inspire their teams, show concern for the well-being, and to become role models for safety compliance (Welles, 2022). In addition, frequent real-time supervision can regulate workers' behaviours by promptly correcting workers' violations (Man et al., 2017). Moreover, closer supervision on the construction sites can guarantee the effective implementation of other incentives, such as fair bonuses and prompt recognition of safe operations by construction workers (Welles, 2022).

4.2.5   Promoting Organizational Safety Policy

Whether it is due governmental, legal or industry requirements, basically all construction enterprises promulgate safety policies, establish safety management systems, and take action plans on health and safety (Di Tecco et al., 2017). Therefore, enacting safety policies is always regarded as one of the common means to incentivize worker safety compliance (Ji et al., 2021). However, different scholars have different opinions on the effectiveness of safety policies. For instance, Zhang et al. (2022) proposed that safety policies have positive but insignificant impacts on safety compliance. They emphasized that safety policies and supervision management could be a passive means to improve workers' safety compliance, but it is impossible to motivate workers to adhere to safety standards at a fundamental level. Others, who hold the opposite view, argue that safety policy is direct and significant in regulating safety compliance for construction workers (Lu & Yang, 2010; Mearns et al., 2003; Wong et al., 2020). These safety policies clearly regulate workers' duties and behaviours and aid in correcting the unsafe behaviour of poorly educated personnel (Lu & Yang, 2010).

4.3 Safety Incentive System (Both Financial and Non-financial)

A safety incentive system/program is a compensation mechanism created by senior managers to reward workers when they achieve certain safety objectives in the workplace (Kim, 2018; Safeopedia, 2018). Safety incentive programs can be divided into rate-based programs and behaviour-based programs. The former is a reward when workers achieve a lower accident rate or injury rate, and the latter is a reward for workers with outstanding safety performance (Safeopedia, 2018). Some scholars have proposed an activity-based program, which rewards workers when they take part in some required safety activities (Choi et al., 2011). In actual engineering, senior management frequently uses incentive systems to motivate workers' safety compliance (Ji et al., 2021; Lu & Yang, 2010). Rewards in the safety incentive programs could be financial or non-financial, so this incentive system is more effective in improving worker safety compliance than a single model of incentives (Lu & Yang, 2010; Zhang et al., 2022). Numerous publications have illustrated that managers and safety practitioners implement safety incentive programs in construction work with the goal of enhancing workers' safety performance (Di Tecco et al., 2017; Wang et al., 2017; Kim et al., 2019; Ji et al., 2021). For instance, Ji et al. (2021) explored a new type of compensation incentive program entitled the tournament mechanism. Workers on construction teams make varied investments in their personal safety due to disparities in their educational backgrounds, professional experiences, and salary. Using competition (e.g., giving different wage levels) and the heterogeneous characteristics of workers (workers' preference for fair income distribution), the tournament mechanism encourages workers to improve their safety behavior in order to receive higher wages. In addition, safety management commitment that gives workers financial or non-financial rewards when they meet specified safety requirements has a strongly positive effect on construction workers' safety compliance (Choi et al., 2011). Central to this incentive is whether senior management makes real safety commitments as workers' enthusiasm for safety compliance decreases if the commitments are frequently not fulfilled (Toole, 2002; Lingard & Rowlinson, 2004). Some scholars, however, have proposed that this effect is indirect because this incentive method is usually to improve the workers' safety compliance by increasing their motivation to spontaneously participate in safety training (Zhang et al., 2022). In their research, the specific safety requirements are often the completion of safety training or participation in some safety activities.

5 Discussion

Prior research has been conducted to study the effectiveness of financial and non-financial incentives as well as various safety incentive programs to incentivize safety compliance (Lu & Yang, 2010; Zhang et al., 2022; Zulkefli. et al., 2014). Researchers have opposite views on the effectiveness of incentives from different perspectives (e.g., managers or workers) and different classification methods of incentives. This section illustrates the opinions from different previous studies on the attributes of different effectiveness factors, such as direct or indirect, substantial or not, and positive or negative consequences. Different researchers have different views on the effectiveness of financial incentives. Financial incentives are regarded as the most common and direct way to motivate workers' safety compliance (Han et al., 2020). For workers, compared to other financial incentives like gift cards or paid vacations, direct increases in wages or cash bonuses are the most effective way to motivate safety compliance (Kim & Kim, 2019). These, however, are not always easy to implement (Teo & Ling, 2009). For instance, from the perspective of leadership, financial incentives have significant risks since workers may conceal risks or hazards in order to earn more rewards (Kim et al., 2019; Michaels, 2010; Zhang et al., 2022). In addition, getting the balance right is difficult. Workers may get weary of the incentive program if they never earn any financial rewards; however, they may not perceive the value in striving to improve their safety behaviour if the threshold is set too low because they are likely to receive a reward every month (Sparer & Dennerlein, 2013). Furthermore, other researchers like Ghasemi et al. (2015) stated that while financial incentives could boost workers' safety performance in a short term, the incentive effects may steadily diminish. Comparing financial incentives to non-financial ones, it was discovered that the former was more effective in terms of recognition (Zulkefli. et al., 2014). However, the core method to incentivizing safety compliance is to cultivate the proper safety attitude/spirit (Hasan & Jha, 2013; Teo & Ling, 2009). Current workers are paying more attention to onsite safety and are willing to undertake safety training as a result of the ongoing increase in their awareness of safety (Ji et al., 2021; Wu et al., 2022). Therefore, meeting their needs through non-economic incentives can more effectively stimulate worker safety compliance (Ji et al., 2021).

The complex conditions on construction sites such as external environment, institutional conditions, and individual characteristics, may affect the incentive effect (Ji et al., 2021). A good working environment and safe atmosphere on the construction site can encourage workers to resist violations spontaneously (Welles, 2022; Wu et al., 2022). Construction workers may have reduced satisfaction if there are non-compliant operations on the work site, and this may make them spontaneously resist violations (Zulkefli. et al., 2014). Workers in a good safe atmosphere are not only reluctant to violate the rules but also are more likely take the initiative and put a stop to others' unsafe actions in order to keep the workplace safe (Wu et al., 2022). Moreover, the effectiveness of using an incentive system to encourage employee safety compliance is positively impacted by a company's ranking in the construction industry (Kim & Kim, 2019). They proposed that the higher the construction company is industry-ranked, the more likely it is to implement a safety incentive system.

The individual attributes of workers refer to human nature, values, degree of education, and intelligence level (Teo & Ling, 2009; Wang et al., 2017). These are the most critical factors affecting the incentive effect because the fundamental goal of both financial and non-financial incentives is to change workers' multiple heterogeneous attributes so their safety compliance can be enhanced (Teo & Ling, 2009; Wu et al., 2022). According to Ji et al. (2021), individual disparities in aptitude may also have an impact on the incentive effect on workers' safety behavior. Workers with higher education or more work experience are more focused on safety compliance (Ji et al., 2021). However, workers easily imitate the unsafe behaviour of other workers once they focus more on the benefits of this 'convenient' behaviour instead of its potential risks (Jiang et al., 2018). On the other hand, safety behaviour could also be spread widely so that more workers spontaneously resist unsafe behaviour on construction sites (Wu et al., 2022). This attribute is common to most workers, especially workers who lack safety education (Ji et al., 2021). Therefore, establishing a positive safety climate can effectively improve the incentive effect (Sparer et al., 2016; Wu et al., 2022). Furthermore, Ji et al. (2021) demonstrated that the efficiency of incentives for safety compliance is significantly impacted by the fairness preference heterogeneity of workers. For instance, the incentive effect and the intensity of rewards can be decreased by workers' preference for equitable income distribution (Dubey et al., 2013). This means that unfair financial incentives can create negative emotions, such as envy or pride, and ultimately have a negative effect on incentivizing safety compliance.

5.1 Future Research Areas

Safety compliance is a complex and important topic for the construction sector. It can be associated with legal, behavioral, psychological, and cultural knowledge. This chapter summarizes most of the ways to incentivize safety compliance and proposes the following three future research areas for this topic that enabling promotion of better safety practice for the construction sector.

- Further develop the research of Hasan and Jha (2013) to examine the applicability of Safety Incentive and Penalty (I/P) provisions/schemes in construction contracts, which can balance the impacts of instrumental and normative approaches to construction practitioners and organizations. Normative control such as legitimacy, morals, justice, and fairness should be adopted (Brown, 1997),
- Revisit the applicability of Deterrence Theory on ensuring safety compliance in construction sector by incorporating social mechanisms (e.g., religion) and demographic factors (e.g., age, gender, work experience and formal training) to examine safety compliance for the sector (Brown, 1997).

- In actual projects, few clients allocate budgets for workers' safety compliance, and this part of the cost is often borne by the contractors themselves (Zahoor et al., 2016). Therefore, it may be difficult for senior management to effectively use incentives to improve workers' safety compliance (Kim & Kim, 2019). Musonda and Pretorius (2015) proposed that it is important to improve the safety compliance of whole projects by using financial incentives so as to affect clients' health and safety performance. This publication aimed to motivate clients to actively participate in health and safety initiatives, which raises the possibility of utilizing safety incentives at the workplace. However, further research on incentive methods other than financial incentives to improve the safety awareness of construction clients still needs to be conducted (Musonda & Pretorius, 2015). There is also a lack of relevant literature on the internal relationship between clients' safety incentives and safety compliance at construction sites.

6 Summary

This chapter discusses various means to incentivize safety compliance for the construction sector, with reference to the two sources of compliance (from normative and instrumental approaches). From normative perspective, compliance is produced by influences through morality and feelings, while in the instrumental approach, underpinned by Deterrence Theory, compliance is produced by legal threat of penalty and prosecution. Governments or authorities routinely adopt the instrumental approach to ensure safety compliance. It is suggested that the applicability of Deterrence Theory should be revisited because effectiveness of this approach on improving health and safety practices is often questioned. A literature review was conducted to summarize various means (other than instrumental approach) to incentivize safety compliance in the construction sector. These cover financial and non-financial incentives. The effectiveness of these incentives is also discussed.

Acknowledgements Special thanks to Mr. Zechen Guan for collecting and reviewing literature for this chapter.

References

Abramovaite, J., Bandyopadhyay, S., Bhattacharya, S., & Cowen, N. (2022). Classical deterrence theory revisited: An empirical analysis of Police Force Areas in England and Wales. *European Journal of Criminology*. https://doi.org/10.1177/14773708211072415

Ahmed, I., & Faheem, A. (2021). How effectively safety incentives work? A randomized experimental investigation. *Safety and Health at Work, 12*(1), 20–27. https://doi.org/10.1016/j.shaw.2020.08.001

Bentham, J. (1789). *An introduction to the principles of morals and legislation*. London: T. Payne and Son.

Biddle, W. C. (1969). A legislative study of the effectiveness of criminal penalties. *Crime & Delinquency, 15*(3), 354–358.

Brown, M. (1997). Compliance with Health and Safety Legislation: Is Deterrence the answer?. In *13th Annual ARCOM Conference* (Vol. 1, pp. 363–371).

Choi, T. N., Chan, D. W., & Chan, A. P. (2011). Perceived benefits of applying Pay for Safety Scheme (PFSS) in construction–A factor analysis approach. *Safety Science, 49*(6), 813–823.

Clarke, S. (2013). Safety leadership: A meta-analytic review of transformational and transactional leadership styles as antecedents of safety behaviours. *Journal of Occupational and Organizational Psychology, 86*(1), 22–49.

Ghasemi, F., Mohammadfam, I., Soltanian, A. R., Mahmoudi, S., & Zarei, E. (2015). Surprising incentive: An instrument for promoting safety performance of construction employees. *Safety and Health at Work, 6*(3), 227–232. https://doi.org/10.1016/j.shaw.2015.02.006

Griffin, M. A., & Hu, X. (2013). How leaders differentially motivate safety compliance and safety participation: The role of monitoring, inspiring, and learning. *Safety Science, 60*, 196–202.

Griffin, M. A., & Neal, A. (2000). Perceptions of safety at work: A framework for linking safety climate to safety performance, knowledge, and motivation. *Journal of Occupational Health Psychology, 5*(3), 347.

Hasan, A., & Jha, K. N. (2013). Safety incentive and penalty provisions in Indian construction projects and their impact on safety performance. *International Journal of Injury Control and Safety Promotion, 20*(1), 3–12. https://doi.org/10.1080/17457300.2011.648676

Herath, T., & Rao, H. R. (2009). Encouraging information security behaviors in organizations: Role of penalties, pressures and perceived effectiveness. *Decision Support Systems, 47*(2), 154–165.

Health and Safety Executive. (2012). *Incentives and rewards for health and safety.* Accessed 20 August 2022, (online) Available at: https://www.hse.gov.uk/construction/lwit/assets/downloads/incentives-and-rewards.pdf.

Hu, Y., Chan, A. P. C., Le, Y., Jiang, W.-P., Xie, L.-L., & Hon, C. H. K. (2012). Improving megasite management performance through incentives: Lessons learned from the shanghai expo construction. *Journal of Management in Engineering, 28*(3), 330–337. https://doi.org/10.1061/(asce)me.1943-5479.0000102

Idoro, G. I. (2008). Health and safety management efforts as correlates of performance in the nigerian construction industry/Sveikatos Ir Saugos Darbe Valdymo Pastangos Nigerijos StatybŲ PramonĖje. *Journal of Civil Engineering and Management, 14*(4), 277–285. https://doi.org/10.3846/1392-3730.2008.14.27

International Labour Organization. (2015). *Construction: a hazardous work.* https://www.ilo.org/safework/areasofwork/hazardous-work/WCMS_356576/lang--en/index.htm

Ji, L., Liu, W., & Zhang, Y. (2021). Research on the tournament incentive mechanism of the safety behavior for construction workers: Considering multiple heterogeneity. *Frontiers in Psychology, 12*, 796295. https://doi.org/10.3389/fpsyg.2021.796295

Jiang, H., et al. (2018). Structural equation model analysis of factors in the spread of unsafe behavior among construction workers. *Information (basel), 9*(2), 39.

Jillian, D., & Kevin, P. (2011). Deterrence and enforcement of occupational health and safety law. *The International Journal of Comparative Labour Law and Industrial Relations, 27*(1), 23–39.https://kluwerlawonline.com/journalarticle/International+Journal+of+Comparative+Labour+Law+and+Industrial+Relations/27.1/IJCL2011003

Khalid, U., Sagoo, A., & Benachir, M. (2021). Safety Management System (SMS) framework development–Mitigating the critical safety factors affecting Health and Safety performance in construction projects. *Safety Science, 143*, 105402.

Kim, G.-H. (2018). Measuring the effectiveness of safety incentives in construction sites in Korea. *Journal of Building Construction and Planning Research, 06*(04), 267–277. https://doi.org/10.4236/jbcpr.2018.64018

Kim, J. D., & Kim, G. H. (2019). Comparison of perceptions of safety motivation factors between construction workers and construction engineers. *Korean Journal of Construction Engineering and Management, 20*(4), 77–85.

Kim, N. K., Rahim, N. F. A., Iranmanesh, M., & Foroughi, B. (2019). The role of the safety climate in the successful implementation of safety management systems. *Safety Science, 118*, 48–56.

Lingard, H., & Rowlinson, S. (2004). *Occupational health and safety in construction project management*. Routledge.

Lipscomb, H. J., Nolan, J., Patterson, D., Sticca, V., & Myers, D. J. (2013). Safety, incentives, and the reporting of work-related injuries among union carpenters You're pretty much screwed if you get hurt at work. *American Journal of Industrial Medicine, 56*(4), 389–399. 11p. https://doi.org/10.1002/ajim.22128

Liu, J., Wang, X., Nie, X., & Lu, R. (2022). Incentive mechanism of construction safety from the perspective of mutual benefit. *Buildings, 12*(5). https://doi.org/10.3390/buildings12050536

Lu, C. S., & Yang, C. S. (2010). Safety leadership and safety behavior in container terminal operations. *Safety Science, 48*(2), 123–134.

Man, S. S., Chan, A. H. S., & Wong, H. M. (2017). Risk-taking behaviors of Hong Kong construction workers—A thematic study. *Safety Science, 98*, 25–36. https://doi.org/10.1016/j.ssci.2017.05.004

McCallum, R., Schofield, T., & Reeve, B. (2012). The role of the judiciary in occupational health and safety prosecutions: Institutional processes and the production of deterrence. *Journal of Industrial Relations, 54*(5), 688–706. https://doi.org/10.1177/0022185612454956

McDermott, V., Zhang, R. P., Hopkins, A., & Hayes, J. (2017). Constructing safety: Investigating senior executive long-term incentive plans and safety objectives in the construction sector. *Construction Management and Economics, 36*(5), 276–290. https://doi.org/10.1080/01446193.2017.1381752

Mearns, K., Whitaker, S. M., & Flin, R. (2003). Safety climate, safety management practice and safety performance in offshore environments. *Safety Science, 41*(8), 641–680.

Meier, R. F., & Johnson, W. T. (1977). Deterrence as social control: The legal and extralegal production of conformity. *American Sociological Review*, 292–304.

Michaels, D. (2010). What to do about safety incentives. *American Society of Safety Engineerings (ASSE), Webinar*.

Mishra, T. (2022). *What is Safety Compliance?—Definition from Safeopedia*. [online] safeopedia.com. Available at: https://www.safeopedia.com/definition/3969/safety-compliance

Musonda, I., & Pretorius, J. H. C. (2015). Effectiveness of economic incentives on clients' participation in health and safety programmes. *Journal of the South African Institution of Civil Engineering, 57*(2), 2–7.

Piquero, A. R., Paternoster, R., Pogarsky, G., & Loughran, T. (2011). Elaborating the individual difference component in deterrence theory. *Annual Review of Law and Social Science, 7*(1), 335–360. https://doi.org/10.1146/annurev-lawsocsci-102510-105404

Pratt, T. C., & Turanovic, J. J. (2018). Celerity and deterrence. In *Deterrence, choice, and crime* (pp. 187–210). Routledge.

Sparer, E. H., Catalano, P. J., Herrick, R. F., & Dennerlein, J. T. (2016). Improving safety climate through a communication and recognition program for construction: A mixed methods study. *Scandinavian Journal of Work, Environment & Health, 42*(4), 329–337. https://doi.org/10.5271/sjweh.3569

Sparer, E., & Dennerlein, J. (2013). Determining safety inspection thresholds for employee incentives programs on construction sites. *Safety Science, 5*(1), 77–84. https://doi.org/10.1016/j.ssci.2012.06.009

Teo, E. A. L., & Ling, F. Y. Y. (2009). Enhancing worksite safety: Impact of personnel characteristics and incentives on safety performance. *International Journal of Construction Management, 9*(2), 103–118. https://doi.org/10.1080/15623599.2009.10773133

Wang, B., Wu, C., Shi, B., & Huang, L. (2017). Evidence-based safety (EBS) management: A new approach to teaching the practice of safety management (SM). *Journal of Safety Research, 63*, 21–28.

Welles, H. (2022). *Incentives That Actually Motivate a Construction Workforce—Construct-Ed*. [online] Construct-Ed. Available at: https://www.construct-ed.com/incentives-that-actually-motivate-a-construction-workforce/. Accessed 8 August 2022.

Wong, T. K. M., Man, S. S., & Chan, A. H. S. (2020). Critical factors for the use or non-use of personal protective equipment amongst construction workers. *Safety Science, 126*.https://doi.org/10.1016/j.ssci.2020.104663

Wu, F., Xu, H., Sun, K.-S., & Hsu, W.-L. (2022). Analysis of behavioral strategies of construction safety subjects based on the evolutionary game theory. *Buildings, 12*(3). https://doi.org/10.3390/buildings12030313

Zahoor, H., Chan, A. P., Masood, R., Choudhry, R. M., Javed, A. A., & Utama, W. P. (2016). Occupational safety and health performance in the Pakistani construction industry: Stakeholders' perspective. *International Journal of Construction Management, 16*(3), 209–219.

Zhang, S., Hua, X., Huang, G., & Shi, X. (2022). How does leadership in safety management affect employees' safety performance? A case study from mining enterprises in China. *International Journal Environment Research Public Health, 19*(10). https://doi.org/10.3390/ijerph19106187

Zulkefli, F. A., Ulang., N. M., & Baharum, F. (2014). Construction health and safety: Effectiveness of safety incentive programme. *SHS Web of Conferences*, Vol. 11, p. 01012. https://doi.org/10.1051/shsconf/20141101012

Dr. Tak Wing Yiu is a professor in Construction Management and Property at the School of Built Environment, University of New South Wales (UNSW), Australia. He is a founding member of the Construction Dispute Resolution Research Unit with which his Ph.D. on behavioural analysis of construction dispute behaviour was completed. His Ph.D. was awarded the 2006 Best Dissertation (Ph.D. Category) by the Hong Kong Institute of Surveyors. Professor Yiu has been delivering construction programmes and conducting research in the field of construction management at Hong Kong and New Zealand universities. Professor Yiu is an associate editor of the ICE Proceedings: Management, Procurement and Law and the ASCE Journal of Legal Affairs and Dispute Resolution in engineering and Construction. Most of his research outputs have been published in reputable construction management journals. He is the Global Accreditation Panel Member of the Chartered Institute of Building and the member of College of Assessors for the Ministry of Business, Innovation & Employment, New Zealand.

Chapter 10
The Role of Incentivization to Mitigate the Negative Impact of COVID-Related Disputes

Peter Shek Pui Wong

Abstract Project delays caused by the COVID outbreak are unprecedented. The associated loss and expenses are supposed to be equitably shared between the client and the contractor. Nonetheless, Standard Forms of Building Contracts in many countries do not consider delay caused by COVID-19 lockdown as a qualifying event for any time and monetary claim. Disagreements and disputes have arisen as a result. In this aspect, incentivization has been advocated as an effective measure to fill the equity gap. But how incentivization can be introduced into construction contracts, and how this may help reduce disputes arising from the COVID-19-associated delay has not yet been explored in prior studies. This chapter presents a study investigating how the claims for COVID-related project delays were managed. Sixteen semi-structured interviews with the contract administration experts were conducted in Melbourne, Australia—a city that experienced the world's most prolonged COVID lockdown in 2020–21. Measures taken to mitigate the consequences of the COVID-related delay were identified. The effect of incentivisation on rebalancing the risk between the client and the contractor was also investigated. The findings reveal that although the existing Standard Forms of Building Contracts cannot be applied flawlessly in managing COVID-related time and monetary claims, interviewees were hesitant to introduce any radical change to the contract provisions. While incentivisation can instigate more active actions towards resolving COVID-related disputes, interviewees preferred the incentive schemes to be developed outside the construction contract regime. Views regarding how incentivisation can be implemented to avoid COVID-related disputes in future projects were sought. The study reported in this chapter illustrates how incentivisation may foster equitable risk sharing between the contracting parties in future contracts.

Keywords COVID-19 · Lockdown · Risks · Equity

P. S. P. Wong (✉)
School of Property, Construction and Project Management, Royal Melbourne Institute of
Technology University, Melbourne, VIC, Australia
e-mail: peterspwong@rmit.edu.au

S. O. Cheung and L. Zhu (eds.), *Construction Incentivization*, Digital Innovations
in Architecture, Engineering and Construction,
https://doi.org/10.1007/978-3-031-28959-0_10

1 Introduction

Construction contracts often enable the project completion date to be extended under the following three conditions:

Condition 1: The delay event should be non-culpable. The contractor should justify that the delay event is not caused by its faults. Delay caused by the client or its agents (such as issuing design change instructions or work suspension orders); and delay caused by factors that are beyond the control of neither contracting party (such as exceptionally adverse weather, strikes, civil commotion, and force majeure); are typical non-culpable events.

Condition 2: Regardless of the nature of the delay, the contractor had used its best endeavour to mitigate the associated negative impact on the project.

Condition 3: The delay event should have disrupted the critical path of the up-to-date program.

   During the extended period, the client's right to charge the contractor liquidated damages (LD hereafter) is forfeited. Furthermore, if the client (or its agents) caused the delay, it is liable for the contractor's loss and expenses (L&E hereafter), the amount would be equivalent to the number of extended days multiply by the pre-agreed daily rate specified in the contract.

   Nonetheless, the above principles can be changed by amending the contract terms. Typical examples of the amendment include deleting 'exceptionally adverse weather' as a reason for claiming Extension of Time (EOT hereafter). Even though the delay caused by the adverse weather is beyond neither contracting party's control, clients genuinely believe that they had given chance to the contractors to 'price on the risk of delay' in their tenders. Contractors can first estimate the additional cost incurred by analysing the historical meteorological records, then reflect such additional cost on the tender price. The allowance is expected to be equal to the estimated delay period, multiply by the daily LD rate.

   As such, risks of delay do not necessarily be equally shared between the contracting parties. Clients can offload their responsibilities for project delays to the contractors by amending the contract terms. They usually see this as 'equitable' for the contractors should have acknowledged the amended terms through their tender prices. Findings from previous studies reveal that such an arrangement shows no impediment to the contract execution, until the COVID-19 outbreak (Chirieac, 2020; Sun & Xu, 2021).

   Clients generally accepted force majeure as a reason for the COVID-19 lockdown-related claims lodged by contractors (Chirieac, 2020; Sun & Xu, 2021). In these cases, force majeure was often interpreted as a 'neutral event'—a delay event that 'prevents performance of a contract, lie outside any of the affected party's control, and cannot be avoided or stopped' (Denison, 2021, p. 89). Under this logic, the contractor is entitled to an EOT. However, the contractor would not be compensated for any L/E associated with the extended work. Furthermore, clients generally do not see other

COVID-19-associated delays, including the disruption of the supply chain, as the 'neutral events'. Under this logic, many EOT and L&E claims were rejected.

The above practice stems from the premise that it is the contractors' own business decision if they didn't make sufficient allowance for undertaking the risk offloaded by the clients. Nonetheless, clients generally ignored the fact that the delays caused by the COVID outbreak are not the regular neutral events that the contractors can predict like inclement weather. Furthermore, long before the pandemic, scholars had already pinpointed the existence of a power relationship between the client and the contractor (Perez et al., 2017; West, 2014). The competitive tendering arrangement, as well as the tender interviews, pressurize the contractors not to reflect their risk-taking on their tender prices.

Findings from recent studies indicate that the clients generally lack compassion for the traumatic loss the contractors suffered from the COVID-19 lockdown (Mosey, 2021). Some contractors went into liquidation as their clients reject to share the risk of COVID-related delays equitably (Mosey, 2021; Larasati et al., 2021). The above reveals that the contracting parties may not easily compromise in settling claims caused by the COVID-19 outbreak. These also triggered debates about the need to review (and update) the standard form of contracts to fill the equity gap between the client and the contractor in a construction contract (Larasati et al., 2021). Equity gap, in this study, refers to the disparities in contract rights and responsibilities between the contracting parties (Zhu & Cheung, 2022).

In this aspect, incentivization has been advocated as an effective measure to fill the equity gap (Zhu & Cheung, 2021). Nonetheless, to the best of our knowledge, no study has explored how incentivization can be introduced into construction contracts. More specifically, how incentivization may help reduce disputes arising from the COVID-19-associated delay has yet to be studied.

In this chapter, we draw on the findings of a qualitative study to explore the role of incentivization to mitigate the negative impact of COVID-related disputes. First, how the construction contract terms were applied to assess the EOT and L&E claims arising from the COVID-related delay are studied. The enforceability of the relevant contract clauses is discussed. Second, measures that were taken to mitigate the consequences of the COVID-related delay are identified. Whether these measures were crucial proof for EOT and L&E claims is studied. Third, the effect of incentivization on rebalancing the risk between the client and the contractor is investigated.

This study is significant because managing construction disputes is not government-driven but a sector-wide practice. The success of dispute resolution usually relies on collaboration among the contracting parties. This study is significant because it examines the effectiveness of collaboration at the corporate level when parties face extraordinary situations like the COVID-19 lockdown. The findings will provide insight into how the existing construction contract paradigm may change through incentivization.

2 Research Methodology

This study adopts a constructivist/interpretivist approach. It analyses the subjective views of the participants in the real-world context (Talabi et al., 2021; Wong et al., 2019). Semi-structured interviews were conducted to investigate how COVID-related delays and claims are managed and how incentivization may help rebalance the risk between the contracting parties. Semi-structured interviews were chosen as this method enables the participants to demonstrate their unique angle on a matter (Madill, 2011). This method helps capture different perspectives of the contracting parties through a set of objective questions (Holdsworth et al., 2019). The interviews were structured in two stages (namely Stage I and Stage II thereafter). At Stage I, interviewees were asked to provide some background information about themselves and their projects. At Stage II, interviewees were requested to base on the background information they provided to respond to the following questions:

1. *Can you explain the rationale behind the decision-making in the relevant EOT and L&E applications in your project?*
2. *What do you think about the current Standard Form of Construction Contract/ or any bespoke contract being used in your project in avoiding disputes arising from the COVID-19-driven delay events?*
3. *What is the role of incentivization in rebalancing the COVID-19-driven delay risk allocation between the client and the contractor?*
4. *Is there a need for amendments to contract provisions in future projects to formalise the incentivization? Why and how?*

Semi-structured interviews were undertaken after approval was obtained from the local Human Research Ethics Committee. The interviews were conducted face to face, digitally audio recorded, and transcribed. Given the exploratory nature of the research, thematic analysis was deemed a suitable method to analyse the collected data. An inductive approach was taken to identify themes from the data (Guest et al., 2006). This was firstly done by the researcher familiarizing himself with the data by reading the transcripts a number of times to develop potential codes along the way. Codes were then used to develop the emergent themes that display the interesting features of answering the four questions asked at Stage II of the interviews. The inductive approach allowed notable themes and patterns to emerge from the transcript themselves (Pablo et al., 2021). Once key themes were identified, data was checked again by the researcher to ensure the reliability of the transcripts.

A purposive sampling strategy was applied. Project managers and contract administrators were the targeted respondents for this study. Potential respondents were identified from two major sources. Firstly, the registered contractors' list, maintained by the Masters Builders Association of Victoria, was utilized. Master Builders is a major building and construction industry association in Australia, and its members represent 95% of all sectors of the Australian building industry. Secondly, potential respondents were searched from general browsing on the official webpages of professional institutes including the Australian Institute of Builders, Australian Institute of

Architects, and Engineers Australia. Interviewees were randomly selected from the above pools of potential respondents.

Researchers in qualitative studies emphasised that there should not be strict guidelines of minimum sample size for the semi-structured interviews (Patton, 2020; Hansen et al., 2020). More importantly, semi-structured interviews should continue until 'the depth of data to reach theoretical saturation—'the point at which no new data emerges to provide additional insights into the research question' (Watkins et al., 2017, pp. 3). Following this approach, sixteen semi-structured interviews were conducted in Melbourne, Australia. Melbourne is considered the best place to conduct this research study because, since March 2020, this city has spent the world most prolonged period (262 days) under COVID-lockdown as any place globally. During this period, the progress of the construction projects was impacted by different levels of site closure and social distancing orders.

3   Interviewee Profile

The demographics are presented in Table 1. Interviewees were assigned reference codes (from A to P). Nine out of the sixteen interviewees of the interviewees are working for the developers and contract administration consultant firms. The rest are from the contractor firms. This sample mix balances contracting parties' views (Saunders et al., 2016; Talabi et al., 2021). All interviewees have more than five years of experience in construction contract administration. Adding strength to the responses, one-third of the interviewees have more than 15 years of contract administration experience. The creditability of the interviewees is indicative of their service to the industry; thus, their responses are believed to be reflective of the industry's views.

Interviewees were first asked whether they had been involved in a construction project in which progress had been affected by the COVID-19 lockdown. All answered 'Yes'; thus, they were invited to continue with the interviews. They were then asked to respond to the interview questions based on their contract administration experience of that specific project. Information including the project nature, contract sum, project duration and the forms of contract being used were collected and presented in Table 2.

Such background information is crucial because this affects how the interviewees respond to the research questions. It's worth noting that the Australian Standard Form of Contract (AS) or amended Australian Standard Form of Contract is being used in the interviewees' projects. Three out of the sixteen projects used AS2124—a standard form for operating lump fixed-price contracts. Fours projects used AS4902 or AS4300, which are intended for design and build projects. The rest of the projects used AS4000, which is suitable for novated design and build projects.

Table 1 Interviewees' profiles

Interviewee	Firm	Role	Experience in contract administration (Years)
A	Developer	Senior contract administrator	6–10
B	Developer's consultant	Senior contract administrator	6–10
C	Developer's consultant	Director	16–20
D	Developer	Senior contract administrator	6–10
E	Contractor	Senior contract administrator	6–10
F	Contractor	Project Manager	16–20
G	Contractor	Contract administrator	11–15
H	Contractor	Contract administrator	6–10
I	Developer	Associate Director	6–10
J	Contractor	Senior cost planner/contract administrator	11–15
K	Contractor	Contract administrator	11–15
L	Developer	Director	30 +
M	Contractor	Senior contract administrator	11–15
N	Developer	Contract administrator	6–10
O	Developer	Director	30 +
P	Developer's consultant	Senior contract administrator	16–20

4 Findings and Discussions

Key findings of the thematic analysis of the elicitation study are presented in this section. From the interviewees' responses, firstly, the rationale behind the decision-making in EOT and L&E claims is introduced. Secondly, the effectiveness of the current Standard Form of Construction Contract/ or any bespoke contract terms in avoiding COVID-19 related disputes is presented. Then, the role of incentivization as understood by interviewees in rebalancing the allocation COVID-19-driven delay risk is articulated. Finally, the need for amendments to contract provisions in future projects to formalise the incentivization is evaluated.

Table 2 Project profiles

Interviewee	Project nature	Project duration (month)	Project sum ($M)	Construction contract used in the project
A	Private—hotel and commercial complex	25	50	Australian Standard Form of Contract (AS) 4902—amended
B	Private—residential	18	60	AS4000—amended
C	Private—residential and commercial complex	24	200	AS4000—amended
D	Public—hospital	18	120	AS4000—amended
E	Private—residential and commercial complex	30	10	AS4000—amended
F	Private—residential and commercial complex	36	190	AS4300
G	Private—residential	18	80	AS4000
H	Private—commercial	17	30	AS4000—amended
I	Public—utilities (civil)	8	17	AS2124
J	Public—hospital	30	480	AS2124
K	Private—clinic	8	13	AS4000
L	Private—residential	16	22	AS4000
M	Private—commercial	18	30	AS4902—amended
N	Public—infrastructure	18	20	AS2124
O	Private—residential	9	10	AS4000—amended
P	Private—residential and commercial complex	24	140	AS4300—amended

4.1   Rationale Behind the Decision-Making in EOT and L&E Claims

The first theme identified under this question is '*government-enforced lockdown was understood as force majeure*'. Standard forms of construction contracts usually have a 'Force majeure' clause that enables the contractor to claim EOT for any qualifying event. However, standard forms of construction contracts rarely articulate what event is qualified as force majeure. Force majeure within the construction field is understood as Acts of God, including natural disasters such as floods, bush fire, tropical cyclones, and earthquakes. Recent case law further extends the applicability of force majeure to manmade effects, including strikes, riots, terrorism, war and cyber-attacks (Denison, 2021). The use of the force majeure clause in managing delay claims associated with the COVID-19 lockdown is a new concept. When the force majeure clause is applied to manage COVID-19 lockdown-related claims, contracting parties may pose different views (Vickery, 2020).

Interviewee A, who is working for the developer in a commercial and hotel complex project responded.

'*We had 4 weeks of state-wide lockdown where no labor is allowed to work on-site. The contractor put in time extensions for that. Obviously, we granted that (EOT) because it was beyond their (the contractor's) control.... But contractually there is nothing there for them (the contractor) to claim loss and expenses. And, without any prejudice, we will entertain something and help you guys (the contractor) out. Most clients are open to hearing and understanding if the contractors are really struggling. People just want to work together to come for the better of the project. it's just good faith, I think.' (Interviewee A, Developer).*

The above view is consistent with the published work of Chirieac (2020) and Sun and Xu (2021) who reported that the construction practitioners often define government-enforced lockdown as force majeure to legitimatise their decisions in granting EOT to the contractor under the existing contract framework. However, the 'good faith' Interviewee A mentioned about might not be shown by all the developers, as the second theme of this question was identified as '*delays irrelated to the government mandated lockdown was not compensated.*'

Interviewee F, who is the senior contract administrator of a first-tier contractor firm says '*Even without the lockdown, the government put in place social distancing measures that limited the number of laborers and reduced the productivity on site... Also, there are lots of additional cost for the disruption of the imported materials supply, sanitization, temperature checking and reporting. All were regarding as the loss and expenses that cannot be predicted during tender submitted. But our client denied our loss and expenses claims.... this is unfair'. (Interviewee F, Contractor).*

Views from the clients can be very different, '*I went back and forward a couple of times and a fair bit of review went into the contract itself. I also sought additional legal advice given the nature of claims lodged by the contractor. Under the contract, the contractor should maintain the risk of any changes to OH&S (Occupational Health and Safety) and safety changes to legislation. And it wasn't provided to them any ground to claim any loss and expenses for complying with the new OH&S requirements.' (Interviewee L, Developer).*

Clients' argument looks sharp and clear. It is the Australian Work, Health, and Safety Act that requires contractors to provide a safe workplace for their employees. With or without the COVID outbreak, the contractors should provide a safe workplace in accordance with the Australian law.

'*But the Law never says we got to employ additional resources to keep those temperature checks and everything in order... the client's denial of sharing the additional expenses is not helping anyone'.* Interviewee E, who is working in the contractor's firm reminded us to consider this matter in another perspective.

As a general principle, it is preferable for the contracting party who has control over an event to assume the delay risk arising from its occurrence. For example, a delay caused by the contractor's suppliers will typically be borne by the contractor. Even though it is not within the contractor's direct control, the contractor should manage the risk through practical steps such as careful selection and monitoring of its suppliers and contingency planning should its preferred source of supply become unavailable. However, there has been no mention of whether the additional cost should be shared.

'Instead of shipping the materials, we got them air-freighted which was 10 to 15 more expensive than estimated… Unfortunately, it was still considered more economical than compensating liquidated damages.' (Interviewee G, Contractor).

Under this logic, clients have every right to reject any EOT and L&E claim caused by COVID-related delay. In the client's perspective, defining government-enforced lockdown as force majeure may have already been a favour they gave to the contractor. Such views may disappoint many governments as they have been encouraging fairer and more responsible contractual arrangements to support the viability of the construction contracts during COVID (United Kingdom Cabinet Office, 2020; Ministry of Law Singapore, 2020).

4.2 Effectiveness of the Existing Contract Terms in Avoiding COVID-19 Related Disputes

Theme 1: Time-related claims under the contract are still possible.

Interviewees who are working for the developers generally perceived that the negotiation of EOT claims under the contract is straightforward:

'….. 'Force majeure' and 'Delays caused by the public/statutory authorities' are valid grounds for EOT claims under the contract. By far I don't aware of any dispute raised by the contractor…. Delay claims caused by COVID lockdown can be settled under the current contract terms. They are just usual EOT claims. Thus, the contractor should justify how the lockdown has affected critical path activities…. It [the contractor] should show its effort in mitigating the delay and check their actual revised critical path…. I need to see its behaviour, not just grant the period it asked for' (Interviewee C, Developer's consultant).

However, interviewees from the contractor firms are generally disappointed as they are accepting the clients' judgments grudgingly:

'Contractor loses most of its flow in the program after the government made it clear that the lockdown would last long enough to make the work impossible to operate in full swing… While the developer pledged to support…., it emailed us saying that any time claim other than the government announced lockdown is not qualified to claim EOT…. You're not getting the same outcome when you are asking your builder to take on more risk. The contractor simply doesn't have the funds to deliver what you may have expected. You can have every mechanism to make every claim impossible, but that might lead you to a court case, not to a good project outcome so.' (Interviewee H, Contractor).

'…clients have not been very receptive to any exclusions or clarifications about the impacts of COVID. All through last year, we are following the contract terms to manage time claims in a way that the client considered as correct. We reluctantly define COVID as Force majeure, Delays caused by the public/statutory authorities

and whatever they like to qualify for EOT claims. But none of them (the conditions) is 100% fit for the nature of COVID-related delay.... And now new contracts kick in, and we will ensure that they will provide a clear definition of COVID delay ... I can see the improvement in avoiding disputes through reading the terms more cautiously' (Interviewee J, Contractor).

The above findings show that disputes caused by COVID-related time-related claims can be effectively eliminated without reforming the construction contract mechanism. Interviewees mainly sought clarification or articulation of the existing contract provisions, and rarely thought of any radical change in the contract administration practice.

Theme 2: Loss and expenses claims are only negotiable outside the contract framework.

Interviewees from the developers and consultant firms conceded that existing contract forms lack relevant contract provisions to deal with COVID-related loss and expenses claims. This made dispute settlement within the existing contractual framework difficult:

'We made our professional judgment on loss and expenses claims, and the contractor has the ability under the contract to dispute the judgment that has been made. That dispute then falls between the developer and the contractor before it gets into further mediation etc. And then they can take it all the way further if the disputes can't be resolved..... I don't see why the client should have to pay loss and expense for the government shutting down their sites, and they're getting nothing out of it in terms of construction and the contract provisions' (Interviewee B, Developer's consultant).

'I think that can become quite difficult dealing with things (COVID-related loss and expenses claims) through the contract, particularly when there isn't any clause there to deal with ... it becomes a matter of interpretation, and those disagreements can escalate. But I think it is quite dangerous to set a precedent to misinterpret or over-interpret the contract term to enable claims within the mechanism too' (Interviewee D, Developer).

Interviewees from the contractor firms shared similar views that the respective loss and expense claims can only be resolved outside the contract framework. The direct loss and expense incurred by COVID outbreak include the additional cost of sanitization of the workplace and machines, the additional government-imposed social distancing measures, late delivery of materials, and the extra labour costs for accelerating the work through night shifts.

'It was difficult to put a valid loss and expenses claim under the contract because catching back the loss of productivity is very difficult to measure. We can genuinely argue that the loss of productivity is caused by the lockdown measures, but clients will ask us for evidence supporting that the direct loss is really caused by the COVID outbreak, not our mismanagement. They knew it is impossible to provide evidence like this before they asked. The messages they conveyed have been very clear. Approving loss and expense claims of this kind under the contract is a no-go zone that the contractors shouldn't reach' (Interviewee K, Contractor).

'*A massive part of our disagreements is, actually, the cost we spent on miti-gating the COVID-related delay…, and the disparity is that under the contract it is not clear who should have the responsibility on such additional cost… COVID is unprecedented. And it is quite obvious that if we only follow the contract provisions, the disputes can never be resolved*' (Interviewee M, Contractor).

Interestingly, if the responsibility for the loss and expense was discussed outside the contract framework, claims become negotiable:

It is understandable that clients don't wanna tarnish the controls or mechanisms or like of the existing contract… but it doesn't mean that we don't wanna help the contractors. At some point, we may suffer more loss if they collapse. We can form a separate agreement, dealing with such claims outside the existing contract mechanism (Interviewee I, Developer).

4.3   The Role of Incentivization in Rebalancing Risk Allocation

Most interviewees believed that incentivization can mitigate some negative impacts of COVID-related disputes.

'*Clients are not always trying to offload risk to the contractors. But it is ironic that they have been quite successful in doing that mainly because they assume the contractors would behave genuinely in tender pricing…..If clients can't expect self-regulation from the contractor, they can incentivize the resolution of COVID-related disputes. The concept is like introducing a contingency or provisional sum in the Bills of Quantities. Clients put upfront the ceiling of their undertakings of the loss and expense caused by the COVID-related delays. As an equal amount of incentive is allowed in tenders for the contractors' to avoid disputes, this enables fairer tender comparison*' (Interviewee P, Developer's consultant).

The major concern should be whether equitable risk allocation can be achieved under the existing procurement and contract mechanism. COVID outbreak reaffirms that the answer is, unfortunately, No. But the industry should admit that the core problem is whether there is COVID that caused delay disputes. The point is that the contractors have long been taking overly aggressive commercial positions, trying to win work and thus taking more risk, reducing their margins, and accepting contract conditions they shouldn't accept. Who's gonna shoot first is the colloquial way of saying it, so some of them do take on more risk to win work and that becomes the status quo and that was. Incentivization, to some extent, relieves such brutal rivalry (Interviewee O, Developer).

Responses from the interviewees with contractor backgrounds focused on the feasibility of incentivizing dispute settlement.

The disruption and the associated loss caused by the COVID-19 outbreak is unprecedented. We find it difficult to execute the existing terms to resolve claims associated with the COVID-19 outbreak anyway. Thinking this outside the box is a good initiative.... Any form of incentivization should be welcomed. At least it shows that the client is willing to resolve this problem through collaboration, not endless finger-pointing.... (Interviewee F, Contractor).

Not only government-enforced lockdown is out of the contractor's direct control, but also the disruption of the global supply chain. Apparently, our client knows no one can address the unprecedented changes in international supply chains. We didn't have any chance to predict such loss during the tendering stage. To me, the incentivization is not merely for unprecedented situations caused by the COVID outbreak, but for any circumstance when risk cannot be reasonably estimated by the contractors. Incentivization is a sign from the clients that they acknowledge this issue ....(Interviewee K, Contractor).

Previous studies have highlighted that the incentivisation can shape cognitive and behavioural change of one party so as to meet the expectation of another (Dix, 2020). Similar findings were revealed in this study. Interviewees found incentivisation as a tool that drives the contractor's behavioural change in tender pricing.

4.4   The Need for Formalising Incentivization in Contracts

Interviewees' responses to the first three questions revealed that the current contract forms they used might have some deficiencies in managing the EOT and L/E claims triggered by the COVID-19 pandemic. It's logical to expect that the interviewees would support formalising incentivization in the contracts. Surprisingly, the findings show that most interviewees opposed it.

'We used to tailor our own set of amended clauses and add them to the Special Conditions of Contract anyway. Any incentive schemes can be enforced even without touching the current version of the Australian Standard (Form of Contract).' (Interviewee N, Developer).

'Everyone's been hurting, both the client and the contractor.... but you can foresee that our appetite to accept risk will change. With or without incentive embedded into the contract, we will pass the additional preliminary costs to the client, then ultimately the end customer. Incentivization, if it won't show how much the client would pay, can make future bidding more complicated. I don't think it's a good idea to show the cost (of incentivization) in the contract.' (Interviewee F, Contractor).

'There is a notable noise asking the developers to be more sympathetic to the contractors' situation. But it remains unclear how the developers can help. To me, incentivization can be operated like an advance payment scheme. If the contractor can justify its cash flow being affected by the pandemic, it can apply for this fund to keep its business afloat. The client will pay the premium of the advanced payment bond for the contractor. I don't think touching the contract terms can help. Touching them may hit the nerves of both parties which can create more disputes. If the intention of incentivization is genuine, let's simplify the procedures to motivate the contractor to get helped quickly.' (Interviewee C, Developer's consultant).

The results indicate that interviewees do recognize incentivization as a tool to mitigate the negative impact of COVID-related disputes. However, formalising incentivization should not involve any radical change to future contract clauses. The findings are in line with the McDonald et al. (2008) who advocated the need a surveillance mechanism to avoid any incentive scheme from creating tensions among the contracting parties.

5   Summary

COVID outbreak has traumatised contracting parties who used to assume time-related risks can be offloaded to the contractors through construction contract terms amendment. Ignoring the contractors' needs to recover the unanticipated losses caused by the COVID outbreak would not only lead to disputes but also contract frustration. COVID-19 exposes the weakness of the Standard Forms of Contract in managing related loss and expenses claims. While interviewees conceded that the contractors suffered an irrecoverable loss in time and cost during the lockdown period, their legal determination constrained their responses.

If the clients are genuinely open to hearing and understanding the contractors' needs, the conversation can start with developing and formalising new measures to avoid disputes arising from COVID-related claims. Interviewees of this study provide valuable suggestions to fair project risk allocation and disputes avoidance.

This study elucidates the role of incentivization in mitigating the negative impact of COVID-related disputes. Surprisingly, the findings of this study do not build a case for reforming the Standard Forms of Contract. Interviewees generally believed that incentivization is enforceable even without amending the existing contract terms. Incentivization helps articulate the client's undertakings of COVID-related delay risk. It rationalises risk assessments, thus promoting reasonable and responsible tender pricing. Incentivization also fosters collaboration in resolving the contractor's cashflow problem which is considered a key motive of claims and disputes.

References

Chirieac, R. M. (2020). Considerations on the positive law institutions that may affect on the execution of construction contracts. *Perspectives of Law and Public Administration, 9*(2), 299–308.

Denison, M. J. (2021). Force majeure clauses in LNG sales and purchase agreements: How do they stand up during the Covid-19 pandemic? *Journal of World Energy Law & Business, 14*(2), 88–99.

Dix, G. (2020). Incentivization: From the current proliferation to the (re)problematization of incentives. *Economy and Society, 49*(4), 642–663.

Guest, G., Bunce, A., & Johnson, L. (2006). How many interviews are enough? An experiment with data saturation and variability. *Field Methods, 18*(1), 59–82.

Hansen, S., Rostiyanti, S. F., & Rif'at, A. (2020). Causes, effects, and mitigations framework of contract change orders: Lessons learned from GBK aquatic stadium project. *Journal of Legal Affairs and Dispute Resolution in Engineering and Construction, 12*(1), 05019008

Holdsworth, S., Sandri, O., Thomas, I., Wong, P., Chester, A., & McLaughlin, P. (2019). The assessment of graduate sustainability attributes in the workplace: Potential advantages of using the theory of planned behaviour (TPB). *Journal of Cleaner Production, 238,* 117929.

Lasrasati, D., Novva, E., Sugeng, T., Aulia, F. M., & Anedya, W. (2021). Impact of the Pandemic COVID-19 on the implementation of construction contracts. *IOP Conference Series: Earth and Environmental Science, 738,* 012075.

Madill, A. (2011). Interaction in the semi-structured interview: A comparative analysis of the use of and response to indirect complaints. *Qualitative Research in Psychology, 8*(4), 333–353.

McDonald, R., Harrison, S., & Checkland, K. (2008). Incentives and control in primary health care: Findings from English pay-for-performance case studies. *Journal of Health Organisation and Management, 22*(1), 48–62.

Ministry of Law Singapore. (2020). *Key Features of Re-Align Framework* [online]. Available from: https://www.mlaw.gov.sg/realign/key-features. Accessed 1 November 2021.

Mosey, D. (2021). What can construction contracting learn from Covid-19? *Construction Law International* [online], 15 (3), Available from: https://www.ibanet.org/article/a6ab6762-65c5-40f9-a990-15081e3e3bcb. Accessed 1 June 2022.

Pablo, Z., London, K., Wong, P. S. P., & Khalfan, M. (2021). Actor-network theory and the evolution of complex adaptive supply networks. *Construction Innovation, 21*(4), 668–684.

Patton, M. Q. (2020). *Qualitative research and evaluation methods* (3rd ed.). Sage Publications.

Perez, D., Gray, J., & Skitmore, M. (2017). Perceptions of risk allocation methods and equitable risk distribution: A study of medium to large Southeast Queensland commercial construction projects. *International Journal of Construction Management, 17*(2), 132–141.

Saunders, M., Lewis, P., & Thornhill, A. (2016). *Research methods for business students* (7th ed.). Pearson.

Sun, C., & Xu, S. (202). Analysis of the impact of the COVID-19 Epidemic on the construction engineering EPC projects and claims. *IOP Conference Series: Earth and Environmental Science, 676,* 012038.

Talabi, S., Koskeia, L., Tzortzopoulos, P., & Kagioglou, M. (2021). Causes of defects associated with tolerances in construction: A case study. *ASCE Journal of Management in Engineering, 37*(4), 05021005.

The United Kingdom Cabinet Office. (2020). *Guidance on responsible contractual behaviour in the performance and enforcement of contracts impacted by the COVID-19* [online]. Available from: https://www.gov.uk/government/publications/guidance-on-responsible-contractual-behaviour-in-the-performance-and-enforcement-of-contracts-impacted-by-the-covid-19-emergency

Vickery, P. (2020). COVID–19: Managing the Impact on Construction Contracts [online]. Available from: https://static1.squarespace.com/static/5aed91101aef1d132050dad6/t/5ecb016440423e1 7efc3172e/1590362470323/20200516-COVID-19-Managing-the-Impact-on-Construction-Contracts-Paper-16-May-2020%281%29.pdf

Watkins, R., Goodwin, V.A., Abbott, R.A., Hall, A., & Tarrant, M. (2017). Exploring residents' experiences of mealtimes in care homes: A qualitative interview study. *BMC Geriatrics* [online], 17:141, Available from: https://doi.org/10.1186/s12877-017-0540-2.

West, J. (2014). Collaborative patterns and power imbalance in strategic alliance networks. *ASCE Journal of Construction Engineering and Management, 140*(6), 04014010.

Wong, S. P., Holdsworth, S., Crameri, L., & Lindsay, A. (2019). Does carbon accounting have an impact on decision-making in building design? *International Journal of Construction Management, 19*(2), 149–161.

Zhu, L. Y., & Cheung, S. O. (2021). Towards an equity-based analysis of construction incentivization. *ASCE Journal of Construction Engineering and Management, 147*(11), 04021148.

Zhu, L. Y., & Cheung, S. O. (2022). Equity gap in construction contracting: Identification and ramifications. *Engineering, Construction and Architectural Management, 29*(1), 262–286.

Dr. Peter Shek Pui Wong is a professor and associate dean (Construction Management) of the School of Property Construction and Project Management, RMIT University, Australia. Professor Wong completed his Ph.D. on learning organisations with the Construction Dispute Resolution Research Unit, Department of Architecture and Civil Engineering of the City University of Hong Kong. His Ph.D. was awarded the 2007 Best Dissertation (Ph.D. Category) by the Hong Kong Institute of Surveyors. His publications overs the topics of construction contract administration, carbon accounting in construction, immersive technology (VR/AR/MR) in construction, prefabrication, and modulation of construction, and organisational change management. He is a Task Group member in 'Sustainable Housing' 'Net-Zero Building' and 'Cities & Infrastructure' of the Australian Sustainable Built Environment Council. He is also an associate editor of the ICE Proceedings: Management, Procurement and Law and editorial board member of the ASCE Journal of Legal Affairs and Dispute Resolution in Engineering and Construction. He is the Education Board Member of the Australian Institute of Quantity Surveyors and the Global Accreditation Panel Member of the Chartered Institute of Building.

Chapter 11
Interweaving Incentives and Disincentives for Construction Dispute Negotiation Settlement

Sen Lin

Abstract What incentives and disincentives motivate negotiators to settle or not in a construction dispute negotiation (CDN)? A thorough literature review is conducted on this subject to identify the antecedents of negotiators' intention to settle (ITS) in CDN. Three relevant constructs are identified: motivation (i.e., prosocial and proself motive), cognition (i.e., justice and power), and psychological bonding (i.e., trust and shared vision). Categorically, this study finds that in the negotiation context, negotiators having a prosocial motive and perceiving justice about the negotiation process and outcome can stimulate negotiators' ITS, which can be seen as incentives; however, the proself motive and perceived power advantage would serve the opposite, thus can be classified as disincentives. In addition, cumulated trust and shared vision during the project collaboration can also play an incentive role in promoting negotiators' intention. As a result, this study develops a link between the incentive/disincentive (I/D) and negotiators' intention to settle through the literature review. A better understanding of these agents of I/D can help explain negotiation conditions and negotiators' decisions whereby appropriate negotiation strategies can be devised.

Keywords Construction dispute negotiation · Disincentive · Incentive · Intention to settle

1 Introduction

1.1 Research Background

It appears that no construction project is free from dispute (Cheung & Yiu, 2006). In this regard, Arcadis (2021) reported three critical causes of dispute: disputing

S. Lin (✉)
Construction Dispute Resolution Research Unit, Department of Architecture and Civil Engineering, City University of Hong Kong, Hong Kong, China
e-mail: slin24-c@my.cityu.edu.hk

parties' misunderstanding of the contractual obligations, changes in scope, and unforeseen events. Another 2021 study published in the World Built Environment Forum involved 1,200 construction and engineering projects across 88 countries as survey respondents. It was reported that the cumulative sum in dispute exceeds US$48.6 billion, and resultant delays amounted to schedule extension by more than 71% of the original (RICS, 2021). Since 2020, the COVID-19 pandemic has been exerting unprecedented pressure on the already strained industry as most projects are inevitably grappled with disputes arising from the lockdown and associated restrictions (Casady & Baxter, 2020). The impact of disputes can be substantive, leading to cost overrun, declining productivity, delays in the delivery of projects, and, potentially, a loss of business viability (Cheung & Pang, 2013; Yiu et al., 2015). Improving dispute resolution has attracted great attention from both the industry and the academia because of the ever presence of construction disputes.

The stair-step chart in Fig. 1 outlines a series of construction dispute resolution methods with reference to the respective hostility and cost. Litigation is the most formal, adversarial, and costly option, commonly regarded as the "last resort"(Jagannathan & Delhi, 2020). The others are collectively termed alternative dispute resolution (ADR) techniques that may save an enormous cost for disputing parties (Cheung, 1999; Yousefi et al., 2010). Furthermore, forms of ADR can also be grouped as binding (formal) and nonbinding (informal). Arbitration offers binding resolution and is commonly specified in most construction contracts. Many projects have arbitration incorporated as the final resolution forum for project disputes. Negotiation, third-party neutral, mediation, mini-trial, and adjudication are the nonbinding options (Cheung et al., 2002; Treacy, 1995). Among all these methods, negotiation is the most cost and time efficient means of resolving disputes. In fact, most disputes are firstly negotiated, thus making negotiation a daily routine for construction practitioners. There are basically no restrictions regarding the form and process of negotiation (Cheung, 1999). Negotiating parties can freely express their will despite having varying goals, expectations, and opinions. Solving differences through negotiation has been regarded as the most commended form of resolution because of resource-saving and relationship-maintenance functions (Lu et al., 2015; Yiu et al., 2018).

A productive negotiation demands cooperative efforts from both parties. In fact, many negotiation studies have informed sufficient advice on best practices in negotiation, including tactics and strategies (Cheung et al., 2009; Yiu et al., 2008), negotiation styles (Cheung et al., 2006; Patton & Balakrishnan, 2010), logrolling and trade-offs (Qu & Cheung, 2012; Tajima & Fraser, 2001), and potential mistakes (Love et al., 2010, 2011; Yiu et al., 2015). These informative materials have contributed to the training of negotiators. Nevertheless, the assumption of these strategic moves is that negotiators are rational and can follow the economic view to take mutual profit maximization as the ultimate goal. In fact, negotiators are not "rational economic men" who would always choose the "right" method and make the "correct" decisions. The social-psychological view reveals that negotiators can deviate from optimality because they are influenced by their personalities, the information available, bounded cognitions, and opportunistic motivations (Bazerman et al., 2000; Caputo,

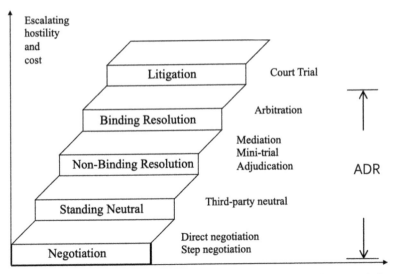

Fig. 1 Hierarchy of methods of construction dispute resolution (Adapted from Cheung (1999))

2013; Tversky & Kahneman, 1985). Furthermore, negotiation is an art of interaction, critically dependent on the relationships between the negotiation parties. Every negotiation move can be influenced by the attitudes of the counterparts and the relationship bonds between the social units (Yiu et al., 2018; Yousefi et al., 2010). The complicated and adversarial habit of construction contracting requires the parties to cooperate and compete simultaneously. In addition, the contracting environment is masked with uncertainties and unexpectedness, thus requiring project participants to make spontaneous decisions and be flexible in response to contingencies (Loosemore, 1999). Above all, prescribing the behavior of negotiators is not practical.

When a negotiator needs to decide how to deal with negotiation, at each critical decision, he needs to consider three options: (1) accept the currently available proposals; (2) continue the negotiation with their counterpart to pursue a better outcome; or (3) leave the negotiating table for lost of interest (Cheung & Chow, 2011; Mitropoulos & Howell, 2001). It is suggested that negotiators' level of intention to settle is the determining factor on which option to take. Table 1 depicts the relationship between intention to settle and the respective negotiating behaviors. It can be expected that despite facing similar negotiation situations, negotiators having different settlement intentions would adopt strategies that would lead to the respective outcomes.

Table 1 State of intention to settle and negotiating behaviors (Adapted from Cheung & Chow, 2011)

Intention to settle	Negotiating behavior	Degree of behavior
Low	Stalemate/Breakdown	Aggressiveness
↓	Irrational argument	↓
	Rational argument	
	Concession	
Strong	Apologies	Cooperativeness

1.2 The Significance of Intention to Settle

Studies on intention have attracted considerable attention in the construction field. Reported studies cover topics such as waste management (Yuan et al., 2018), construction labor productivity (Johari & Jha, 2020), construction insurance (Liu et al., 2018), and failure learning (Liu et al., 2017). Intention can be regarded as an aspiration for a specific outcome that will motivate people to set goals and plans (Hagger et al., 2002). According to behavioral school, intention is a key behavioral attribute that can affect the way in which people act (Johari & Jha, 2020). The theory of planned behavior (TPB) is one of the classical "intention–behavior" models that highlight the relevance of intention to predict the practice of certain behaviors (Ajzen, 1991). If one develops a high level of intention toward a certain object, there will be a higher chance that he/she will conduct purposive behaviors to achieve that object. Furthermore, Ajzen (2011) reviewed the applications of TPB and suggested that intention has better predictive validity when one has actual control over behaviors and the time intervals for the intention to take effect is not that long. Considering the characteristics of daily routine and voluntariness of CDN, negotiators' intention is paramount. In addition to these theoretical predictions, field observations from the Global Construction Disputes Report also identified that parties' intention to settle is the most crucial element for early resolution (Arcadis, 2021).

Lin and Cheung (2021) described intention to settle (ITS) in construction dispute negotiation (CDN) as the state of favorably engaging in ending a dispute through negotiation. Three forms of intention (i.e., technique-based, relationship-based, and cognition-based intention) were examined that respectively represent the willing-to-settle negotiators' subjective perceptions toward negotiation issues, the counterpart, and themselves. Considering the pivotal role of intention to settle in bringing negotiated settlement, it is invaluable to understand their formation. With this aim, a systematic literature review was conducted to identify what would facilitate (i.e., incentives) or impede (i.e., disincentives) negotiators' intentions.

1.3 Sources of Incentives and Disincentives

Studies on the influence of negotiation behaviors and outcomes were reviewed to identify the sources of incentives and disincentives. For example, Thompson (1990) summarized the impact of individual differences, motivational, and cognitive approaches in negotiation, and asserted that the role of personality and individual differences is minimal. Thompson et al. (2010) further proposed five levels of negotiation behaviors: intrapersonal, interpersonal, group, organizational, and virtual. Brett (2000) put forward a dyadic negotiation model with two key concepts: (1) interests and priorities; and (2) negotiation strategies. The model asserts that negotiators' interests and priorities would affect the outcome potential, and strategies would influence the negotiation process; their combined effects result in different negotiation outcomes (Brett & Thompson, 2016). Even though these negotiation models are not focused on CDNs, they provide valuable insight into incentives and disincentives of settlement intentions.

Unlike two-person negotiation, CDN is a two-party negotiation whereby organizational factors are involved and hence the settlement intention. To operationalize in the negotiation context, two dimensions of antecedents are included: (i) social motive at the intra-organizational level (i.e., prosocial motive and proself motive); (ii) relational cognition at the inter-organizational level (i.e., justice and power). This is in line with the negotiation model proposed by Brett (2000) and Brett and Thompson (2016). It can be explained that in CDNs, negotiators are representatives of their organizations. Their expectations and interests in the distribution of negotiation outcomes are aligned with those of their organizations, ultimately determining their social motive for the settlement. Moreover, their negotiation interactions make them form different perceptions about the negotiation situation and relationship, which indicate their relational cognition and complicate their settlement intention. In addition, construction projects usually face long-term work periods, complex technical requirements, and high uncertainty environments, which provide room to cultivate relationships and make negotiations more complex than those in general business operations. In such a scenario, the connecting mechanism formed during the project cooperation would also take effect in negotiations, thereby influencing negotiators' settlement intention. Considering the social network in construction projects, this study proposes psychological bonding at the project level (i.e., trust and shared vision) that should exert a "high-level" influence to regulate the negotiation trends and affect negotiators' intention to settle.

The first two levels have more to do with the negotiation context, while the third level is more subjective and pertains to the degree of psychological connection between the two parties at project level. Thus, the social motive, relational cognition, and psychological bonding together determine the negotiating parties' intention to settle, as illustrated in Fig. 2.

Specifically, the three levels of factors are classified as incentives and disincentives on intention to settle (Fig. 3). In the negotiation context, proself motive and power are identified as disincentives, while prosocial motive and justice are incentives. In the

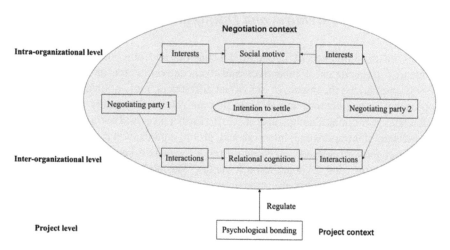

Fig. 2 A model of intention to settle in CDNs

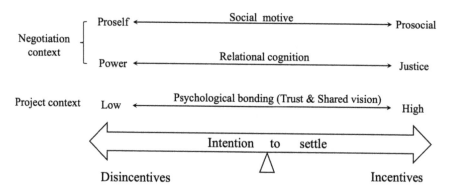

Fig. 3 Incentives and disincentives on intention to settle

project context, a low psychological bonding mechanism will impede negotiators' settlement. However, high psychological bonding will serve the opposite. Each factor playing the role of incentive/disincentive (I/D) is addressed in the following sections.

2 Social Motive (Intra-Organizational Level)

Disputes arise when individuals, groups, or organizations have different viewpoints on certain issues. Negotiating parties have different levels of aspiration in achieving their goals (De Dreu et al., 2000). Intuitively, one might feel that negotiators would strive to maximize their gains. In fact, negotiators are not always aiming to maximize their own profits while minimizing those of their counterparts. Many negotiation situations are mixed-motive, with the interests of negotiating parties neither completely opposite nor fundamentally compatible, leaving ample room for negotiation (Beersma & De Dreu, 2002). This leads to one of the main theoretical variables in negotiation—social motive.

2.1 The Anatomy of Social Motive in Negotiation

Social motive refers to the preference for a distribution of negotiation outcomes for the self as well as these of the others (Butt & Choi, 2006). A number of social motives have been identified. For example, McClintock (1977) classified four types of motive: altruistic, competitive, individualistic, and cooperative. Competitive and individualistic motives are commonly grouped as self-centered. Trötschel and Gollwitzer (2007) identified two types of motive as prosocial and egoistic. Carnevale and Lawler (1986) distinguished individualistic and cooperative orientation. This study adopts the widely used classification of prosocial motive and proself motive. Prosocial motives involve more cooperative and altruistic goals, whereas proself motives would drive competitive and individualistic goals. Negotiators with prosocial motives desire to maximize both parties' profits, and they try to maintain a fair and harmonious game environment. In contrast, proself-motivated negotiators, who are more egoistical, would ignore their counterparts' outcomes and pursue their maximum gain. Social motive embraces the intrinsic triggers of negotiation behaviors, whether constructively or destructively. Social motive has been widely explored in negotiation studies (Beersma & De Dreu, 2005; De Dreu et al., 1998; Li et al., 2021; Trötschel & Gollwitzer, 2007).

Social motive can be viewed as a trait variable reflecting negotiators' differences in "social value orientation"; it can also act as a state variable of "motivational orientation" (De Dreu et al., 2000). This is because social motive in negotiation can be induced or affected by both individual differences and situational elements in negotiation. Scholars have conducted two types of studies to identify the origins of social motives. Regarding individual differences, Antonioni (1998) tested the influence of the big five personalities and showed that the personality of extroversion, conscientiousness, openness, and agreeableness could induce a more integrative and cooperative motive. Social motives are also highly related to social value orientations. Carnevale and Probst (1998) confirmed the positive relationship between individualism and proself motive, and the positive correlations between allocentrism and prosocial motive. In addition, a large range of work has attempted to reveal the influence of situational factors on social motives, especially in CDNs. For example, Li et al. (2021) found that trust can facilitate prosocial motive while conflict event criticality can induce proself motive. Wei and Luo (2012) revealed the significance of interactive effects of power and social motive on problem-solving behaviors. De Dreu

(2004) also summarized the rooted influence of superiors, culture, reward structures, and social relationships on motivation.

De Dreu et al. (2000) suggested that the application of social motive is functionally equivalent and produces similar effects on negotiation, regardless of whether acting as a trait or state variable. In this study, the social motive is treated as a state variable that negotiators with different "motivational orientations" can stimulate or inhibit negotiators' intention and behavior.

2.2   The Incentive and Disincentive of Social Motive

Before discussing how social motive incentivizes (or disincentivizes) negotiators' intentions or behaviors, it is useful to draw on the underlying theory—dual concern theory. Dual concern theory advocates that the choice of negotiation strategies is affected by negotiators' relative degree of concern for self and concern for others, or in other words, social motive (Pruitt, 1983; Rahim, 1983). As shown in Fig. 4, there are five types of negotiation strategies that negotiators may take, depending on the level and object of concern: problem-solving, yielding, forcing, avoiding, and compromising. It is argued that the function of social motives can predict negotiating behaviors; more specifically, prosocial rather than proself motive (i.e., strong rather than weak concern for others) can drive more problem-solving, less contentious behaviors, and ultimately achieve more integrative outcomes (Li et al., 2021). The central message of dual concern theory is that negotiators with a high degree of self-concern and other-concern can achieve more amicable outcomes than negotiators who only care about their own outcomes and negotiators who only care about the other party's outcomes. This view is also supported by the theory of cooperation and competition (Deutsch, 1949).

In addition to strategy choice, social motive also explains how negotiators process negotiation information (De Dreu & Carnevale, 2003). Carnevale and Probst (1998) suggested that prosocial negotiators are more flexible and inclusive. Gelfand and Christakopoulou (1999) added that negotiators with collective backgrounds could better understand the priorities and preferences of both parties, thus fostering better trade-offs and effective logrolling. It is suggested that when processing negotiation information, proself negotiators are likely to fall prey to motivational biases and strengthen their selfishness. Prosocial negotiators, however, are more inclusive and thereby more likely to achieve joint gains.

This study focuses on the influence of prosocial and proself motives on negotiators' intention to settle. Considering the effect of social motives on strategy choice and information process, this study argues that prosocial as opposed to proself motivated negotiating parties can better facilitate the intention to settle. That is, prosocial motive can be an incentivizing agent, while proself motive is a disincentivizing agent.

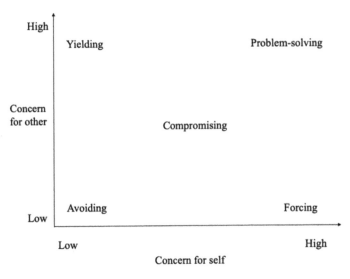

Fig. 4 Dual concern model (adapted from Rahim, 1983)

(1) Prosocial motive as incentivizing agent

Prosocial motive is closely related to concern for others. Negotiators with prosocial motives consider negotiations more like a cooperative game for which the relationship, fairness, and joint outcomes are prominent. Prosocial negotiators have a better chance to form positive attitudes, put themselves in others' shoes, build trust, carefully listen, and commit to constructive information exchange (Beersma & De Dreu, 2005; De Dreu et al., 2000). As a result, integrative behaviors, such as problem-solving or trade-offs, that represent a high level of settlement intention are more likely to happen. In this regard, the prosocial motive can be one of the significant antecedents that incentivize negotiators' intention to settle.

(2) Proself motive as disincentivizing agent

In contrast to prosocial motive, proself motive is more associated with self-concern, which is also akin to "toughness," "resistance to yielding," and "intransigence to concession making." Selfish negotiators commonly have higher aspirations of the negotiation outcomes; thus, they will see negotiation as a competitive game. Negotiators with proself motive are more likely to develop aggressive attitudes, distrust, and negative perceptions of negotiation situations (De Dreu et al., 2000). In this case, they tend to use protracted approaches, demanding proposals, threats, or even coercion, which will impede the intention to settle. Accordingly, this study posits that proself motive can be the disincentive antecedent to negotiators' intention to settle.

3 Cognition Against the Counterpart (Inter-Organizational Level)

The other main line in negotiation is cognitive perspective during the interactions. Starting from the 1980s, socio-psychological scholars applied to cognition approach to examine structural or situational variables that shaped behavioral decision research in negotiation as opposed to game theory or rational mathematical analysis (Bazerman et al., 2000). Cognition studies cover topics such as framing effect (Tversky & Kahneman, 1985, 1986), perceived problem-solving feasibility (Pruitt, 1983), prescriptive perspective (Raiffa, 1982), and attributional perspectives (Shaver, 2012). To gain a deeper understanding of how negotiators form their intention to settle, it is necessary to know how they describe or perceive their opponents and negotiation situations. Therefore, this study focuses on negotiators' cognition against their counterparts (i.e., power and justice) in order to better understand their negotiation decisions. The weight of evidence seems to support that justice can incentivize negotiators' intention to settle, while power may serve the opposite.

3.1 Justice

Justice or fairness has been identified as a major issue in managing construction projects. According to Lind and Tyler (1988), organizational justice is related to whether people are treated in a fair manner, whether the pay is fair, whether the adopted procedures are not biased, and whether people's responses to outcomes are justified. Justice is of prime importance in construction dispute management, verified by a series of studies. For example, Maqsoom et al. (2020) explained that prospective disputes could be avoided if contractors perceive fairness. Tatum and Eberlin (2008) found that justice is highly related to the conflict management style, and people who are careless about justice would be more inclined to competitive or dominating resolution tactics. Zhang et al. (2021) highlighted the role of justice on claimants' satisfaction. Justice is a complex concept that contains just interactions as well as just outcomes (Druckman & Wagner, 2016). According to the holistic review by Colquitt et al. (2013), there are four main trends of study on organizational justice:

(1) the distributive justice trend (from the 1950s to 1970s) that focuses on the allocation norms and just outcomes with the theory of equity;
(2) the procedural justice trend (from the 1970s to 1990s) that focuses on the development of just rules to promote a sense of process fairness;
(3) the interactional justice trend (from the 1980s to 2000s) that attends more to interpersonal interactions and treatment as a unique form of justice; and.
(4) the integrative trend (from the 1980s to the 2000s) that put justice as an integral element of an organization.

This study argues that multidimensional justice (i.e., distributive justice, procedural justice, and interactional justice) is positively related to negotiators' intention to settle. The role of each dimension in negotiation is explained in turn.

3.1.1 Distributive Justice

Distributive justice refers to the perceived fairness of the decision outcomes (Blau, 2017). Distributive justice is regarded as the beginning of organizational justice, which can trace back to the equity theory (Adams, 1965). According to equity theory, results are usually measured as a ratio of inputs to outputs and compared to a reference standard to determine whether the distribution of resources is equitable or not. The reference standard is adopted according to the preference of the individual and is usually derived from another individual or organization which can be comparable to himself/herself. Three possible results may occur: (1) Equity-People are assumed to be satisfied if they find their input-outcome ratio is equal to the reference standard. This situation is ideal or optimal to be achieved. (2) Underpayment inequity-People will feel under-benefited if their input-outcome ratio is less than the comparative standard. In this case, retaliation may occur to restore the equity, such as asking for a raise in compensation or cutting corners. Adams (1965) further found that the feeling of underpayment will impede the organizational commitment or even drive people to "leave the field." (3) Overpayment inequity-People may feel guilty when they find their input-outcome ratio is greater than the reference standard; this will motivate people to do more to match the outcomes and address this inequity (Zhang et al., 2021). The equity theory provides the basis of distributive justice with the principles of equality, proportionality, compensation, and need, which are considered appliable to negotiation (Druckman & Wagner, 2016).

Based on the classic definition of equity (Adams, 1965) and considering the context of CDNs, distributive justice can be defined as how negotiators perceive the fairness of their counterparts' offers regarding the dispute being negotiated (Lu et al., 2017). The dispute can be monetary (e.g., additional costs) or nonmonetary (e.g., the extension of time or technical criteria) (Maqsoom et al., 2020). Negotiators will form their expectations or judgments about offers based on the work they have done, which can then serve as a reference standard for the distribution outcome (Folger & Konovsky, 1989). Lu et al. (2017) revealed that when negotiators perceive that the offers are fair, they are more likely to take cooperative behaviors. On the other hand, the fear of being exploited will make negotiators cautious about the offers and impede any possible reciprocation (Fisher et al., 2011; Zhang & Han, 2007). Furthermore, Youngblood et al. (1992) found that perceived unjust decisions can be the reason for conflict escalation. As such, it can be assumed that negotiators will compare the offers with their expectations and will be willing to resolve disputes only when the distribution appears reasonably fair.

3.1.2 Procedural Justice

Procedural justice is defined as the perceived fairness of the procedures that regulate the process and decide outcomes (Colquitt et al., 2013). In contrast to distributive justice, which is primarily concerned with the satisfaction of outcomes, procedural justice focuses more on the fairness of the rules. Thibaut and Walker (1975) proposed the psychology of procedural justice by distinguishing between two types of control: process control and decision control. Thibaut and Walker (1975) compared the Anglo-American adversarial legal system, in which the judge controls the decision but not the evidence process, and the European Inquisitorial legal system, in which the judge controls both the decision and process. Through the response of participants, they found that people consider the Anglo-American adversarial legal system fairer as it allows them some control over the process, even though the final decisions may not favor them. The psychology of procedural justice explains the significance of impartiality of process as people care about their direct or indirect control over decisions (Tyler, 1989). Procedural issues, including the impartial rules, unbiased process, opportunity to express, grounds for decisions, and authority of decision making, are considered crucial to enhance the perception of procedure justice (Al-Zu'bi 2010; Bayles, 2012; Lind and Tyler, 1988).

Procedural justice in negotiation relates to the perceived fairness of procedures and criteria adopted by negotiating parties in proposing offers (Lu et al., 2017; Luo, 2007). To be procedurally fair, negotiators should follow certain rules and provide evidence that proves their offer is legitimate. More specifically, negotiators are suggested to abide by the explicit and implicit terms of the contract, give sufficient explanation about why their offer should be accepted, leave room for the other side to express views, and timely adjust mistakes or unreasonable problems (Luo, 2007). Lind et al. (1993) found that decision to accept or reject the offer is highly correlated with negotiators' judgment of procedural justice. With unfair procedures, negotiators would think there is little chance to recover their losses even if they are right (Zhang et al., 2021). Thus, they will be less incentivized to settle the problems. It can be summarized that the fairer the negotiation process, the more likely negotiators will collaboratively work forwards a settlement (Druckman & Wagner, 2016; Lu et al., 2017).

3.1.3 Interactional Justice

Interactional justice is the extent to which people perceive fairness based on the "quality of interpersonal treatment received during the execution of a procedure" (Bies, 1986). Interactional justice also relates to the "fair process" and plays a complementary role to procedural justice. Procedural justice focuses more on the formal process that is applied to make decisions, while interactional justice concerns more person-to-person interactions. Bies (1986) delineated the attributes of interactional justice as truthfulness, justification, respect, and propriety, which can help to distinguish interactional justice from the rules and criteria of procedural justice. In some studies, interactional justice is also called the quality of treatment (Aibinu et al., 2011;

Maqsoom et al., 2020). Tyler and Blader (2013) stated that negative feelings about the quality of treatment could lead to the devaluation of a group and dissatisfaction with organizational decisions. No matter interactional justice or quality of treatment, it places more on social sensitivity and involves actions showing people's kindness and goodwill, such as careful listening, open discussion, and attentive communication.

CDN typically involves interactions among negotiators from different parties. The perception of whether negotiators are treated fairly during their interpersonal and informational exchange can be considered interactional justice (Luo, 2007). Compared to distributive and procedural justice, interactional justice is informal and relates more to the relationship between negotiating parties. Interactional justice is suggested to affect negotiators' cognitive, affective, and, ultimately, behavioral reactions (Tyler & Bies, 2015). Macfarlane (2001) found that if negotiators feel offended during a negotiation, they will likely harden their position. Kadefors (2005) added that the possible reactions are anger, loss of motivation, or even resentment if negotiators are unfairly treated. The feeling of being fairly treated is the minimum requirement for a voluntary resolution. Negotiators who are satisfied with the just interactions will be more inclined to communicate and give positive feedback, thus creating a harmonious negotiation atmosphere that is conducive to settling their differences (Rupp & Spencer, 2006).

For distributive justice, the greater negotiators perceive the favorability of the outcomes, the higher chances that they will accept the decisions. Procedural and interactional justice, which emphasize the procedures and interactions during the decision-making process, are also proved to be reasonable concepts to determine negotiators' settlement intention. Negotiators will voluntarily come to the negotiating table when they feel fair about the process and results of negotiations; otherwise, negotiations will be frozen in endless blame-shifting and competition.

3.2 Power

3.2.1 Definition of Power

Power in negotiation is mainly analyzed in two distinct ways. The economic view of power is determined by the presence and quality of alternatives, named BATNA (Best Alternative to a Negotiated Agreement) (Fisher et al., 2011). Ideally, the more appealing one's alternatives, the less dependent on the other party, and the greater one's power. In this genre, power was manipulated by comparing negotiating parties' value of BATNA (Wei & Luo, 2012), or the number of alternatives (Van Kleef et al., 2006). However, this form of operationalization requires estimation of the parties' BATNA, which can only be applied in laboratory or simulated experiments. Moreover, Brett and Thompson (2016) reported that the effects of BATNA were influenced by the context of the negotiation. Only when the bargaining zone is small and certain can the BATNA exert a clear impact on negotiators' decisions (Kim & Fragale, 2005).

This leads to a more relational way of conceptualizing power in negotiation. In inter-personal negotiations, negotiators' power is more related to their status, which is esteem and respect accorded by others. Higher-status negotiators can enjoy the obedience of others but are also expected to take responsibility for the welfare of others (Magee & Galinsky, 2008). For group- or organizational-level negotiations, the definition of power is more about the degree of influence, resource dependency, and the capacity to achieve outcomes (Christen, 2004; Deutsch, 1973; Keltner et al., 2003; Lu et al., 2015). Accordingly, Bacharach and Lawler (1981) gave an overall description of power as the ability to exert influence on others (i.e., people or parties). Keltner et al. (2003) defined power as "an individual's relative capacity to modify others' states by providing or withholding resources or administering punishments". Lu et al. (2015) analyzed power in the context of CDN and defined it as "negotiators' ability to achieve their desired outcomes". This study takes the organizational level perspective to explore the impact of power on negotiators' intention to settle.

3.2.2 Power-Related Theory

Numerous theoretical perspectives have been developed to study power in negotiations, including bilateral deterrence theory (Lawler, 1986), conflict spirals theory (Lawler, 1986), approach/inhibition theory of power (Keltner et al., 2003), and power-dependence theory (Emerson, 1962, 1964). These power theories have different perspectives and their application in CDN is discussed.

(1) Bilateral deterrence theory and conflict spirals theory

The two theories were proposed by Lawler (1986) to explain how negotiators decide to use coercive tactics with different power levels, but the findings suggest opposing outcomes. By the bilateral deterrence theory, unequal power produces more coercive behavior than power parity. It is stated that when power is unequal, the powerful party does not fear retaliation and is, therefore, more inclined to take punitive measures, while the less powerful party is also motivated to take punitive actions as a signal of unwillingness to submit. The bilateral deterrence theory predicts that equal power will reduce the frequency of punitive moves as parties can perceive the coercive capability of their counterparts. However, conflict spiral theory posits the opposite conclusion, which states that equal power relationships produce more coercive behavior. This theory believes that the guarantee of security or the protection of profits is determined by their power. In the situation of equal power, coercive tactics can help portray a tough image and maintain their relative advantage, avoiding looking weak or being attacked from the other side.

(2) Approach/inhibition theory

Approach/inhibition theory argues that power is developed as a relative ability to change others' states by applying resources or imposing punishments (Keltner et al., 2003). This theory deliberates how power influences human behavior and compares the affective, cognitive, and behavioral tendencies between the power-advantaged and power-disadvantaged. The findings show that the powerful side tends to take the system of behavioral approach with which they will form positive affect, pay more attention to rewards that can satisfy personal goals, take automatic information processing, and apply stimulative behavior. On the contrary, reduced power will lead to negative affect, attention to threat and punishment, concern for others' interests and goals, controlled information processing, and inhibited social behavior (Jordan et al., 2011; Keltner et al., 2003; Tost, 2015).

(3) Power-dependence theory

Power-dependence theory rests on the social exchange relationship and the dependence between each other (Emerson, 1962; Molm, 2015). This theory has a social premise, in which individuals/parties control resources valued by each other. For example, Party A has power over Party B if A controls resources that B highly values. Power is thus derived from the other side's dependence. Moreover, the exchange interdependence asymmetry determines which party has greater power and is considered to be the reverse of dependence (Cuevas et al., 2015). Several studies have indicated that the asymmetry of power allows the power-advantaged party to exert influence on the less powerful party to act in ways they are not willing to, determine the relationship process, and direct outcomes to their favor (Caniëls & Gelderman, 2007; Piskorski & Casciaro, 2005).

This study tends to focus on the power-dependence theory that describes power as stemming from the degree of dependence in an exchange relationship. Compared to traditional buyer–supplier relationships, project organizations seem to be more complex and highly characterized by exchange interdependence (Senescu et al., 2013). Different interdependent units (e.g., relations among owner, contractor, subcontractor, or consultant) in construction projects need to maintain their exchange relationship with each other to gain resources and achieve their desired outcomes (Bankvall et al., 2010; Chinowsky et al., 2011; Guo et al., 2021). Moreover, exchange interdependence is often asymmetric in the construction industry and will thus result in power asymmetry. For example, if there are many contractors of the same type, it will create a buyer's market, and the owner is regarded as having a power advantage. In other cases, the owner will not have relative power if the contractor has irreplaceable techniques. Once there exists asymmetry, it can be expected that both parties' attitudes and behaviors in negotiation will be affected. This study applies the power-dependence theory in CDN. The other three power theories are supplemental and assist in understanding negotiators' intention to settle.

3.2.3 The Role of Power in Negotiation

Empirical evidence of the influence of power on negotiation outcomes is mixed, especially at the dyadic level. Some studies found that parties of equal power achieve higher joint gains than unequal dyads (Mannix & Neale, 1993). Wolfe and McGinn (2005) also supported this point of view, they found that negotiating parties with small perceived power differentials can reach better agreements than parties with higher power differentials. However, other studies have the opposite results. For example, Sondak and Bazerman (1991) argued that unequal power between dyad members would improve the negotiated outcomes. Wei and Luo (2012) supported that negotiators with unequal power achieved higher joint gains; more specifically, the high-low power dyads and high-high power dyads could get better results than the low-low power dyads. The relationship between power and negotiation outcomes can be contingent and complex, as it also relates to other elements, such as negotiators' affects (Anderson & Thompson, 2004) or their aspiration levels (Mannix & Neale, 1993).

Even though the mixed effects on negotiation outcomes, the influence of power on negotiators' settlement intention and behaviors should be clear. This study argues that, in CDNs, the higher the negotiators perceive their power, the less motivation for them to consider a negotiated settlement (i.e., less intention to settle) (Christen, 2004). In other words, power can disincentivize the intention to settle. This viewpoint is in line with the power-dependence theory. The high-power negotiators who have overwhelming resources commonly set higher goals (Wong, 2014), demand more but concede less (De Dreu, 1995), behave in more aggressive actions (e.g., threats and bluffs) (Brett & Thompson, 2016), less concern about the other side's emotions and act at will without serious considerations (Van Kleef et al., 2006), which are all considered as barriers to the negotiated settlement. In contrast, to maintain a harmonious exchange relationship, low-power negotiators have the impetus to make a good impression. In this regard, they are inclined to care about the potential retaliation, pay more attention to the other party, have a higher chance to concede, and be more vigilant to deal with the negotiated issues (Lawler et al., 1988; Piskorski & Casciaro, 2005). Moreover, Van Kleef et al. (2006) found that negotiators with lower power are motivated to foster information processing so that they can have an accurate understanding of both the negotiation situation and their counterparts. Compared to the high- and low-power negotiators, it is suggested that negotiators' intention to settle may decrease as their perceived power increases, thereby showing power is a disincentive to settlement.

4   Psychological Bonding Mechanism (Project Level)

As negotiations are embedded in the context of construction projects, this study suggests that project-level factors should also be taken into account in analyzing negotiators' intention to settle. Construction project is a network-based unit in which all stakeholders have a mutual goal of completing a project spanning years or even decades. The project members of different specialties are supposed to accomplish contractual obligations together. They need to cope with project emergencies arising from changes in the environment (Koh & Rowlinson, 2012). It is suggested that through long-term cooperation and interaction, parties are able to cultivate certain level of social capital. Social capital refers to a set of relational resources embedded in a network that allows individuals/parties to work effectively in achieving a common purpose (Gulati et al., 2000; Nahapiet & Goshal, 1998). Nahapiet and Goshal (1998) specified three aspects of social capital: structural dimension (represented by network ties), relational dimension (represented by trust), and cognitive dimension (represented by shared vision). Among these dimensions, trust and shared vision can be considered interrelated yet different aspects of psychological resources in an organization (Li, 2005). Rodríguez and Wilson (2002) called these psychological factors social bonds of relationships. In this study, we take trust and shared vision as the psychological bonding mechanisms that emerge in the collaboration of the project and can play the role of incentive in negotiations.

4.1 Trust

4.1.1 Definition and Dimensions of Trust

Trust is the positive expectation or faith that others will act in a mutually acceptable manner, considering all parties' well-being, and avoiding opportunistic behaviors even when there is a chance (Das & Teng, 2001). McAllister (1995) defined interpersonal trust as "the extent to which a person is confident in and willing to act on the basis of, the words, actions, and decisions of another." Many scholars have indicated the central importance of trust in developing and strengthening relationships and considered it a prominent mechanism for partnership sustainability (Jagosh et al., 2015; Wong & Cheung, 2004; Wu et al., 2012). The formation of trust is affected by individuals' chronic disposition, contextual factors during interactions, and historical experiences with others (Kleiman et al., 2015; Lewicki et al., 2011). The traditional saying, "it takes twenty years to build trust but five minutes to ruin it," aptly describes the fragile nature of trust. Trusting is always risk-taking as the trustee can exploit the trustor. In this study, considering the definition of organizational trust proposed by Mayer et al. (1995), trust is defined as a temporary state reflecting one party's willingness to be vulnerable to the other party's action.

Trust is a complex construct containing multiple dimensions that vary in different domains. For example, Ring (1996) took the economic views and divided two distinct forms of trust as fragile and resilient. Ganesan and Hess (1997) distinguished interpersonal and organizational types of trust. Two trust dimensions: credibility and benevolence, are used. On the other hand, Rousseau et al. (1998) focused on organizational trust and proposed four types of trust: deterrence-based, calculus-based, relational, and institution-based. Cheung et al. (2011) developed a trust inventory that includes system-based, cognition-based, and affect-based trust in construction contracting. These perspectives are inspiring and provide sufficient sources for trust conceptualization. In this study, we adopted a commonly used classification in organizational management studies, which considers both the cognitive approach and the facet of emotion, named competence trust (i.e., cognitive trust) and goodwill trust (i.e., affective trust) (Dirks & Ferrin, 2002; Mayer et al., 1995; Zhang et al., 2016). The two forms of trust are derived from different theoretical views, as competence trust focuses more on the character-based perspective while goodwill trust is more on the relationship-based perspective.

(1) Competence trust

Competence trust is a relatively rational evaluation of the other party about whether they can fulfill the required work. It develops based on past successful interactions, similar working experiences, or qualifications of the trustee (Mayer et al., 1995). Parties holding competence trust believe the trustors can accomplish the tasks considering their reliability of the resources and capabilities (Newell et al., 2019). This is essential to construction projects as there are always large scales of investment and high uncertainties (Pinto et al., 2009). The existence of competence trust can ease trustors' anxiety about the potential obstacles and believe in the success of the project. Nevertheless, competence trust will erode if they do not feel that the other parties possess the resources or ability to fulfill their commitments.

(2) Goodwill trust

Goodwill trust refers to one's perception of good faith or integrity from the trustee. It addresses the question, "will the other party consistently concern for my interests?" Goodwill trust is developed from the other party's ethical behavior and their unwillingness to take advantage of another person (Hartman, 2002). It is suggested that goodwill trust can reduce transaction costs and improve performance as both parties believe that the other party will act in the interest of their partnership (Dyer & Chu, 2003). When there are contingencies that go beyond the contractual obligations, goodwill trust can increase work flexibility and allow quick decisions to solve problems. The friendly communication, elimination of competitive behaviors, and voluntariness to contribute will come when goodwill trust is built (Pinto et al., 2009).

4.1.2 The Role of Trust in Construction Projects and Negotiations

Reported studies covering topics of leadership (Tan & Tan, 2000), communication (Cheung et al., 2013), commitment (Kwon & Suh, 2005), project-stakeholder relationship (Karlsen et al., 2008), and dispute resolution (Zhang et al., 2016) have indicated the critical role of trust in organizational management. The presence of trust is desirable as it can foster information exchange, reduce the transaction costs of monitoring, avoid opportunism, reinforce willingness to overcome risk/uncertainty, and cultivate a more harmonious working environment (Dyer & Chu, 2003; Woolthuis et al., 2005). Zhang et al. (2016) highlighted the role of trust as it can be a complementary means of contractual governance. Mutual trust allows parties to capture the "hearts and minds" of each other and motivates them to "go that extra mile" if it is necessary (Newell et al., 2019). It is therefore suggested that trust helps bond the members of the construction team (Wong & Cheung, 2004).

These merits of trust can also come into play in negotiations. This study argues that trust is one of the psychological bonding mechanisms that would foster cooperation and positively influence negotiators' intention to settle. Given the inherent information asymmetry between the negotiating parties, they may take an opportunistic lens to assess the proposals from their counterparts (Cheung et al., 2011). Moreover, the act of openly asking and answering questions would expose parties' priorities and interests. Distrust is common in construction and would lead to endless disputes. Building trust can ease skeptical thoughts, facilitate negotiators' information sharing, and suppress negotiators' opportunistic behaviors (Brett & Thompson, 2016; Gunia et al., 2011). In addition, trust can also affect the choice of negotiation tactics and strategies. Zhang et al. (2016) found that goodwill trust would encourage negotiators to spend more time and even go beyond their obligations to do the other a favor; thus, negotiators with goodwill trust are more likely to take integrative instead of distributive approaches. Negotiators with competence trust were found to be highly committed to the relationship, which leads them to put more effort into seeking mutually acceptable solutions (Zhang et al., 2016). Even though trust-building is always challenging in construction, it is undeniable that trust is the glue between negotiating parties and can facilitate negotiators' choice to settle the differences.

4.2 Shared Vision

During project interactions, parties may develop shared vision. Shared vision means members having shared values, mutual goals, and mental models in a relationship (Li, 2005). In this regard, Nanus (1992) demonstrated shared vision as "a mental model of a future state of a process, a group, or an organization." Thoms and Greenberger (1995) indicated that shared vision is the psychological basis of an organization's motivation and work plans to achieve an ultimate goal. Pearce and Ensley (2004) described shared vision as "a common mental model of the future state of the team or its tasks that provides the orientation for action within the team." Chi et al. (2022) gave a definition in the context of construction projects as "a shared understanding of collaboration and appropriate methods of cooperation for all collective goals and directions as formed by stakeholders involved in projects." The function of shared vision is to delineate the expectations of task performance (Converse et al., 1993). The shared vision is regarded as a "top-level" psychological concept, and two parts of connotation can be summarized from its definition: shared understanding and action norms (Wang et al., 2021).

The role of shared vision has been widely recognized in the organizational arena. Wang et al. (2021) argued that shared vision could facilitate the adoption of specific behaviors (i.e., behavioral integration) that are valuable to the groups' goals. Parkhill et al. (2015) found that a high shared vision creates organizational principles, explains why certain futures are expected, and leads to related social actions. Chi et al. (2022) suggested the critical role of shared vision on value co-creation. Jacobson and Choi (2008) found that high commitment and shared vision among clients, contractors, and architects are key to project success. As a top-level psychological factor, shared vision can bring parties together and facilitate a series of variables conducive to project cooperation and performance. As part of the significant work in construction projects, negotiation settlement should also be governed by the shared vision.

Due to the diversity of contradictions, the highly dynamic nature of the construction process, and the incompleteness of the contract, negotiators may have a fear of being exploited in negotiations, which leads them to be short-term vision and unwilling to cooperate with each other (Cheung & Pang, 2013; Wang et al., 2021). In this case, if parties have a well-developed shared vision, it would remind negotiators to have better coordination and work amicably to achieve their overall goal of the project (Shenhar & Holzmann, 2017). Specifically, shared vision, on the one hand, increases negotiators' mutual understanding of each other and the negotiated issues, thus reducing potential prejudice and friction during the negotiation process. On the other hand, their shared goal provides a good foundation for negotiators to communicate and exchange information, contributing to the negotiation efficiency. Therefore, shared vision can be regarded as the other psychological bonding mechanism for triggering negotiators' settlement intention in CDNs, which indicates that shared vision is also an incentive of intention to settle.

5 The Integrative Incentive/Disincentive Influence on Intention to Settle

Prior studies have provided valuable sources for identifying the I/D on negotiators' intention to settle. Based on the above discussion in this chapter, an integrative I/D framework on negotiators' intention to settle is portrayed in Fig. 5. Disputing issues to be negotiated and the negotiating parties are the two primary inputs of a CDN. During the negotiation, three aspects of I/D, including motivation, cognition, and psychological bonding, are having influence on the negotiators' intention to settle.

More specifically, the operation mechanism of the three levels of I/D is summarized as follows:

(1) In the negotiation context, negotiators' motivation mainly describes how they value their interests. This is the intrinsic trigger at the intra-organizational level. Negotiation is a goal-directed process with negotiating parties having

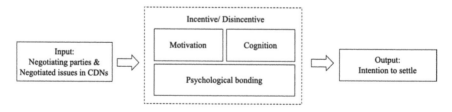

Fig. 5 The integrative influence of incentive/disincentive on intention to settle

different preferences, interests, or priorities. Their expectation of the negotiation outcomes informs their pursuit of prosocial or proself motives, and hence their settlement intention: prosocial motive encourages intention to settle while proself motive disincentives it.

(2) In addition to the intrinsic motivation, how well the negotiation outcome can materialize the potential value of joint gains depends on the negotiation interactions (Brett and Thompson 2016). Relational cognition attributes (i.e., justice and power) that capture at the inter-organizational level underpin negotiators' position. It is believed that how negotiators view each other and value their relationship will directly affect their settlement intention. In terms of justice, negotiators' feeling of being fairly treated in the negotiation process and outcome can motivate negotiators to settle. However, if negotiators feel that they have a power advantage over their counterpart, they would take advantage of that and adopt a more competitive manner. The chance of settlement will thereby be impeded.

(3) Except for the negotiation-related factors, negotiations are embedded in constriction projects and should also be regulated by the project-level factors. It is proposed that trust and shared vision are the psychological bonding mechanisms to be developed for long-run cooperation that are not limited to dispute negotiation. These can be seen as the superior psychological determinants that have enduring effect on negotiation behaviors. High levels of trust and shared vision are conducive to the strengthening of the parties' relationships, reducing tension during negotiations, and promoting cooperation, hence explaining how psychological bonding regulates the intention of negotiators to reach a settlement.

6 Summary

Negotiation has been recognized as the most efficient method to resolve construction disputes. However, analyzing negotiating behavior is not practical without understanding the underlying settlement intention. This study argues that negotiators' intention to settle is indispensable for negotiated settlement. It is thus crucial to explain why negotiators intend to settle or not. In this regard, a thorough literature review is conducted to identify what facilitates (i.e., incentive) or impedes (i.e., disincentive) negotiators' intention to settle. Three levels of agents are summarized: (i) social motive (i.e., prosocial and proself motive) at the intra-organizational level; (ii) relational cognition (i.e., justice and power) at the inter-organizational level; and (iii) psychological bonding (i.e., trust and shared vision) at the project level. Incentives/disincentives explain why and how negotiators develop settlement intention. Creating an environment conducive to settlement can therefore be deployed by fostering incentives and suppressing disincentives.

References

Adams, J. S. (1965). Inequity in social exchange. *Advances in Experiment Social Psychology, 2*(C), 267–299. https://doi.org/10.1016/S0065-2601(08)60108-2.

Aibinu, A. A., Ling, F. Y. Y., & Ofori, G. (2011). Structural equation modelling of organizational justice and cooperative behaviour in the construction project claims process: Contractors' perspectives. *Construction Management and Economics, 29*(5), 463–481. https://doi.org/10.1080/01446193.2011.564195

Ajzen, I. (1991). The theory of planned behavior. *Organizational Behavior and Human Decision Processes, 50*(2), 179–211. https://doi.org/10.4135/9781446249215.n22

Ajzen, I. (2011). The theory of planned behaviour: Reactions and reflections. *Psychology and Health, 26*(9), 1113–1127. https://doi.org/10.1080/08870446.2011.613995

Al-Zu'bi, H. A. (2010). A Study of relationship between organizational justice and job satisfaction. *Internationa of Journal Business Management, 5*(12), 102–109.https://doi.org/10.5539/ijbm.v5n12p102

Anderson, C., & Thompson, L. L. (2004). Affect from the top down: How powerful individuals' positive affect shapes negotiations. *Organizational Behavior and Human Decision Processes, 95*(2), 125–139. https://doi.org/10.1016/j.obhdp.2004.05.002

Antonioni, D. (1998). Relationship between the big five personality factors and conflict management styles. *International Journal of Conflict Management, 9*(4), 336–355. https://doi.org/10.1108/eb022814

Arcadis. (2021). *2021 Global Construction Disputes Report The road to early resolution.*

Bacharach, S. B., & Lawler, E. J. (1981). *Bargaining: Power, tactics, and outcomes.*

Bankvall, L., Bygballe, L. E., Dubois, A., & Jahre, M. (2010). Interdependence in supply chains and projects in construction. *Supply Chain Management, 15*(5), 385–393. https://doi.org/10.1108/13598541011068314

Bayles, M. E. (2012). *Procedural justice: Allocating to individuals.* Springer Science & Business Media.

Bazerman, M. H., Curhan, J. R., Moore, D. A., & Valley, K. L. (2000). Negotiation. *Annual Review of Psychology, 51*(1), 279–314.

Beersma, B., & De Dreu, C. K. W. 2(002). Integrative and distributive negotiation in small groups: Effects of task structure, decision rule, and social motive. *Organization Behaviour and Human Decision Processes, 87*(2), 227–252. Academic Press Inc. https://doi.org/10.1006/obhd.2001.2964.

Beersma, B., & De Dreu, C. K. W. (2005). Conflict's consequences: Effects of social motives on post-negotiation creative and convergent group functioning and performance. *Journal of Personality and Social Psychology, 89*(3), 358–374. https://doi.org/10.1037/0022-3514.89.3.358

Bies, R. J. (1986). Interactional justice: Communication criteria of fairness. *Res. Negot. Organ., 1*, 43–55.

Blau, P. (2017). *Exchange and power in social life.* Routledge.

Brett, J. M. (2000). Culture and negotiation. *International Journal of Psychology, 35*(2), 97–104. https://doi.org/10.1108/JBIM-11-2015-0230

Brett, J., & Thompson, L. (2016). Negotiation. *Organizational Behavior and Human Decision Processes, 136*, 68–79. https://doi.org/10.1016/j.obhdp.2016.06.003

Butt, A. N., & Choi, J. N. (2006). The effects of cognitive appraisal and emotion on social motive and negotiation behavior: The critical role of agency of negotiator emotion. *Human Performance, 19*(4), 305–325. https://doi.org/10.1207/s15327043hup1904_1

Caniëls, M. C. J., & Gelderman, C. J. (2007). Power and interdependence in buyer supplier relationships: A purchasing portfolio approach. *Industrial Marketing Management, 36*(2), 219–229. https://doi.org/10.1016/j.indmarman.2005.08.012

Caputo, A. (2013). A literature review of cognitive biases in negotiation processes. *International Journal of Conflict Management.*

Carnevale, P. J., & Lawler, E. J. (1986). Time pressure and the development of integrative agreements in bilateral negotiations. *Journal of Conflict Resolution, 30*(4), 636–659.

Carnevale, P. J., & Probst, T. M. (1998). Social values and social conflict in creative problem solving and categorization. *Journal of Personality and Social Psychology, 74*(5), 1300–1309.

Casady, C. B., & Baxter, D. (2020). Pandemics, public-private partnerships (PPPs), and force majeure | COVID-19 expectations and implications. *Construction Managament Economics, 38*(12), 1077–1085. Routledge. https://doi.org/10.1080/01446193.2020.1817516.

Cheung, S. (1999). Critical factors affecting the use of alternative dispute resolution processes in construction. *International Journal of Project Management, 17*(3), 189–194. https://doi.org/10.1016/S0263-7863(98)00027-1

Cheung, S. O., & Chow, P. T. (2011). Withdrawal in construction project dispute negotiation. *Journal of Construction Engineering and Management, 137*(12), 1071–1079. https://doi.org/10.1061/(ASCE)CO.1943-7862.0000388

Cheung, S. O., Chow, P. T., & Yiu, T. W. (2009). Contingent use of negotiators' tactics in construction dispute negotiation. *Journal of Construction Engineering and Management, 135*(6), 466–476. https://doi.org/10.1061/(asce)0733-9364(2009)135:6(466)

Cheung, S. O., & Pang, K. H. Y. (2013). Anatomy of construction disputes. *Journal of Construction Engineering and Management, 139*(1), 15–23. https://doi.org/10.1061/(ASCE)CO.1943-7862.0000532

Cheung, S. O., Suen, H. C. H., & Lam, T. I. (2002). Fundamentals of alternative dispute resolution processes in construction. *Journal of Construction Engineering and Management, 128*(5), 409–417. https://doi.org/10.1061/(ASCE)0733-9364(2002)128:5(409)

Cheung, S. O., Wong, W. K., Yiu, T. W., & Pang, H. Y. (2011). Developing a trust inventory for construction contracting. *International Journal Project Management, 29*(2): 184–196. Elsevier Ltd and IPMA. https://doi.org/10.1016/j.ijproman.2010.02.007.

Cheung, S. O., & Yiu, T. W. (2006). Are construction disputes inevitable? *IEEE Transactions on Engineering Management, 53*(3), 456–470. https://doi.org/10.1109/TEM.2006.877445

Cheung, S. O., Yiu, T. W., & Lam, M. C. (2013). Interweaving trust and communication with project performance. *Journal of Construction Engineering and Management, 139*(8), 941–950. https://doi.org/10.1061/(ASCE)CO.1943-7862.0000681

Cheung, S. O., Yiu, T. W., & Yeung, S. F. (2006). A study of styles and outcomes in construction dispute negotiation. *Journal of Construction Engineering and Management, 132*(8), 871–881. https://doi.org/10.1061/(ASCE)0733-9364(2006)132

Chi, M., Chong, H.-Y., & Xu, Y. (2022). The effects of shared vision on value co-creation in megaprojects: A multigroup analysis between clients and main contractors. *International Journal Project Management*, (January). Elsevier Ltd. https://doi.org/10.1016/j.ijproman.2022.01.008.

Chinowsky, P., Taylor, J. E., & Di Marco, M. (2011). Project network interdependency alignment: New approach to assessing project effectiveness. *Journal of Management in Engineering, 27*(3), 170–178. https://doi.org/10.1061/(asce)me.1943-5479.0000048

Christen, C. T. (2004). Predicting willingness to negotiate: The effects of perceived power and trustworthiness in a model of strategic public relations. *Journal Public Relations Research, 16*(3), 243–267. https://doi.org/10.1080/1532-754X.2004.11925129

Colquitt, J. A., Greenberg, J., & Zapata-Phelan, C. P. (2013). What is organizational justice? A historical overview. *Handbook of Organizational Justice*, 3–56. Psychology Press.

Converse, S., Cannon-Bowers, J. A., & Salas, E. (1993). Shared mental models in expert team decision making. *Individual Group Decision Market Current Issues, 221*, 221–246.

Cuevas, J. M., Julkunen, S., & Gabrielsson, M. (2015). "Power symmetry and the development of trust in interdependent relationships: The mediating role of goal congruence. *Industrial Market Management, 48*, 149–159. Elsevier Inc. https://doi.org/10.1016/j.indmarman.2015.03.015.

Das, T. K., & Teng, B. S. (2001). Trust, control, and risk in strategic alliances: An integrated framework. *Organization Studies, 22*(2), 251–283.

Deutsch, M. (1949). A theory of co-operation and competition. *Hum. Relations, 2*(2), 129–152.

Deutsch, M. (1973). *The resolution of conflict: Constructive and destructive processes.* Yale University Press.

Dirks, K. T., & Ferrin, D. L. (2002). Trust in leadership: Meta-analytic findings and implications for research and practice. *Journal of Applied Psychology, 87*(4), 611–628. https://doi.org/10.1037/0021-9010.87.4.611

De Dreu, C. K., Weingart, L. R., & Kwon, S. (2000). Influence of social motives on integrative negotiation: A meta-analytic review and test of two theories. *Journal of Personality and Social Psychology, 78*(5), 889–905. https://doi.org/10.1037//0022-3514.78.5.889

De Dreu, C. K. W. (1995). Coercive power and concession making in bilateral negotiation. *Journal of Conflict Resolution, 39*(4), 646–670.

De Dreu, C. K. W. (2004). Motivation in negotiation: A social psychological analysis. *Handbook Negotiation Culture*, 114–135.

De Dreu, C. K. W., & Carnevale, P. J. (2003). Motivational bases of information processing and strategy in conflict and negotiation. *Advances in Experimental Social Psychology, 35*, 235–291. https://doi.org/10.1016/S0378-1097(99)00465-6

De Dreu, C. K. W., Giebels, E., & Van De Vliert, E. (1998). Social motives and trust in integrative negotiation: The disruptive effects of punitive capability. *Journal of Applied Psychology, 83*(3), 408–422. https://doi.org/10.1037/0021-9010.83.3.408

Druckman, D., & Wagner, L. M. (2016). Justice and negotiation. *Annual Review of Psychology, 67*(August), 387–413. https://doi.org/10.1146/annurev-psych-122414-033308

Dyer, J. H., & Chu, W. (2003). The role of trustworthiness in reducing transaction costs and improving performance: Empirical evidence from the United States, Japan, and Korea. *Organization Science, 14*(1), 57–68. https://doi.org/10.1287/orsc.14.1.57.12806

Emerson, R. M. (1962). Power-dependency relations. *American Sociological Review, 27*(1), 31–41.

Emerson, R. M. (1964). Power-dependence relations: Two experiments. *Sociometry*, 282–298.

Fisher, R., Ury, W. L., & Patton, B. (2011). *Getting to yes: Negotiating agreement without giving in.* Penguin.

Folger, R., & Konovsky, M. A. (1989). Effects of Procedural and Distributive Justice on Reactions to Pay Raise Decisions. *Academy of Management Journal, 32*(1), 115–130. https://doi.org/10.5465/256422

Ganesan, S., & Hess, R. (1997). Dimensions and Levels of Trust: For Implications to a Relationship Commitment. *Marketing Letters, 8*(4), 439–448.

Gelfand, M. J., & Christakopoulou, S. (1999). Culture and negotiator cognition: Judgment accuracy and negotiation processes in individualistic and collectivistic cultures. *Organizational Behavior and Human Decision Processes, 79*(3), 248–269. https://doi.org/10.1006/obhd.1999.2845

Gulati, R., Nohria, N., & Zaheer, A. (2000). Strategic networks. *Strategic Management Journal, 21*(3), 203–215. https://doi.org/10.1002/(SICI)1097-0266(200003)21:3%3c203::AID-SMJ102%3e3.0.CO;2-K

Gunia, B. C., Brett, J. M., Nandkeolyar, A. K., & Kamdar, D. (2011). Paying a price: Culture, trust, and negotiation consequences. *Journal Application Psychology, 96* (4), 5–24. https://doi.org/10.1037/2Fa0021986.

Guo, W., Lu, W., Hao, L., & Gao, X. (2021). Interdependence and information exchange between conflicting parties: The role of interorganizational trust. *IEEE Transactions on Engineering Management*, 1–17.https://doi.org/10.1109/TEM.2020.3047499

Hagger, M. S., Chatzisarantis, N. L. D., & Biddle, S. J. H. (2002). A meta-analytic review of the theories of reasoned action and planned behavior in physical activity: An examination of predictive validity and the contribution of additional variables. *Journal Sport Exercise Psychology, 24*(1), 3–32. 20.500.11937/10206.

Hartman, F. T. (2002). The role of trust in project management. *Front. Project Management Research Square, Pennsylvania, PMI*, 225–235.

Jacobson, C., & Choi, S. O. (2008). Success factors: Public works and public-private partnerships. *International Journal of Public Sector Management, 21*(6), 637–657. https://doi.org/10.1108/09513550810896514

Jagannathan, M., & Delhi, V. S. K. (2020). Litigation in construction contracts: Literature review. *Journal of Legal Affairs and Dispute Resolution in Engineering and Construction, 12*(1), 03119001.

Jagosh, J., Bush, P. L., Salsberg, J., Macaulay, A. C., Greenhalgh, T., Wong, G., Cargo, M., Green, L. W., Herbert, C. P., & Pluye, P. (2015). A realist evaluation of community-based participatory research: Partnership synergy, trust building and related ripple effects. *BMC Public Health, 15*(1), 1–11. BMC Public Health. https://doi.org/10.1186/s12889-015-1949-1.

Johari, S., & Jha, K. N. (2020). Interrelationship among belief, intention, attitude, behavior, and performance of construction workers. *Journal of Management in Engineering, 36*(6), 1–14. https://doi.org/10.1061/(ASCE)ME.1943-5479.0000851

Jordan, J., Sivanathan, N., & Galinsky, A. D. (2011). Something to lose and nothing to gain: The role of stress in the interactive effect of power and stability on risk taking. *Administrative Science Quarterly, 56*(4), 530–558. https://doi.org/10.1177/0001839212441928

Kadefors, A. (2005). Fairness in interorganizational project relations: Norms and strategies. *Construction Management and Economics, 23*(8), 871–878. https://doi.org/10.1080/014461905 00184238

Karlsen, J. T., Grae, K., & Massaoud, M. J. (2008). The role of trust in project-stakeholder relationships: A study of a construction project. *International Journal Project Organization Management, 1*(1), 105–118. https://doi.org/10.1504/IJPOM.2008.020031

Keltner, D., Gruenfeld, D. H., & Anderson, C. (2003). Power, approach and inibition. *Psychological Review, 110*(2), 265.

Kim, P. H., & Fragale, A. R. (2005). Choosing the path to bargaining power: An empirical comparison of BATNAs and contributions in negotiation. *Journal of Applied Psychology, 90*(2), 373–381. https://doi.org/10.1037/0021-9010.90.2.373

Van Kleef, G. A., De Dreu, C. K. W., Pietroni, D., & Manstead, A. S. R. (2006). Power and emotion in negotiation: Power moderates the interpersonal effects of anger and happiness on concession making. *European Journal of Social Psychology, 36*(4), 557–581. https://doi.org/10.1002/ejs p.320

Kleiman, T., Sher, N., Elster, A., & Mayo, R.(2015). Accessibility is a matter of trust: Dispositional and contextual distrust blocks accessibility effects. *Cognition, 142*, 333–344. Elsevier B.V. https://doi.org/10.1016/j.cognition.2015.06.001.

Koh, T. Y., & Rowlinson, S. (2012). Relational approach in managing construction project safety: A social capital perspective. *Accident Analysis and Prevention, 48*, 134–144. Elsevier Ltd. https://doi.org/10.1016/j.aap.2011.03.020.

Kwon, I. W. G., & Suh, T. (2005). Trust, commitment and relationships in supply chain management: A path analysis. *Supply Chain Management, 10*(1), 26–33. https://doi.org/10.1108/135985405 10578351

Lawler, E. J. (1986). Bilateral deterrence and conflict spiral: A theoretical analysis. *Graphical Processes, 3*, 107–130. https://doi.org/10.1108/s0882-6145_2014_0000031014

Lawler, E. J., Ford, R. S., & Blegen, M. A. (1988). Coercive capability in conflict: A test of bilateral deterrence versus conflict spiral theory. *Social Psychology Quarterly, 51*(2), 93–107.

Lewicki, R. J., Saunders, D. M., Minton, J. W., Roy, J., & Lewicki, N. (2011). *Essentials of negotiation.* McGraw-Hill/Irwin.

Li, B., Gao, Y., Zhang, S., & Wang, C. (2021). Understanding the effects of trust and conflict event criticality on conflict resolution behavior in construction projects: Mediating role of social motives. *Journal Management Engineering, 37*(6): 04021066. American Society of Civil Engineers (ASCE). https://doi.org/10.1061/(asce)me.1943-5479.0000962.

Li, L. (2005). The effects of trust and shared vision on inward knowledge transfer in subsidiaries' intra- and inter-organizational relationships. *International Business Review, 14*(1), 77–95. https://doi.org/10.1016/j.ibusrev.2004.12.005

Lin, S., & Cheung, S. O. (2021). Analyzing intention to settle in construction dispute negotiation. *Creation Construction Conference 2021.* Opatija.

Lind, E. A., Ambrose, M., & Park, M. V. D. V. (1993). Individual and corporate dispute resolution : using procedural fairness as a decision heuristic. *Administrative Science Quarterly*, 224–251.

Lind, E. A., & Tyler, T. R. (1988). *The social psychology of procedural justice*. Springer Science & Business Media.

Liu, J., Geng, L., Xia, B., & Bridge, A. (2017). Never let a good crisis go to waste: Exploring the effects of psychological distance of project failure on learning intention. *Journal of Management in Engineering, 33*(4), 1–7. https://doi.org/10.1061/(ASCE)ME.1943-5479.0000513

Liu, J., Lin, S., & Feng, Y. (2018). Understanding why Chinese contractors are not willing to purchase construction insurance. *Engineering Construction and Architectural Management, 25*(2), 257–272. https://doi.org/10.1108/ECAM-08-2016-0186

Loosemore, M. (1999). A grounded theory of construction crisis management. *Construction Management and Economics, 17*(1), 9–19. https://doi.org/10.1080/014461999371781

Love, P., Davis, P., Ellis, J., & Cheung, S. O. (2010). Dispute causation: Identification of pathogenic influences in construction. *Engineering Construction and Architectural Management, 17*(4), 404–423. https://doi.org/10.1108/09699981011056592

Love, P. E. D., Davis, P. R., Cheung, S. O., & Irani, Z. (2011). Causal discovery and inference of project disputes. *IEEE Transactions Engineering Management, 58*(3), 400–411. IEEE. https://doi.org/10.1109/TEM.2010.2048907.

Lu, W., Li, Z., & Wang, S. (2017). The role of justice for cooperation and contract's moderating effect in construction dispute negotiation. *Engineering Construction Architcture Management, 24* (1): 133–153. Emerald Group Publishing Ltd. https://doi.org/10.1108/ECAM-01-2015-0002.

Lu, W., Zhang, L., & Li, Z. (2015). Influence of negotiation risk attitude and power on behaviors and outcomes when negotiating construction claims. *Journal of Construction Engineering and Management, 141*(2), 1–11. https://doi.org/10.1061/(ASCE)CO.1943-7862.0000927

Luo, Y. (2007). The independent and interactive roles of procedural, distributive, and interactional justice in strategic alliances. *Academy of Management Journal, 50*(3), 644–664. https://doi.org/10.5465/AMJ.2007.25526452

Macfarlane, J. (2001). Why do people settle? *McGill Law J., 46*(3), 663–711.

Magee, J. C., & Galinsky, A. D. (2008). The self-reinforcing nature of social hierarchy: Origins and consequences of power and status. *IACM 21st Annual Conference Paper.*

Mannix, E. A., & Neale, M. A. (1993). Power imbalance and the pattern of exchange in dyadic negotiation. *Group Decision and Negotiation, 2*(2), 119–133. https://doi.org/10.1007/BF01884767

Maqsoom, A., Wazir, S. J., Choudhry, R. M., Thaheem, M. J., & Zahoor, H. (2020). Influence of perceived fairness on contractors' potential to dispute: moderating effect of engineering ethics. *Journal Construction Engineering Management, 146*(1): 04019090. American Society of Civil Engineers (ASCE). https://doi.org/10.1061/(asce)co.1943-7862.0001740.

Mayer, R. C., Davis, J. H., & Schoorman, F. D. (1995). An integrative model of organizational trust. *Academy of Management Review, 20*(3), 709–734.

McAllister, D. J. (1995). Affect- and cognition-based trust as foundations for interpersonal cooperation in organizations. *Academy of Management Journal, 38*(1), 24–59. https://doi.org/10.5465/256727

McClintock, C. (1977). Social motives in settings of outcome interdependence. *Negotional Social Psychology Perspective*, 49–77.

Mitropoulos, P., & Howell, G. (2001). Model for understanding, preventing and resolving project disputes. *Journal of Construction Engineering and Management, 127*(3), 223–231. https://doi.org/10.1061/(ASCE)0733-9364(2001)127:3(223)

Molm, L. D. (2015). Power-dependence theory. *Blackwell Encyclogy Sociology*. https://doi.org/10.1002/9781405165518.wbeosp082.pub2.

Nahapiet, J., & Goshal, S. (1998). Social capital, intellectual capital, and the organizational advantage. *Academy of Management Review, 23*(2), 242–266.

Nanus, B. (1992). *Visionary leadership: Creating a compelling sense of direction for your organization*. Jossey-Bass.

Newell, W. J., Ellegaard, C., & Esbjerg, L. (2019). The effects of goodwill and competence trust on strategic information sharing in buyer–supplier relationships. *The Journal of Business and Industrial Marketing, 34*(2), 389–400. https://doi.org/10.1108/JBIM-02-2017-0035

Parkhill, K. A., Shirani, F., Butler, C., Henwood, K. L., Groves, C., & Pidgeon, N. F. (2015). We are a community [but] that takes a certain amount of energy': Exploring shared visions, social action, and resilience in place-based community-led energy initiatives. *Environment Science Policy, 53*, 60–69. Elsevier Ltd. https://doi.org/10.1016/j.envsci.2015.05.014.

Patton, C., & Balakrishnan, P. V. S. (2010). The impact of expectation of future negotiation interaction on bargaining processes and outcomes. *Journal of Business Research, 63*(8), 809–816. https://doi.org/10.1016/j.jbusres.2009.07.002.

Pearce, C. L., & Ensley, M. D. (2004). A reciprocal and longitudinal investigation of the innovation process: The central role of shared vision in product and process innovation teams (PPITs). *Journal of Organizational Behavior, 25*(2), 259–278. https://doi.org/10.1002/job.235

Pinto, J. K., Slevin, D. P., & English, B. (2009). Trust in projects: An empirical assessment of owner/contractor relationships. *International Journal Project Management, 27*(6): 638–648. Elsevier Ltd and IPMA. https://doi.org/10.1016/j.ijproman.2008.09.010.

Piskorski, M. J., & Casciaro, T. (2005). Power imbalance, mutual dependence, and constraint absorption: A closer look at resource dependence theory. *Administrative Science Quarterly, 50*(2), 167–199.

Pruitt, D. G. (1983). Strategic choice in negotiation. *American Behavioral Scientist, 27*(2), 167–194.

Qu, Y., & Cheung, S. O. (2012). Logrolling 'win–win' settlement in construction dispute mediation. *Automation in Construction, 24*, 61–71. https://doi.org/10.1016/j.autcon.2012.02.010

Rahim, M. A. (1983). A measure of styles of handling interpersonal conflict. *Acadmics Journal Management, 26*(2), 368–376.

Raiffa, H. (1982). *The art and science of negotiation.* Harvard University Press.

RICS. (2021). More common, more costly: Is COVID-19 causing conflict in the construction sector? https://www.rics.org/es/wbef/megatrends/markets-geopolitics/more-common-more-costly-is-covid-19-causing-conflict-in-the-construction-sector/.

Ring, P. S. (1996). Fragile and resilient trust and their roles in economic exchange. *Business & Society, 35*(2), 148–175.

Rodríguez, C. M., & Wilson, D. T. (2002). Relationship bonding and trust as a foundation for commitment in U.S.-Mexican strategic alliances: A structural equation modeling approach. *Journal of International Marketing, 10*(4), 53–76. https://doi.org/10.1509/jimk.10.4.53.19553

Rousseau, D. M., Sitkin, S. B., Burt, R. S., & Camerer, C. (1998). Not so different after all: A cross-discipline view of trust. *Academy of Management Review, 23*(3), 393–404.

Rupp, D. E., & Spencer, S. (2006). When customers lash out: The effects of customer interactional injustice on emotional labor and the mediating role of discrete emotions. *Journal of Applied Psychology, 91*(4), 971.

Senescu, R. R., Aranda-Mena, G., & Haymaker, J. R. (2013). Relationships between project complexity and communication. *Journal of Management in Engineering, 29*(2), 183–197.

Shaver, K. G. (2012). *The attribution of blame: Causality, responsibility, and blameworthiness.* Springer Science & Business Media.

Shenhar, A., & Holzmann, V. (2017). The three secrets of megaproject success: Clear strategic vision, total alignment, and adapting to complexity. *Project Management Journal, 48*(6), 29–46. https://doi.org/10.1177/875697281704800604

Sondak, H., & Bazerman, M. H. (1991). Power balance and the rationality of outcomes in matching markets. *Organizational Behavior and Human Decision Processes, 50*(1), 1–23. https://doi.org/10.1016/0749-5978(91)90031-N

Tajima, M., & Fraser, N. M. (2001). Logrolling procedure for multi-issue negotiation. *Graphical Decision Negotiation, 10*(3), 217–235. https://doi.org/10.1023/A:1011262625052

Tan, H. H., & Tan, C. S. (2000). Toward the differentiation of trust in supervisor and trust in organization. *Genetic, Social, and General Psychology Monographs, 126*, 241–260.

Tatum, B. C., & Eberlin, R. J. (2008). The relationship between organizational justice and conflict style. *Business Strategy Series, 9*(6), 297–305. https://doi.org/10.1108/17515630810923603

Thibaut, J. W., & Walker, L. (1975). *Procedural justice: A psychological analysis.* L. Erlbaum Associates.

Thompson, L. (1990). Negotiation behavior and outcomes: Empirical evidence and theoretical issues. *Psychological Bulletin, 108*(3), 515–532. https://doi.org/10.1037/0033-2909.108.3.515

Thompson, L. L., Wang, J., & Gunia, B. C. (2010). Negotiation. *Annual Review of Psychology, 61*, 491–515. https://doi.org/10.1146/annurev.psych.093008.100458

Thoms, P., & Greenberger, D. B. (1995). Training business leaders to create positive organizational visions of the future: Is it successful? *Acad. Manag., 1*, 212–216.

Tost, L. P. (2015). When, why, and how do powerholders 'feel the power'? Examining the links between structural and psychological power and reviving the connection between power and responsibility. *Research Organization Behaviour, 35*, 29–56. Elsevier Ltd. https://doi.org/10.1016/j.riob.2015.10.004.

Treacy, T. B. (1995). Use of alternative dispute resolution in construction. *Journal of Management in Engineering, 11*(1), 58–63. https://doi.org/10.1201/9780203859926.ch125

Trötschel, R., & Gollwitzer, P. M. (2007). Implementation intentions and the willful pursuit of prosocial goals in negotiations. *Journal of Experimental Social Psychology, 43*(4), 579–598. https://doi.org/10.1016/j.jesp.2006.06.002

Tversky, A., & Kahneman, D. (1985). The framing of decisions and the psychology of choice. *Behaviour Decision Market*, 25–41. Spring, Boston.

Tversky, A., & Kahneman, D. (1986). Rational choice and the framing of decisions. *Journal of Business, 59*(4), 251–278. https://doi.org/10.1086/296365

Tyler, T., & Blader, S. (2013). *Cooperation in groups: Procedural justice, social identity, and behavioral engagement.* Routledge.

Tyler, T. R. (1989). The psychology of procedural justice: A test of the group-value model. *Journal of Personality and Social Psychology, 57*(5), 830–838. https://doi.org/10.1037/0022-3514.57.5.830

Tyler, T. R., & Bies, R. J. (2015). Beyond formal procedures: The interpersonal context of procedural justice. *Application Social Psychology Organization settings*, 77–98. Psychology Press.

Wang, G., Locatelli, G., Wan, J., Li, Y., & Le, Y. (2021). Governing behavioral integration of top management team in megaprojects: A social capital perspective. *International Journal Project Management, 39*(4), 365–376. Elsevier Ltd. https://doi.org/10.1016/j.ijproman.2020.11.005.

Wei, Q., & Luo, X. (2012). The impact of power differential and social motivation on negotiation behavior and outcome. *Public Personal Management, 41*(5), 47–58.

Wolfe, R. J., & McGinn, K. L. (2005). Perceived relative power and its influence on negotiations. *Group Decision Negotiation, 14*(1), 3–20. https://doi.org/10.1007/s10726-005-3873-8

Wong, P. S. P., & Cheung, S. O. (2004). Trust in construction partnering: Views from parties of the partnering dance. *International Journal of Project Management, 22*(6), 437–446. https://doi.org/10.1016/j.ijproman.2004.01.001

Wong, R. S. (2014). Same power but different goals: how does knowledge of opponents' power affect negotiators' aspiration in power-asymmetric negotiations? *Global Journal Business Research, 8*(3), 77–89.

Woolthuis, R. K., Hillebrand, B., & Nooteboom, B. (2005). Trust, contract and relationship development. *Organization Studies, 26*(6), 813–840. https://doi.org/10.1177/0170840606050554594

Wu, M. Y., Weng, Y. C., & Huang, I. C. (2012). A study of supply chain partnerships based on the commitment-trust theory. *Asia Pacific Journal Market Logistics, 24*(4), 690–707. https://doi.org/10.1108/13555851211259098

Yiu, T. W., Cheung, S. O., & Chow, P. T. (2008). Logistic regression modeling of construction negotiation outcomes. *IEEE Transactions on Engineering Management, 55*(3), 468–478. https://doi.org/10.1109/TEM.2008.922630

Yiu, T. W., Cheung, S. O., & Lok, C. L. (2015). A fuzzy fault tree framework of construction dispute negotiation failure. *IEEE Transactions on Engineering Management, 62*(2), 171–183. https://doi.org/10.1109/TEM.2015.2407369

Yiu, T. W., Liu, T., & Kwok, L. C. (2018). Explicating the role of relationship in construction claim negotiations. *Journal of Construction Engineering and Management, 144*(2), 1–9. https://doi.org/10.1061/(ASCE)CO.1943-7862.0001431

Youngblood, S. A., Trevino, L. K., & Favia, M. (1992). Reactions to unjust dismissal and third-party dispute resolution: A justice framework. *Employee Responsibilities and Rights Journal, 5*(4), 283–307. https://doi.org/10.1007/BF01388306

Yousefi, S., Hipel, K. W., & Hegazy, T. (2010). Attitude-based negotiation methodology for the management of construction disputes. *Journal of Management in Engineering, 26*(3), 114–122. https://doi.org/10.1061/(asce)me.1943-5479.0000013

Yuan, H., Wu, H., & Zuo, J. (2018). Understanding factors influencing project managers' behavioral intentions to reduce waste in construction projects. *Journal of Management in Engineering, 34*(6), 04018031. https://doi.org/10.1061/(asce)me.1943-5479.0000642

Zhang, L., Fu, Y., & Lu, W. (2021). Contract enforcement for claimants' satisfaction with construction dispute resolution: moderating role of shadow of the future, fairness perception, and trust. *Journal Construction Engineering Management, 147*(2), 04020168. American Society of Civil Engineers (ASCE). https://doi.org/10.1061/(asce)co.1943-7862.0001981.

Zhang, S. B., Fu, Y. F., Gao, Y., & Zheng, X. D. (2016). Influence of trust and contract on dispute negotiation behavioral strategy in construction subcontracting. *Journal of Management in Engineering, 32*(4), 04016001. https://doi.org/10.1061/(asce)me.1943-5479.0000427

Zhang, Z. X., & Han, Y. L. (2007). The effects of reciprocation wariness on negotiation behavior and outcomes. *Group Decision Negotiation, 16*(6), 507–525. https://doi.org/10.1007/s10726-006-9070-6

Miss Sen Lin is a Ph.D. candidate of the Department of Architecture and Civil Engineering, City University of Hong Kong. Her doctoral study covers conceptualizing negotiators' intention to settle (ITS), investigating the barriers and enablers of negotiators' ITS, and suggesting ways to facilitate a better negotiation environment for settlement. She is a member of the Construction Dispute Resolution Research Unit that supports her Ph.D. program. Her current research interests are construction dispute resolution, contracting behaviour, construction project management, and construction risk management. Miss Lin has published journal and conference papers, as well as book chapters in these topics.

Chapter 12
Voluntary Participation as an Incentive of Construction Dispute Mediation—A Reality Check

Nan Cao

Abstract In recent years, the use of mediation as an alternative to arbitration/litigation has gathered momentum at both industry and national levels. One characterising feature of the mediation movement is keeping voluntary participation as one of the core design features of mediation arrangements. The use of mediation in construction has started in the mid-eighties in Hong Kong. Despite the concerted efforts of the Hong Kong Government and the mediation services providers, its adoption had soon flattened off after an initial rise. A slight decline in usage has in fact been recorded recently. Use of mediation to resolve construction disputes has not been as promising as expected. From a pragmatic point of view, this study identified four potential mismatches between contracting arrangements with the voluntary participation. These are (i) principal-agent relationship; (ii) power asymmetry between the parties; (iii) quasi-imposed adoption; and (iv) biases of the disputing parties on the process. It is concluded that voluntary participation may not directly lead to the adoption of construction dispute mediation.

Keywords Construction dispute mediation · Reality check · Voluntary participation · Compulsion · Indifference

1 Introduction

Construction disputes are likely to increase because of the disruption caused by COVID-19 to construction projects (Kim et al., 2021). Dispute should be resolved early with negotiation being the most used method (Cheung et al., 2000). However, negotiation is not always successful in reaching a settlement and the dispute will need the service of other more formal dispute resolution processes (Chong & Zin, 2012). Currently, the most designated methods of construction dispute resolution

N. Cao (✉)
Construction Dispute Resolution Research Unit, Department of Architecture and Civil Engineering, City University of Hong Kong, Hong Kong, China
e-mail: nancao7-c@my.cityu.edu.hk

© The Author(s), under exclusive license to Springer Nature Switzerland AG 2023
S. O. Cheung and L. Zhu (eds.), *Construction Incentivization*, Digital Innovations in Architecture, Engineering and Construction,
https://doi.org/10.1007/978-3-031-28959-0_12

are mediation and arbitration (Chan & Suen, 2005). Arbitration clause is included in most of the standard forms of contract and was originally introduced as a less notorious to litigation. Moreover, it has been developed as a replicate of litigation due to the adversarial process the arbitral procedures have adopted (Harmon, 2003). Furthermore, most contract dispute resolution clauses specify that arbitration cannot be commenced until the construction work has reached substantial completion or the contract is terminated. Thus, the two parties may be stuck in a sour relationship for a long time, especially if the dispute occurs early in the project (Chau, 2007). Furthermore, typical construction contract dispute resolution provisions are multi-tiered, with mediation incorporated as an intermediate step before arbitration. The attempt of mediation is often served as a condition precedent to arbitration.

Using mediation to resolve dispute has a long history. Knowing the broad concept of mediation in both local and international contexts would help understanding better the design of the process. The Hong Kong judiciary (2020) defined mediation as a voluntary process in which a trained and impartial third person, the mediator, helps the parties in dispute to reach an amicable settlement that meets their needs. The American Arbitration Association (2004) defined mediation as a process in which an impartial third party facilitates communication and negotiation and promotes voluntary decision making by the parties to settle the dispute across different forms and contexts. The European Union Directive (2008) on "Certain Aspects of Media-tion in Civil and Commercial Matters" defined mediation as "a structured process, however named or referred to, whereby two or more parties to a dispute attempt by themselves, on a voluntary basis, to reach an agreement to settle their dispute with the assistance of a mediator. This process may be initiated by the parties or suggested/ordered by a court. Therefore, the definitions of mediation in different countries and regions share a common design that mediation should be regarded as a voluntary dispute resolution process and to be assisted by an impartial mediator. In Australia, Canada and the United Kingdom where mandatory mediation is implemented, it is found that there is no significant difference in the settlement rate between voluntary and mandatory mediation (Quek, 2009). In fact, there is little concern over manda-tory use if the outcome is self-determined. Moreover, involuntary use may create the debate of denial to justice that has always been viewed as a constitutional right (Boettger, 2004; Wissler, 1997). There will not be a simple black or white answer to make voluntary participation a divine design principle of mediation. Voluntariness is a multidimensional concept and encapsulates the idea of participation at one's own will. To arouse the interest of the disputing parties, voluntary participation offers the attraction that parties have nothing to lose in attempting mediation. This study looks into the viability of voluntary participation as a bait for use with reference to prevailing construction contracting practice. In this connection, a thorough review of construction dispute resolution is conducted to unveil the potential incompatibil-ities. This study therefore works as a reality check of the following: can voluntary participation be an incentive of construction dispute mediation?

2   Construction Dispute Resolution

The literature review on construction dispute resolution covers the following topics:

- Nature of construction dispute
- Approaches in resolution
- Resolution methods
- Use of mediation in the Hong Kong.
- Voluntary participation of CDM.

2.1   Nature of Construction Dispute

Disputes in the construction industry can result from a variety of reasons; contractual, environmental, and behavioural. Typical construction projects last for several years, during which many changes may happen. Furthermore, physical, and environmental conditions may also prove to be materially different from those envisaged at tender. It has often been proved in vain for efforts to exhaust all contingencies. Disputes frequently occur when there is no provision to deal with unanticipated happenings. Cheung and Pang (2014) suggested that construction disputes have three primary contributing factors: task, contract incompleteness, and people (Cheung, & Pang, 2014).

Task Factor

It is imperative to do a thorough risk assessment during the tendering process. The time to do a risk assessment at tender is frequently very short. There are many examples of projects that take far longer to complete than that time anticipated and agreed upon because the risks of the project were not sufficiently evaluated ex ante. Inevitably, delay would incur increased expenses by the contractor. The owner's potential to sue for delay damages would made the contracting environment extremely acrimonious. Project risks could have an impact on the project's potential.

Uncertainty is the discrepancy between the information needed to complete the task and the information already available (Klir, 2006). The complexity of the task and the performance requirements determine the amount of time and cost required. Uncertainty means that not all project elements can be planned out before work begins (Marti et al., 2010). When uncertainty is high, initial designs and specifications will inevitably be insufficient. If disputes happen, project participants will have to work together to find solutions.

Contractual Incompleteness

Every construction contract dispute must have a contractual basis (Totterdill, 1991). Standard contract forms explicitly set out the risks and responsibilities which contracting parties have agreed to undertake. Moreover, this drafting objective may not be achieved for transactions of long duration and with works to be executed in uncertain environment. When customised contracts or amended standard forms are used, inconsistencies or unintended misunderstandings will lead to disputes. In extreme circumstances, contradictory provisions are resulted. As such, incompleteness, omissions, errors etc., may cause disagreement over risk ownership by the contracting parties.

People Factor

While incomplete contracts create minefields, opportunism behaviours exploited by contracting parties can take the form of commitment violations, forced renegotiations, responsibility evasion, and refusals to adapt (Wathne & Heide, 2000). Since contracts cannot account for every eventuality, when a problem surfaces, one party may wish to take advantages of as much as possible. The counterpart may pretend to be ignorance and avoid taking responsibilities. The parties may also have different interpretations of the happening. It is also common for the parties to find their expectations being miles apart from the outcome. Another sting of the situation is when the project team members are having personal conflict among them (Mitropoulos & Howell, 2001). The emotion involved frequently intensifies the conflict and prevents the parties from taking rational decisions.

2.2   *Approaches in Resolution*

Construction problems are prevalent, which suggests that there is a need to identify suitable ways to manage them before being blown out as major disputes. In fact, there are quite a number of methods that have been put into practice. Most construction contracts would specify several methods, usually in a tiered arrangement, to resolve disputes arising from the project. Moreover, tiered dispute resolution is far often being deployed as separate independent options. Construction contract drafters frequently overlook the fact that dispute prevention and dispute resolution techniques can be integrated to maximize the chance of disposing the disputes. The following three resolution approaches are commonly used.

Dispute Avoidance

The construction industry has made significant strides in creating more effective dispute resolution procedures over the past few decades. In fact, experts usually named the construction sector as offering cutting edge of innovation. However, it appears that the construction industry has not given enough thoughts to prevent dispute. Dispute prevention techniques are routinely overlooked in the design of dispute resolution clauses in the construction contracts. One notable exception is the use of Dispute Resolution Advisor (DRA) in the HKSAR government works contracts. DRA aims to facilitate early resolution of problems that arise during the construction before these crystalize into dispute.

It is crucial to comprehend the individual project specificities to avoid disputes in the construction projects. In this sense, it might be wise to work with a DRA or another impartial third party during the construction stage. By aiding the parties to create appropriate dispute prevention strategies, the value of DRA can be influential.

DRA is currently used for projects in Hong Kong with the Architectural Services Department and the Housing Authority. The DRA is tasked with preventing disputes at the main contract and nominated subcontract levels. DRA is jointly appointed by both the employer and the main contractor.

The fundamental idea behind a Dispute Resolution Advisor (DRA) is the use of an impartial third-party neutral who counsels the parties to a prospective dispute and offers viable solutions to settle it. From the start of the contract through its conclusion, the employer, and the contractor jointly appoint the DRA. The primary responsibility of the DRA is to help the parties identify possible solutions to the issue and assist in settling those conflicts before they become official disputes. The DRA does not have any decision-making authority, and his/her role is to encourage parties to collaborate and complete the works in line with the contract.

Negotiated Resolution

Most disputes are settled by inter party negotiation without outside assistance. The Contract shall permit the provision of various options for dispute resolution techniques to enable the contractual matters of different opinions be negotiated before triggering the more formal procedures. Construction contracts typically resolve disputes through arbitration or litigation if the interparty negotiations fail. However, not every disagreement can be settled by the parties themselves. In those circumstances, intervention by a third party would become necessary. Nonetheless, it is advisable to use more flexible ways first because litigation and arbitration are expensive and time-consuming.

At this point, the participation of a third-party neutral adds value by facilitating exchanges to aid the parties in resolving the issue quickly and effectively before it worsens to the point where it has a significant negative impact on the project. A skilful third-party neutral can help the disputing parties to exchange more focused proposals by providing advice. Expedited settlement and low costs are the obvious advantages of negotiation. Additionally, facilitated discussions help to maintain good relations between the parties and do not typically disrupt the project.

Binding Resolution

Not all disagreements can be settled by negotiation. To address these issues, a complete dispute resolution framework must include options like mediation, adjudication, arbitration, or litigation. However, comprehension of the construction business and construction conflicts is essential for providing advice and making decisions in such disputes, particularly in complicated construction disputes. In litigation, the parties may select a counsel with relevant knowledge, but they have no right to choose the judge. Arbitration does allow the appointment of construction-related arbitrators by the parties. As a result, using arbitration as a dispute resolution method rather than litigation may give the parties a more construction appropriate outcome. Post-completion arbitration is the general arrangement in most construction contracts in Hong Kong. Moreover, in recent years adjudication and voluntary mediation have been introduced as pre-arbitration intermediate proceedings. Additionally, to obtain expedited resolution, contracting parties should be required to pursue alternative dispute resolution if one of them causes the event. This would eliminate the necessity for a mutual consent. Nevertheless, the Guidelines on Dispute Resolution (HKCIC, 2010), states that alternate dispute resolution may not be adequate to address the various sorts of issues that may arise throughout the course of the contract.

2.3   Resolution Methods

It is impossible to resolve every issue and account for every possibility at the pre-contract stage simply because of the unpredictability and complexity faced by every construction project. The reality is that unanticipated risks may surface after the project commencement. When the responsibility for the parties is unclear, dispute arises. In such situation, the contracting parties would first try to resolve it amicably since this is probably the quickest and most cost-effective course of action. Prompt mediation would allow the project to go without interruption and preserving strong working relationships. If this isn't an option, it could be essential to look for a third party to assist settlement. However, going to court to resolve a dispute can be costly, complicated, contentious, and time-consuming.

Speedy settlement of disputes is always preferred so that the work can move forward. A close working relationship would also be helpful. Unfortunately, this is seldom the situation. As a result, provisions for use of alternative dispute resolution (ADR) processes are planned and incorporated into the contract dispute resolution procedure. ADR processes are usually less formal than the court proceedings and are intended to be quicker, cheaper, less time-consuming, and allow the parties to preserve the relationship.

It is advised that the contracting parties be given the option to select any or a combination of the following dispute resolution procedures: 1. Mediation, 2. Adjudication, 3. Arbitration, 4. Litigation. The contracting parties may opt for post-completion arbitration as the final means of dispute resolution. An account of each of the commonly used resolution method is given here follows.

Negotiation

Almost all dispute resolution commences with negotiation. It has been well documented that negotiation is the most resource efficient resolution method. In fact, before triggering the dispute resolution clause of the contracts, the parties having differences are likely to have exchanged their positions. Categorically, dispute negotiations follow a stepped approach. The frontline project personnel are usually the initial negotiators. However, it is found that the success rate is not very high at this round of negotiation. Chow et al., (2015) opined that this is likely because they are the people making the decisions that are in dispute. This makes it very hard for both sides to back down without a fight. There are many research studies to champion negotiation in providing prescriptive advice (Fisher et al., 2011), quantitative tools Zaden

(1965) and behavioural deliberations (Afzalur and Bonoma (1979). Conventional wisdom suggests that most disputes were settled through negotiation. While there is no doubt about the versatility of negotiation. The quantum of major construction disputes makes negotiation relatively less successful. In fact, almost all construction contracts would not rely solely on negotiation, instead more formal dispute resolution provisions like arbitration are incorporated.

Mediation

In mediation, a neutral third party assists to reconcile the conflicts between the disputing parties through a mutually agreed procedure. In the facilitative form of mediation, a neutral third party, the mediator, helps the disputing parties to negotiate a settlement. A mediator does not have the authority to determine the core issues. For the past 20 years, Hong Kong has accumulated a reasonable amount of mediation experience in the building sector. Although post-completion arbitration is still predominantly used for major dispute, especially for public works projects.

The following are some characterizing features of mediation. The mediator can meet with a party in private. The parties may end the mediation at any time and the mediator does not render a decision or opinion. If the mediation succeeds, a settlement agreement that is a contract supplemental to the construction contract will be signed. If the mediation fails, the dispute will advance to the next tier of resolution.

Adjudication

An impartial third person, the adjudicator, is tasked with making decision on the disputes referred to him. After the parties have presented their evidence and made their written and/or spoken arguments, the adjudicator renders a decision. The decision is binding unless being challenged. In post-completion arbitration, the adjudicator's decision can be reviewed by the arbitrator. The decision of an adjudicator therefore is often described as interim binding only. In Hong Kong, adjudication has not been utilized much. Several adjudications were conducted under the Airport Core Programme (ACP) in the 1990s. The Government has adopted the DRA system in conjunction with voluntary adjudication for major engineering works contracts valued at more than HK$200 million. More recently, the Hong Kong Government has introduced the Security of Payment provisions for the use in Government projects. Adjudication is used to provide quick decisions on payment related disputes, at least on an interim basis. How popular adjudication will become is yet to be seen. It is anticipated adjudication will be a strong competitor of mediation.

Arbitration

Arbitration is a private yet highly regulated method to resolve disputes. Most construction contracts in the world designate arbitration as the final resolution forum. In some arbitration friendly jurisdictions, such as Hong Kong and the United Kingdom, arbitration can be agreed by the contracting parties as the ultimate form of dispute resolution. In simple terms, all disputes arising from the project will be resolved by arbitration and the right to litigation is limited. In essence, a dispute will be settled by an arbitrator or arbitration panel chosen by the disputing parties. Disputes are decided based on appropriate legal precedents and evidence on the facts. An arbitrator is appointed by the disputing parties or nominated by the designated appointing authority to conduct the arbitration, according to any applicable contractual clauses and the relevant statutory regulatory framework. The costs of the arbitration shall be borne by the losing party. Arbitration has evolved to follow many civil court procedures. In general, only a handful of cases are eligible for appeal as a very strict rule on appeal are imposed under most arbitration law.

Litigation

Litigation is the practice of pursuing or opposing legal action in court to settle disputes. The party's rights or responsibilities may be enforced or determined by the court. Even with arbitration friendly jurisdictions, there remains a fair amount of construction disputes decided by the court. Conflicts in the construction industry can result from a mix of factual and legal issues. Defective contract documents are one of the prime sources of legal disputes. To maintain fair proceedings, litigation procedures are suspectable to the tactics that may end with protracted proceedings with significant cost implications. It may take years to get a judgment from the court and the drawn-out process has proved to be a nightmare for less resourceful litigants. Litigation is open to the public and may generate undesirable publicity.

2.4 Use of Mediation in Hong Kong

Mediation has been promoted for use in the Hong Kong construction industry as an alternative to costly arbitration and litigation (Chau, 2007). Construction mediation was introduced in about mid 1980s. Since then, its use has been part of the mediation movement in Hong Kong. With the Hong Kong Government aiming to promote Hong Kong as a regional dispute resolution services hub, facilitations in the forms of legislation and revisions of court practice direction, mediation has been well placed as the mainstream alternative dispute resolution (ADR) regime in Hong Kong. The mediation movement reached its peak in 2009 when the Hong Kong Civil Justice Reform (CJR hereafter) came into effect. With these policies driven efforts, the use of mediation is expected to rise. Nevertheless, the adoption of mediation has not been particularly impressive. For example, the success rate of building management disputes has fluctuated in recent years. The number of building management mediation cases has shown a gradual decline since 2015 (The Hong Kong Judiciary, 2021a). The averaged data for 2008–2013 and annual data for 2013–2020 for the building management cases are shown in Fig. 1.

Another record also portrays a similar decline in use. The mediation reports filed with the Court of First Instance from 2011 to 2021, the number of mediations conducted in 2020 underwent a sharp decline (The Hong Kong Judiciary, 2021b). As shown in Table 1, the number of mediation certificates increased from 2012 to 2015. Since 2015, there has been no growth in use. Instead, a graduate decline is observed.

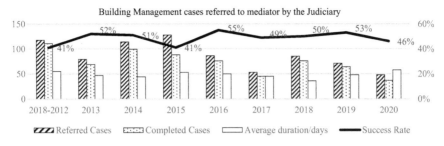

Fig. 1 Building Management cases referred to mediators by the Hong Kong Judiciary

Table 1 Number of mediation related documents filed in the court of first instance*

	2012	2013	2014	2015	2016	2017	2018	2019	2020	2021
Mediation certificate	2,977	2,878	3,271	3,668	3,623	3,716	3,590	2,138	1,793	1,703
Mediation notice	1,146	1,164	1,223	1,381	1,380	1,399	1,248	958	627	642
Mediation response	1,062	1,031	1,078	1,258	1,181	1,249	1,140	876	553	550
Mediation minutes	508	541	602	652	666	663	634	478	266	303
Settlement rate	38%	45%	48%	46%	48%	48%	51%	51%	47%	42%

* It only includes cases commenced by the 5 CJR related case types in the Court of First Instance, i.e., Civil Action (HCA), Admiralty Action (HCAJ), Commercial Action (HCCL), Construction and Arbitration List (HCCT)

To examine the use of mediation in major construction disputes, the following summaries are collected. Table 2 presents the number of Construction and Arbitration Proceedings (HCCT)-related cases (Legal Reference System, 2022).

The Hong Kong International Arbitration Centre (HKIAC hereafter) is the main dispute resolution provider in Hong Kong, their record of mediation shall be useful reference of the practice pattern. Tables 3 and 4 summarise the number of disputes handled by the HKIAC It can be observed that there has been no increase in the use of mediation for construction disputes.

Table 2 Number of construction and arbitration cases filed in the high court

	2011	2012	2013	2014	2015	2016	2017	2018	2019	2020	2021
HCCT	22	18	21	9	16	14	20	26	30	32	28

Table 3 Number of disputes involving HKIAC in recent years

	2014	2015	2016	2017	2018	2019	2020	2021
Arbitration	252	271	262	297	265	308	318	277
Mediation	24	22	15	15	21	12	16	12
Adjudication	0	0	0	0	0	1	0	0

Table 4 Ratio of construction disputes involving HKIAC

	2015	2016	2017	2018	2019	2020	2021
Construction dispute	22.2%	19.2%	19.2%	13.7%	14.8%	10.7%	9.4%

2.5 Voluntary Participation of CDM

Most mediation research are about understanding the mediation process. Mediation engages parties and the mediator in moving through a sequence of developmental stages. The general sequence of a mediation is outlined in Fig. 2 (Moore, 2014). The potential activities and moves of the mediator, mainly two broad categories of stages: (1) those conducted before the formal problem-solving sessions begin, often with the intermediary meeting and working with parties individually to better understand the conflict and develop possible mediation strategies; and (2) those conducted in stages after the mediator has entered into formal discussions with multiple disputants, either in a joint session or by shuttling between them, and has commenced some aspect of problem-solving.

A rise in use of mediation happened after the Civil Justice Reform came to effect in 2009. However, the rise was not sustained. Instead, a plateauing off soon surfaced then followed by a slight drop in the last five years. This happening is unexpected and quite disheartening to the mediation advocates.

In search for an explanation of this, the design and practice of mediation are first reviewed. The main attractions of mediation include privacy and flexibility. Voluntary participation is identified as the characterising feature to exemplify the

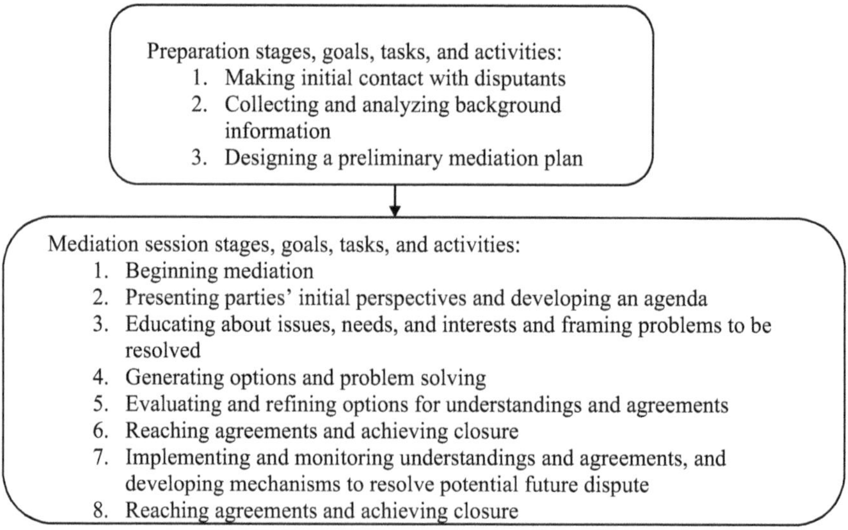

Preparation stages, goals, tasks, and activities:
1. Making initial contact with disputants
2. Collecting and analyzing background information
3. Designing a preliminary mediation plan

Mediation session stages, goals, tasks, and activities:
1. Beginning mediation
2. Presenting parties' initial perspectives and developing an agenda
3. Educating about issues, needs, and interests and framing problems to be resolved
4. Generating options and problem solving
5. Evaluating and refining options for understandings and agreements
6. Reaching agreements and achieving closure
7. Implementing and monitoring understandings and agreements, and developing mechanisms to resolve potential future dispute
8. Reaching agreements and achieving closure

Fig. 2 The mediation process roadmap (Moore, 2014)

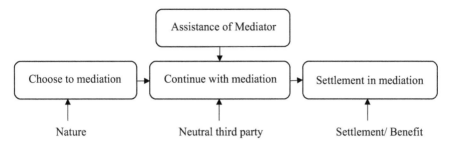

Fig. 3 Critical attributes of the mediation process

self-determination nature of the proceeding. This is supported by not conferring any decision power to the mediator. The freedom of exit at any time is attractive to disputants who are new to the process. Coercion runs against the voluntariness and can have three facets: coercion to mediate, coercion to continue and coercion to settle. Cheung et al. (2002) examined mediation from four aspects: nature, the neutral third party, settlement, and benefits. Those critical attributes are shown in Fig. 3. The first aspect normally serves as the main reason for choosing mediation, and the second aspect is usually used to justify continuing mediation. The last two aspects are mostly related to the willingness to settle.

Nature

Apart from voluntariness, confidentiality and enforceability are other major attractive attributes of mediation (Cheung, 1999). Confidentiality is very important to organisations, especially for listed companies. The potential of having long drawn litigation is detrimental to their share price. Thus, this form of loss is to be avoided. It is now well recognised the importance of keeping the whole mediation proceeding private and confidential. In some mediation rules also specifically state that the discussion during mediation proceedings shall not be used in subsequent resolution forums should the mediation fails to achieve a settlement. Nevertheless, how to invoke the parties' desire to mediate remains a challenge. Good faith provisions have been used to augment the value of mediation. Nevertheless, such provisions are vulnerable in common law courts.

Neutral Third Party

Mediation is a form of assisted negotiation; thus, the input of the mediator can have pivotal influence. For disputes over technical issues, it is useful to have mediator having the relevant technical background. Mediation services providers should enlist mediators of different backgrounds. Nevertheless, mediators shall discharge their function impartially. It can therefore be expected that the qualifications for mediators to be regulated especially for substantive disputes. At the moment, it seems that mediation does not have attraction for major disputes.

Settlement/Benefit

Only parties who are having the willingness to settle can lead to a sensible resolution. Aggression is therefore not the appropriate strategy for non-adversarial mediation process. Instead, identifying common interests would pave a path to settlement. As for settlement, mediation has the advantage of offering a wider range of remedies than formal proceedings. Non-monetary terms of settlement are possible as the settlement is in the form of contract rather than arbitration ward/court judgement. Very often remedy that ease the tension between the parties can help structuring a settlement agreement. Apology is the most obvious example. Lateral thinking is vital in considering settlement options. The average duration of mediation is within several days at the most. The associated benefits are self-evident.

The tension between voluntary participation as an underlying core value of mediation and quasi-imposition cannot be underestimated. Although some studies reported that parties can still benefit from using ADR even though their participation is not voluntary (Sherman, 1992). The prominence of self-determination in many mediation guidelines demonstrates the significance of voluntary participation in mediation (Hedeen, 2005). Some researchers have remarked that voluntary action in mediation is part of the "magic of mediation" that would lead to better outcomes: higher satisfaction with process, higher rates of settlement, and greater adherence to settlement terms (Shack, 2003, Wissler, 2004). And it is commonly held that mediators are expected to protect and nurture parties' self-determination and to facilitate the parties to be ultimate decision-makers (Welsh, 2001). If mediation is forced upon unwilling parties, the likely consequence is making the process perfunctory (Smith, 1998).

Nevertheless, consensus mediation as a key success factor remains non-conclusive. Welsh (2001) notes that many speak of "self-determination" but that they understand the term quite differently. The "dialogue of solidarities" perspective of voluntary participation (Petrzelka & Bell, 2000) reminded that individuals are embedded not only in ties of interest, but also in sentimental affection and normative commitment. Some researchers emphasise party's self-determination as participating freely at all stages of the process of the mediation (Merry, 1989). Self-determination theory (SDT hereafter) identifies three innate needs of competence, relatedness, and autonomy (Ryan & Deci, 2000). Competence refers to the capabilities to understand the possible results and effectiveness for the decision (Harter, 1978). Relatedness is based on the instinct to interact with others and considers whether the decision has an opportunity to interact with others (Baumeister & Leary, 2017). Autonomy refers to whether the motivation is from the heart and whether the behaviour is self-decided and not influenced by others (Deci & Vansteenkiste, 2003). Furthermore, SDT can be used to distinguish intrinsic and extrinsic motivation. Intrinsic motivation is the natural, inherent drive to seek out challenges and new possibilities that SDT associates with cognitive and social development while extrinsic motivation comes from external sources (Deci & Ryan, 1985; Vallerand, 1997).

Apart from Self-determination theory, there are other theoretical explanations that could be used to study the willingness to mediate (Pederson et al., 2007, Esses &

Dovidio, 2002). Social exchange theory, transaction cost economics theory, transactional value theory, social cognitive theory and planned behaviour theory are notable examples (Martins & Monroe, 1994, Wu, et al, 2014, Williamson, 1993, Zajac et al., 1993, Bandura, 1986, Schifter & Ajzen, 1985).

Transaction cost economics (TCE hereafter) explains how transactions are organized with the aim of minimizing transaction costs (Williamson, 1979). TCE suggests that each type of transaction produces coordination costs of monitoring, controlling, and managing transactions, in which cost is the primary determinant of such a decision (Williamson, 1979). The spectrum of transaction costs is extended to include any mechanism for coordinating the actions of individuals, which includes the costs of deciding, planning, arranging, and negotiating actions and the terms of exchange between two or more parties (Williamson, 1993). Zajac et al. (1993) examined inter-organisational strategies from a transactional value rather than transaction cost perspective to maximize joint value and create value.

Social cognitive theory (SCT hereafter) states that when people observe a model performing a behaviour and the consequences of that behaviour, they remember and use this information to guide the subsequent behaviours (Bandura, 1986). SCT can be applied to interpret the willingness to participate in the mediation and mainly aims to explain the influence of past mediation participation experience on current mediation participation.

Planned behaviour theory (PBT hereafter) is an improved model based on the rational behaviour theory that explains individuals' attitudes, subject norms, and perceived behavioural control in the light of their intention (Schifter & Ajzen, 1985). Although TPB was originally developed for the study of individual behaviours, it has been extended to understand organizational behaviour in recent years (Cheng, 2016).

The literature on mediation is growing, but the anchoring voluntary participation in mediation has not been scrutinised and in fact seems has been taken for granted. Voluntary participation embraces the implicit assumptions of intentional action, absence of controlling influences and no-role restriction. Based on the literature review conducted for this study, Cao (2021) summarised a list of manifestations of voluntary participation as presented in Table 5. Nelson et. al. (2011) focused on intention and the absence of coercion and manipulation. Appelbaum et al. (2009) on the other hand looked into inducement and persuasion to indicate the exertion of external forces. The approach of Brunk (1979) centered on manipulation. The societal stance of Kamuya et al. (2011) is thought provoking yet lacking construction perspective. As voluntariness is best identified by the parties' own initiative, no compulsion, and no indifference proposed by Poitras (2005) are adopted in this study.

Table 5 Identifications of voluntariness (Cao, 2021)

Dimensions of voluntariness	Manifestations	References
Intentional action	The party in the performance of actions uses intentional action	Nelson et al. (2011)
The absence of persuasion	No side persuading another side believes something through the merit of reasons proposed	
The absence of coercion	No side intentionally forces another side or uses credible and severe threats of harm to control another side	
The absence of information manipulation	There is no use of non-persuasive means to alter a side's understanding of a situation	
The absence of reward manipulation	No side motivates another side to do what the agent of influence intends	
Inducement	No offer to provide incentives are made	Appelbaum et al. (2009)
persuasion	No application of interpersonal pressure or by an exhortation to self-interest or community norms is applied	
Force	No enforcement by non-consensual intervention or the issuance of threats is used	
Diminished capacity	Supply or funding chains are disrupted	Brunk (1979)
Goals	There is a willingness to mediate the dispute to achieve a mutual goal	
Manipulation	The choice of action is free from constraints imposed by other persons or social institutions	
Understanding of the proposed program	Potential participants have an adequate understanding of specific aspects of the proposed program or even of the program in general	Kamuya et al. (2011)
Social norms	No side considers decision making by the other side as the social norm	
Social relations	Cross-cutting interpersonal and contextual domains do not make it difficult to say no	
Value	There is a willingness to mediate the dispute for shared value	

(continued)

Table 5 (continued)

Dimensions of voluntariness	Manifestations	References
Inducements	The voluntariness of the disputants is undermined by "inducements" or "offers" designed to encourage the parties to enter mediation	

3 Reality Checks on Construction Contracting and Mediation

Mediation is one of the most common means of ADR used in Hong Kong to resolve construction disputes. However, every means has her own advantages and shortcomings. There is a need to review the practicality of the design assumptions. This section serves as realty checks on the conventional construction dispute mediation design.

3.1 Principal-Agent Relationship

One of the earliest and most popular theories to explain how organisations interact is the principal-agent theory (PAT hereafter). In a typical principal-agent arrangement, the principal delegates authority to the agent, to act and make decisions on his behalf. Moreover, the principal and the agent may have inherent conflicting interest (Jensen & Meckling, 2019; Meckling & Jensen, 1976). Eisenhardt (1989) also employed an agency perspective to study the problems of cooperation within coalition of members of diverse background. The typical principal-agent relationships in construction are those between developer and contractor, between the main contractor and the subcontractor. Although it is typically expected that all parties will work together to effectively complete a construction project, the principle-agent theory suggests that because each party is motivated by their own interests, inherent conflict between them is inevitable. Additionally, due to the characteristics of construction projects, such as their high level of asset specificity and uncertainty (Zhu & Cheung, 2021), their relationship may change as far as interdependency is concerned. The special characteristics are illustrated as follows:

First, Williamson (1983) proposed six dimensions of asset specificity: (i) human asset specificity; (ii) physical asset specificity; (iii) site specificity; (iv) dedicated asset specificity, (v) capital specificity; (vi) temporal specificity (Masten et al., 1991; Riordan & Williamson, 1985). Among them, construction projects highlight dedicated asset specificity and temporal specificity. Dedicated assets are those that have been specifically made for a particular transactional. It is therefore built on the anticipation of a long-term partnership. In contrast to a standard buyer–seller contractual relationship, termination of a construction contract, particularly in the middle and later stages of the project, could have serious financial implications. According to

Masten et al. (1991), temporal specificity relates to the significance of timing and coordination needed for a transactional relationship. When it comes to construction projects, on-time delivery becomes essential to avoid costly delays (Chang & Ive, 2007). Furthermore, there has been much discussion about how asset specificity relates to the performance of inter-firm relationships. According to the TCE, asset specificity raises the risks associated with opportunism (Heide & Stump, 1995). Nevertheless, the relational exchange theory (RET hereafter) is more focused on the resource efficiency of the firms (Pitelis, 2007). Lui et al. (2009) found that the development of intangible, relation-specific assets and trust-based collaborative behaviour would improve inter-firm relationship.

Second, every project is subject to the uncertainty of, among others, scope, environment, and human decisions (Ward & Chapman, 2003). It is not a random act but methodically related to the elements of people's workplace, activities, and resources (Williamson, 1979; Love et al., 2020). Uncertainty in construction projects can come from five different sources: (i) program: bad weather and environmental approvals; (ii) quality: substandard work, non-conformance, and inadequate quality assurance; (iii) management: supervisors and engineers' behaviour, stakeholder relations; (iv) design: contract variations, errors and omissions in documentation, approval; (v) safety: safety culture, fatigue, competency of construction plant operators. Because each construction project is unique, they inherently carry new risk. This uncertainty is exacerbated by the intrinsic complexity and ambiguity of the construction process, which is also rendered worse by process variability (Cheung & Yiu, 2006; Love et al., 2016).

3.2   Power Asymmetry

The discussion of asset specificity in construction projects in Sect. 3.1 has brought out the issue of power between the contracting parties. Empirical evidence is available to confirm that power differential does exist between contracting parties (Gaski, 1984; Dwyer et al., 1981; McAlister et al., 1986).

Emerson (1962) contends that power is a function of resource availability and criticality: power increases when a particular organization's resource is in higher demand and less generally available. Likewise, asymmetric information may also be a function of power whereby one party is having better information (Schieg, 2008). For instance, the contractor will be in an information advantaged position when the developer is not able to observe the performance of the contractor. Equity theory (ET hereafter) explains the negative impact of imbalanced distribution of output in relation to the respective input (Adams, 1963). Emerson's (1962) power-dependence theory (PDT hereafter) also points out the use of benchmarks to determine if the outcome is a gain or not. Power, according to negotiation professionals (Fisher et al., 2011), is a function of what alternatives one has instead of reaching a settlement. Power asymmetry permeates basically every phase of mediation.

Power asymmetry combines both the substantive and relational aspects of mediation by comparing with the other party to create a valuable perspective. The extent to which the parties have outside options, substantial resources, or other means of preserving some level of independence from their counterpart and expected risks and returns could vary greatly. In this way, it unites issues and people rather than separating them from issues.

3.3   Quasi Imposition of Construction Mediation

The high level of uncertainty in construction projects coupled with bounded rationality renders the construction contract incomplete. The level of contract completeness influences the type of conflict whereby different resolution approaches would deemed appropriate (Lumineau & Malhotra, 2011). Although ADR usage patterns vary, Hong Kong has been actively developing and promoting the use of mediation as an efficient means of dispute resolution. There are several reasons for this preference, but the main ones include the time and cost savings, allowing parties to maintain control over the issue resolution, and flexible structuring of settlement options (Stipanowich & Lamare, 2014).

Mediation can be used at any stage of a dispute, and voluntary participation is one of the central designs (Nolan-Haley, 2012). A voluntary mediation process is one freely chosen by the participants. that is freely chosen by the participants; voluntarily made agreements; parties are not forced to mediate and settle by an internal or external party to a dispute (Moore, 2014). Stulberg (1996) notes that "there is no legal liability to any party refusing to participate in a mediation process. Since a mediator has no authority to impose a decision on the parties, he/she cannot threaten the recalcitrant party with a judgment." During a mediation, every party is free to suggest options. The alternative to a negotiated agreement is considered the outcome is accruing to a party from not interacting with the other party and not agreeing to participate in the mediation with them. If the party's mediation outcomes are equal to or exceed their alternative outcomes, they will continue with the mediation. On the other hand, if the alternative outcomes exceed the one on the table, they will take steps to improve, and if failed, they will withdraw from the mediation (Wall, 1981). However, voluntary participation does not mean that they may not be 'compelled' to try mediation (Moore, 2014). Other disputants or external figures, such as friends, colleagues, constituents, authoritative leaders, or judges, may put pressure on a party to attempt mediation. In addition, some courts on family and civil cases in the United States require parties to participate in mediation and make good faith efforts for a settlement before the court will be willing to hear the case. Apart from that, attempting mediation does not mean that the participants must settle. Apparently, forced negotiation may not provide the necessary conducive platform for genuine attempts to settle (Trakman & Sharma, 2014). Moreover, there has been a call for the mandatory use of mediation to accelerate its adoption (Nolan-Haley, 2011; Quek, 2009). As far as the practice of mediation is concerned, any form of imposition has been viewed as a departure of voluntary participation. Nevertheless, in construction contracting, certain effort to overcome the inertia is necessary. In this connection, By analyzing the current arrangements related to construction mediation, the mediation rules, contractual use, court encouragement and court-connected, an analysis of voluntary participation is illustrated.

Mediation Rules

The Hong Kong International Arbitration Centre (HKIAC) is the leading dispute resolution services provider in Hong Kong. The rules of the HKIAC are most representative of the conduct of mediation in Hong Kong. According to the HKIAC mediation rule, a failure by any party to reply within 14 days shall be treated as a refusal to mediate. Thus, mediation can only be conducted if all parties agree to mediate. This echoes the conceptual approach that mediation should be participated voluntarily (Katz, 1993). Delay tactics are less likely to be pursued if the disputants decide to mediate at their own will. The parties are much more likely to make meaningful contribution, such as negotiating in good faith. In addition, in fully voluntary mediation. Afterall, the parties are free to leave at any time.

Contractual Use of Mediation

Most construction contracts have adopted a tiered- resolution procedures that include mediaton as an intermediate step before binding resolution forums. Cheung (2016) summarised the construction mediation landscape as follows. Typically, a three-tiered dispute resolution procedure is used. According to the HKG General Conditions of Contract for Building Works/Civil Engineering Works/Design and Build Contracts Clause 86 and General Conditions of Contract for Term Contract for Building Works Clause 92/Civil Engineering Works Clause 89, when a dispute arises, it shall be reported to and settled by the designated contract administrator. If either party is dissatisfied with the decision made, they can refer the matter to mediation within 28 days of the decision. If the matter cannot or does not need to be resolved by mediation, any reference to arbitration shall be made in accordance with the Arbitration Ordinance within 90 days. A similar design is also adopted in the private building projects form of contract. More recently, the New Engineering Contract (NEC) has been used extensively for public works projects in Hong Kong. The 2017 NEC4 Dispute Resolution Service Contract (DRSC) offers three dispute resolution options (W1, W2, and W3), and Z-clauses that provide bespoke additional contract conditions can be added, allowing unique requirements for local dispute resolution practices. W1 and W2 under NEC4 use adjudication as the primary means of dispute resolution, W3 uses dispute avoidance, while mediation can be added to the Z-clauses as a construction dispute resolution tool in the NEC, such as adjudication and arbitration. This contractual use of mediation is quite different from its mandatory use because voluntary participation is retained. To ensure the validity of ADR clauses, it is important that mediation provision should be specific enough so that objective criteria can be deduced to determine compliance or otherwise. As such, a mediation clause should specify the model and rules to be used. In addition, a clear time frame for its implementation, the nominating authority and the minimum amount of participation are essential items to be incorporated to develop an enforceable mediation clause. Besides, contractual mediation clauses are often found in construction contracts which mandate mediation when a dispute arises, like a statutory mandate. However, the initial decision to mandate mediation for disagreements is made by the parties themselves and leaves intact voluntariness of the agreement (Nelle, 1991).

Court Encouraged Mediation

The legislation and judicial pronouncements appear to be pushing the parties to mediate. Some researchers worried that mediation de facto would lose its voluntary nature (Ahmed, 2012). However, the initial decision to mediate their disagreements is made by the parties themselves, and as such voluntariness is maintained (Nelle, 1991). According to Section F of Hong Kong High Court Practice Direction 6.1, construction cases reaching the Hong Kong High Court are encouraged to attempt mediation. Accordingly, upon receiving the Mediation Notice, the Respondent should respond to the Applicant in writing within 14 days, although he has the right to refuse to mediate. The principal way to encourage an attempt to mediate to involves the imposition of cost sanctions where a party unreasonably refuses to attempt. However, if a party (1) has engaged in mediation to the minimum level of expected participation agreed upon by the parties beforehand or as determined by the court or (2) has a reasonable explanation for nonparticipation, he should not suffer any adverse costs order.

Court-Connected Mediation

The dilemma of compeling parties to voluntary mediation is a paradox in itself (Cao, 2021). The debate over imposition of mediation will never (Cheung, 2016; Hilmer, 2013; Leung, 2014; Meggitt, 2018). In Canada, Australia, the United Kingdom and Singapore, the courts are more open about the use of compulsory ADR. Court-annexed mediation is the most direct way to ensure attempts of mediation. The Civil Justice Reform's (CJR's) Working Party has proposed court-annexed mediation; in its Interim in 2000. However, the proposal was finally rejected and no court-annexed mediation is put to practice (Cheung, 2016; Hilmer, 2013; Meggitt, 2018). Statutory use denotes that disputes will be automatically directed for mediation, irrespective of its nature. The downside is the parties only mediate perfunctorily (Leung, 2014). Parties being forced to mediate are unlikely mediate in good faith (Meggitt, 2018). Court-annexed mediation undermines the voluntary nature of mediation (Cheung, 2016). If parties are forced to mediate, particaption may merely be taken to satisfy the mandatory requirements. Rules of law and justice may not even be on the agenda, which may address commercial issues in a way that lacks clarity and certainty (Hilmer, 2013).

3.4 Disputants' Perceptions of the Used of Mediation

Biases in construction projects can be in the following forms: strategic misrepresentation and normalization of deviation (Pinto, 2013); opportunistic decision criteria and value perception (Brewer & Runeson, 2009); heuristics and organizational learning (Winch & Kelsey, 2005); anchoring, overconfidence, self-serving, hindsight, and confirmation (Cheung & Li, 2019). Previous studies mainly focused on bias a mediator holds toward the disputants, as an impartial role in mediation procedures. Anchoring and confirmation, both forms of bias were found to be more likely to occur in construction dispute (Cheung et al., 2019; Izumi, 2010). Moreover, unintentional biases may also affect disputants' judgment who need to be rational to achieve better outcomes. The selective accessibility model argues that anchoring involves estimating the target closes to the anchor that may not be realistic or rational (Strack et al., 2016). Disputant who is influenced by anchoring bias is more likely to make judgments based on the first set of information they have. It is challenging to make appropriate judgments on a final assessment in this situation that can differ significantly from the disputants' preconceptions. Confirmation bias describes the tendency to search or interpret information that conform to those already held views, expectations, and situational context. (Kassin et al., 2013; Nickerson, 1998). With the presence of confirmation bias, it can be expected that disputants would make decisions based on the information in possession. It can therefore be assumed that biases may affect the disputing parties if they have certain pre-occupation on the process.

4 Voluntary Mediation and the Reality

Table 5 gives the dimensions of voluntariness. It is advocated that in the context of and is advocated that these dimensions have two characteristics: no compulsion and no indifference. This section discusses the compatibility between reality and the use of voluntary participation as an incentive for the use of mediation. Table 6 summarises the initial evaluations. Apparently, the reality may not render voluntary mediation a pragmatic option. In addition, the actual practice also suggested that pure voluntary participation is not easy to come by.

The following two sub-sections argue that the reality checks of the use of construction mediation in Hong Kong do not neatly meet the "no compulsion" and "no indifference" views of voluntariness.

Table 6 Reality check and elements of voluntariness

Principal agent relationship (*Characterized by self-interested principal and work-averse agent*)	Power asymmetry (*Characterized by differential in resource, information, and expectation*)	Quasi-imposition (*Characterized by use of adverse sanction on non-compliance*)	Disputants' perception (Characterised by anchoring and confirmation)
No Compulsion? No Indifference?			

4.1 Power Asymmetry

Principal agency theory highlights the inherent conflict of interest between the principal and the agent (PA hereafter). In construction contracting, although the relationship between the employer and the contractor can be largely identified as one of principal and agent. Moreover, the increasing involvement of the employer during construction has significantly made the relationship with the contractor as one between collaborators. Nevertheless, the asymmetric conditions inherited from a P-A relationship remain. Apparently, mediation is a kind of assisted negotiation and assumes parties enjoy free negotiation as formulated in most negotiation theories. With a P-A relationship the power differentials can be in the form of resources and information. From a behavioural perspective, the process assumes open negotiation with the disputants having no worries about the consequences if the mediation fails. Moreover, mediation is only one of the steps of a multi-tiered arrangement (Li & Cheung, 2018). The caveat that mediation is being used as a rehearsal of planned arbitration and even litigation cannot be overruled. It is therefore advocated that power asymmetry can deter the use of mediation due to the inevitable power asymmetry. The degree of power asymmetry between the parties appears to be directly related to the initiation of mediation (Richmond, 1998). Not much research on the influence of asymmetric conditions between the disputants in construction dispute mediation have been identified. It is crucial to further this study by developing the constructs of power asymmetry and investigate the implications of voluntary participation of mediation.

4.2 Quasi-Imposition

The use of mediation to resolve construction dispute is largely contractual based. Contractual use of mediation was first introduced in Hong Kong in the 80's with the Hong Kong Government Architectural Services Department taking the lead. Voluntary mediation was enabled for allowing proceed to next tier of resolution when a referral to mediation is not met with positive response. This approach has been taken in most contracts used in Hong Kong. When a construction case reaches to the High

Court of Hong Kong, the provisions of minimum participation of mediation under PD 6.1 shall apply. To avoid adverse cost order, it is necessary to attempt mediation though reasonable refusal is allowed. Thus, the immediate question is: "under these conditions can voluntary participation still be claimed as far as no compulsion?". According to mediation theories, like the Harvard concept (Fisher et al., 2011), developing as many options as possible and with no fixed position are the underlying principles of successful mediation. Human judgment biases always affect disputants' decisions since they are having varied experiences and certain degrees of irrationality. Anchoring and confirmation biases are sources of entrenched positions which should be properly managed. Mediators need to facilitate the disputants to eliminate biases, therefore discovering a better solution, or achieving a better outcome than they would have done without such a mechanism.

However, the procedure has inherent default because the disputants are somehow steered by the 'imposed' procedures. For example, one's willingness to make concessions is simply not be imposed. During a mediation, various factors have an impact upon the willingness to negotiate. The general definition of a mediation is that the disputing parties voluntarily come together to try and discover a better solution or accomplish a better result than they would have done without such a mechanism. As for "no indifference", it is not unheard of the perfunctory attitude taken by the participating parties when the mediation is quasi-imposed. The caveat of using mediation as rehearsal equally applies in this situation.

5 Voluntary Participation as an Incentive for Use of Construction Dispute Mediation

Sections 4.1 and 4.2 highlight the possible incompatible conditions against voluntary participation in construction dispute mediation. Power asymmetry between disputants and the quasi-imposition are identified as two interesting inherent obstacles that would marginalise the incentivising function of voluntary participation.

The principal–agent problem refers to conflict of interest arises when one entity (the "agent") acts on behalf of another entity (the "principal") (Grossman & Hart, 1992). The approach/inhibition theory of power examines how power influences individual's psychological states and transform their behaviour (Keltner er al., 2003). The principle-agent problem frequently happens between construction project parties due to the high level of asset specificity and unpredictability. In dyadic relationships, the more powerful partner is linked to positive affect, attention to rewards, automatic information processing, and unrestrained conduct. Conversely, a weaker party is linked to negative affect, attention to threat, restrained information processing, and social inhibition. This study applies principal–agent problem and approach/inhibition theory of power to explain whether and how power asymmetry affects the voluntary

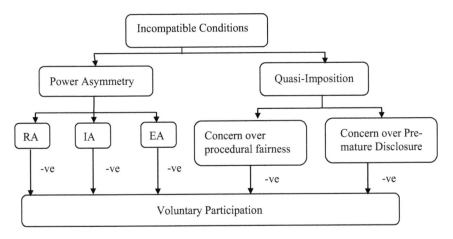

Fig. 4 Obstacles against voluntary participation

to participate in mediation. A principal agent relationship is characterized by a self-interested principal and a work-averse agent, and power asymmetry is characterized by differences in resource, information, and expectation.

Consider current arrangements related to construction mediation, contractual use, court encouragement and court-connected mediation, this study argues that quasi-imposition negates parties' willingness to mediate. Perception of fairness has been an important variable when studying behavioral willingness (Maxwell, 2002). And bias which has been found to be pervasive in negotiation obviously has an impact on perceived fairness (Gelfand et al., 2002).

This study points out that concerns of disputant may over the procedural fairness, and the way mediation is installed would negate the willingness to mediate. Apart from the mediation at the table, there still have concerns about the opportunistic motive in conducting the mediation. Such as treating this as a rehearsal of arbitration. It is pointed out that voluntary participation can be a sweetener of mediation adoption. However, due to the prevailing contracting arrangement and practice, willingness to mediate can be marginalised by the structural issue of power asymmetry as well as the implementation issue of quasi-imposition. Figure 4 summarises the key arguments put forward by this study.

6   Summary

This chapter first reviews the landscape of construction dispute resolution including nature of dispute, approaches to resolution, forms of resolution. It is also found that mediation has been identified as the preferred alternative dispute resolution method for civil disputes by the Hong Kong government. In construction, several governmental initiatives have been taken to promote its use. These promotional efforts are to upbring potential users' knowledge on the benefits of using mediation. Reported usage of mediation has dropped recently after promising initial uptake after the 2009 Civil Justice Reform. Voluntary participation has been one of the key attributes promoted by mediation service providers. The control of the proceedings by the disputing parties is considered as a very appealing feature. Thus, voluntary participation has been the core design of mediation procedures whether it is contractual or court-encouraged. A reality check on contracting practice revealed two note-worthying incompatible contracting conditions that would negate willingness to mediate. This study serves a timely reminder of the limitations of construction dispute mediation through a reality check on the prevailing construction contracting practice.

References

Adams, J. S. (1963). Towards an understanding of inequity. *The Journal of Abnormal and Social Psychology, 67*(5), 422.

Afzalur Rahim, M., & Bonoma, T. V. (1979). Managing organisational conflict: A model for diagnosis and intervention. *Psychological Reports, 44*, 1323–1344.

American Arbitration Association, American Bar Association, and the Association for Conflict Resolution. (2004). *Model standards of conduct for mediators.* Washington, DC: American Arbitration Association, American Bar Association, and the Association for Conflict Resolution.

Appelbaum, P. S., Lidz, C. W., & Klitzman, R. (2009). Voluntariness of consent to research: A conceptual model. *Hastings Center Report, 39*(1), 30–39. https://doi.org/10.1353/hcr.0.0103

Bandura, A. (1986). The explanatory and predictive scope of self-efficacy theory. *Journal of Social and Clinical Psychology, 4*(3), 359.

Baumeister, R. F., & Leary, M. R. (2017). The need to belong: Desire for interpersonal attachments as a fundamental human motivation. Interpersonal development, 57–89.

Brewer, G., & Runeson, G. (2009). Innovation and attitude: Mapping the profile of ICT decision-makers in architectural, engineering and construction firms. *International Journal of Managing Projects in Business.*

Brunk, C. G. (1979). The problem of voluntariness and coercion in the negotiated plea. *Law and Society Review*, 527–553.

Boettger, U. (2004). Efficiency versus party empowerment-against a good-faith requirement in mandatory mediation. *The Review of Litigation, 23*, 1.

Cao, N. (2021). The paradox of power asymmetry and voluntary participation in construction dispute mediation. In *Construction Dispute Ressearch Expanded* (Edr: Cheung S.O.), Springer International.

Chan, E. H., & Suen, H. C. (2005). Disputes and dispute resolution systems in Sino-foreign joint venture construction projects in China. *Journal of Professional Issues in Engineering Education and Practice, 131*(2), 141–148.

Chang, C. Y., & Ive, G. (2007). The hold-up problem in the management of construction projects: A case study of the Channel Tunnel. *International Journal of Project Management, 25*(4), 394–404.

Chau, K. W. (2007). Insight into resolving construction disputes by mediation/adjudication in Hong Kong. *Journal of Professional Issues in Engineering Education and Practice, 133*(2), 143–147.

Cheung, S. O. (2016). 4 court-connected mediation in Hong Kong. In *Court-Connected Construction Mediation Practice: A Comparative International Review* (pp. 55–77).

Cheung, S. O., & Li, K. (2019). *Biases in construction project dispute resolution.* Engineering, Construction and Architectural Management.

Cheung, S. O., Li, K., & Levina, B. (2019). Paradox of bias and impartiality in facilitating construction dispute resolution. *Journal of Legal Affairs and Dispute Resolution in Engineering and Construction, 11*(3), 04519007.

Cheung, S. O., Tam, C. M., Ndekugri, I., & Harris, F. C. (2000). Factors affecting clients' project dispute resolution satisfaction in Hong Kong. *Construction Management & Economics, 18*(3), 281–294.

Cheung, S. O., & Pang, H. Y. (2014). Conceptualising construction disputes. In *Construction Dispute Research* (pp. 19–37). Springer, Cham. Routledge.

Cheung, S. O., & Yiu, T. W. (2006). Are construction disputes inevitable? *IEEE Transactions on Engineering Management, 53*(3), 456–470.

Chong, H. Y., & Zin, R. M. (2012). Selection of dispute resolution methods: Factor analysis approach. *Engineering, Construction and Architectural Management.*

Chow, P. T., Cheung, S. O., Young, C. Y., & Wah, C. K. (2015). The roles of withdrawal in the negotiator personality-tactic relationship. *Journal of Business Economics and Management, 16*(4), 808–821.

Deci, E. L., & Ryan, R. M. (1985). The general causality orientations scale: Self-determination in personality. *Journal of Research in Personality, 19*(2), 109–134.

Deci, E. L., & Vansteenkiste, M. (2003). Self-determination theory and basic need satisfaction: Understanding human development in positive psychology.

Dwyer, F. R., & Walker, O. C., Jr. (1981). Bargaining in an asymmetrical power structure. *Journal of Marketing, 45*(1), 104–115.

Emerson, R. M. (1962). Power-dependence relations. *American Sociological Review, 27*(1), 31–41.

Eisenhardt, K. M. (1989). Agency theory: An assessment and review. *Academy of Management Review, 14*(1), 57–74.

Fisher, R., Ury, W. L., & Patton, B. (2011). Getting to yes: Negotiating agreement without giving in. Penguin.

Gaski, J. F. (1984). The theory of power and conflict in channels of distribution. *Journal of Marketing, 48*(3), 9–29.

Gelfand, M. J., Higgins, M., Nishii, L. H., Raver, J. L., Dominguez, A., Murakami, F., ... & Toyama, M. (2002). Culture and egocentric perceptions of fairness in conflict and negotiation. *Journal of Applied Psychology, 87*(5), 833.

Grossman, S. J., & Hart, O. D. (1992). An analysis of the principal-agent problem. In Foundations of insurance economics (pp. 302–340). Springer, Dordrecht.

Harmon, K. M. (2003). Resolution of construction disputes: A review of current methodologies. *Leadership and Management in Engineering, 3*(4), 187–201.

Harter, S. (1978). Effective motivation reconsidered. *Toward a Developmental Model. Human Development, 21*(1), 34–64.

Hedeen, T. (2005). Coercion and self-determination in court-connected mediation: All mediations are voluntary, but some are more voluntary than others. *Justice System Journal, 26*(3), 273–291.

Heide, J. B., & Stump, R. L. (1995). Performance implications of buyer-supplier relationships in industrial markets: A transaction cost explanation. *Journal of Business Research, 32*(1), 57–66.

Hilmer, S. E. (2013). Mandatory mediation in Hong Kong: A workable solution based on Australian experiences. *China-EU Law Journal, 1*(3–4), 61–96. https://doi.org/10.1007/s12689-012-0016-y

Izumi, C. (2010). Implicit bias and the illusion of mediator neutrality. *Wash. UJL & Pol'y, 34*, 71.

Jensen, M. C., & Meckling, W. H. (2019). Theory of the firm: Managerial behavior, agency costs and ownership structure. In Corporate Governance (pp. 77–132). Gower.

Kamuya, D., Marsh, V., & Molyneux, S. (2011). What we learned about voluntariness and consent: Incorporating "background situations" and understanding into analyses. *The American Journal of Bioethics, 11*(8), 31–33.

Kassin, S. M., Dror, I. E., & Kukucka, J. (2013). The forensic confirmation bias: Problems, perspectives, and proposed solutions. *Journal of Applied Research in Memory and Cognition, 2*(1), 42–52.

Katz, L. V. (1993). Compulsory alternative dispute resolution and voluntarism: Two-headed monster or two sides of the coin. *Journal Dispute Resolution, 1.*

Keltner, D., Gruenfeld, D. H., & Anderson, C. (2003). Power, approach, and inhibition. *Psychological Review, 110*(2), 265.

Kim, S., Kong, M., Choi, J., Han, S., Baek, H., & Hong, T. (2021). Feasibility analysis of COVID-19 response guidelines at construction sites in south Korea using CYCLONE in terms of cost and time. *Journal of Management in Engineering, 37*(5), 04021048.

Klir, G. J. (2006). Uncertainty and information. Foundations of Generalised Information Theory.

Legal Reference System. (2022). Cases reached Construction and Arbitration Proceedings in the high court of Hong Kong special administrative region. Retrieved from https://legalref.judiciary.hk/lrs/common/ju/judgment.jsp?L1=HC&L2=CT&AR=14#A14.

Leung, R. H. M. (2014). *Hong Kong mediation handbook.* Sweet & Maxwell.

Li, K., & Cheung, S. O. (2018). Bias measurement scale in repeated dispute evaluations. *Journal of Management in Engineering, 34*(4): https://doi.org/10.1061/(ASCE)ME.1943-5479.0000617.

Love, P. E., Edwards, D. J., & Smith, J. (2016). Rework causation: Emergent theoretical insights and implications for research. *Journal of Construction Engineering and Management, 142*(6), 04016010–04016011.

Love, P. E., & Matthews, J. (2020). Quality, requisite imagination and resilience: Managing risk and uncertainty in construction. *Reliability Engineering & System Safety, 204*, 107172.

Lui, S. S., Wong, Y. Y., & Liu, W. (2009). Asset specificity roles in interfirm cooperation: Reducing opportunistic behavior or increasing cooperative behavior? *Journal of Business Research, 62*(11), 1214–1219.

Lumineau, F., & Malhotra, D. (2011). Shadow of the contract: How contract structure shapes interfirm dispute resolution. *Strategic Management Journal, 32*(5), 532–555.

Marti, K., Ermoliev, Y., Makowski, M., & Pflug, G. (2010). *Coping with uncertainty.* Springer.

Masten, S. E., Meehan, J. W., Jr., & Snyder, E. A. (1991). The costs of organisation. *Journal Economics and Orgnization, 7*, 1.

Martins, M., & Monroe, K. B. (1994). Perceived price fairness: A new look at an old construct. *ACR North American Advances.*

Maxwell, S. (2002). Rule-based price fairness and its effect on willingness to purchase. *Journal of Economic Psychology, 23*(2), 191–212.

Meggitt, G. (2018). The cases for (and against) compulsory court-connected mediation in Hong Kong. *5th Asian Law Institute Conference*, 22–23.

Meckling, W. H., & Jensen, M. C. (1976). Theory of the firm: Managerial behavior, agency costs and ownership structure. *Journal of Financial Economics, 3*(4), 305–360.

Merry, S. E. (1989). Myth and practice in the mediation process. Mediation and Criminal Justice: Victims, Offenders, and Community. London: Sage, 239–250.

Mitropoulos, P., & Howell, G. (2001). Model for understanding, preventing, and resolving project disputes. *Journal of Construction Engineering and Management, 127*(3), 223–231.

Moore, C. W. (2014). *The mediation process: Practical strategies for resolving conflict.* Wiley.

Nelle, A. (1991). Making mediation mandatory: A proposed framework. *Ohio State Journal on Dispute Resolution, 7*, 287.

Nelson, R. M., Beauchamp, T., Miller, V. A., Reynolds, W., Ittenbach, R. F., & Luce, M. F. (2011). The concept of voluntary consent. *The American Journal of Bioethics, 11*(8), 6–16. https://doi.org/10.1080/15265161.2011.583318

Nickerson, R. S. (1998). Confirmation bias: A ubiquitous phenomenon in many guises. *Review of General Psychology, 2*(2), 175–220.

Nolan-Haley, J. M. (2011). Is Europe headed down the primrose path with mandatory mediation. *NCJ International Com. Reg., 37*, 981.

Nolan-Haley, J. (2012). Mediation: The new arbitration. *Harvard Negotiation Review, 17*, 61.

Petrzelka, P., & Bell, M. (2000). Rationality and solidarities: The social organization of common property resources in the Imdrhas Valley of Morocco. *Human Organization, 59*(3), 343–352.

Pitelis, C. N. (2007). A behavioral resource-based view of the firm: The synergy of Cyert and March (1963) and Penrose (1959). *Organization Science, 18*(3), 478–490.

Poitras, J. (2005). A study of the emergence of cooperation in mediation. *Negotiation Journal, 21*(2), 281–300.

Pinto, J. K. (2013). Lies, damned lies, and project plans: Recurring human errors that can ruin the project planning process. *Business Horizons, 56*(5), 643–653.

Quek, D. (2009). Mandatory mediation: An oxymoron-examining the feasibility of implementing a court-mandated mediation program. *Cardozo Journal Conflict Resolution, 11*, 479.

Riordan, M. H., & Williamson, O. E. (1985). Asset specificity and economic organisation. *International Journal of Industrial Organization, 3*(4), 365–378.

Richmond, O. (1998). Devious objectives and the disputants' view of international mediation: A theoretical framework. *Journal of Peace Research, 35*(6), 707–722.

Ryan, R. M., & Deci, E. L. (2000). Self-determination theory and the facilitation of intrinsic motivation, social development, and well-being. *American Psychologist, 55*(1), 68.

Schieg, M. (2008). Strategies for avoiding asymmetric information in construction project management. *Journal of Business Economics and Management, 9*(1), 47–51.

Schifter, D. E., & Ajzen, I. (1985). Intention, perceived control, and weight loss: An application of the theory of planned behavior. *Journal of Personality and Social Psychology, 49*(3), 843.

Sherman, E. F. (1992). Court-mandated alternative dispute resolution: what form of participation should be required. *SMUL Review, 46*, 2079.

Smith, G. (1998). Unwilling actors: Why voluntary mediation works, why mandatory mediation might not. *Osgoode Hall LJ, 36*, 847.

Stipanowich, T. J., & Lamare, J. R. (2014). Living with ADR: Evolving perceptions and use of mediation, arbitration, and conflict management in fortune 1000 corporations. *Harvard Negotiation l. Review, 19*, 1.

Strack, F., Bahník, Š, & Mussweiler, T. (2016). Anchoring: Accessibility as a cause of judgmental assimilation. *Current Opinion in Psychology, 12*, 67–70.

Stulberg, J. B. (1996). Facilitative versus evaluative mediator orientations: Piercing the grid lock. *Fla. St. UL Review, 24*, 985.

The Hong Kong judiciary. (2020). What is mediation? Retrieved from https://mediation.judiciary.hk/en/what_is_mediation.html

The Hong Kong Judiciary. (2021a). Mediation Statistics for Building Management Cases. Retrieved from https://mediation.judiciary.hk/en/figures and statistics.html

The Hong Kong Judiciary. (2021b). Mediation Statistics for Civil Justice Reform related cases. Retrieved from https://mediation.judiciary.hk/en/figures_and _statistics.html

The Hong Kong Construction Industry Council. (2010). Guidelines on Dispute Resolution. Retrieved from https://www.cic.hk/cic_data/pdf/about_cic/publications/eng/V10_6_e_V00_Guidelines_DisputeResolution.pdf

The European Union Directive. (2008). The European Parliament and of the Council of 21 May 2008 on certain aspects of mediation in civil and commercial matters OJ L 136, 24.5.2008, p. 3–8 (BG, ES, CS, DA, DE, ET, EL, EN, FR, IT, LV, LT, HU, MT, NL, PL, PT, RO, SK, SL, FI, SV) Special edition in Croatian: Chapter 19 Volume 009 P. 281–286.

Totterdill, B. W. (1991). Does the construction industry need alternative dispute resolution? The opinion of an engineer. *Construction Law Journal, 7*(3), 189–199.

Trakman, L. E., & Sharma, K. (2014). The binding force of agreements to negotiate in good faith. *The Cambridge Law Journal, 73*(3), 598–628.

Wathne, K. H., & Heide, J. B. (2000). Opportunism in interfirm relationships: Forms, outcomes, and solutions. *Journal of Marketing, 64*(4), 36–51.

Vallerand, R. J. (1997). Toward a hierarchical model of intrinsic and extrinsic motivation. In Advances in experimental social psychology (Vol. 29, pp. 271–360). Academic Press.

Ward, S., & Chapman, C. (2003). Transforming project risk management into project uncertainty management. *International Journal of Project Management, 21*(2), 97–105.

Welsh, N. A. (2001). Making deals in court-connected mediation: What's justice got to do with it. *Wash. ULQ, 79*, 787.

Winch, G. M., & Kelsey, J. (2005). What do construction project planners do? *International Journal of Project Management, 23*(2), 141–149.

Wissler, R. L. (1997). The effects of mandatory mediation: Empirical research on the experience of small claims and common pleas courts. *Willamette l. Rev., 33*, 565.

Williamson, O. E. (1979). Transaction-cost economics: The governance of contractual relations. *The Journal of Law and Economics, 22*(2), 233–261.

Williamson, O. E. (1983). Credible commitments: Using hostages to support exchange. *The American Economic Review, 73*(4), 519–540.

Williamson, O. E. (1993). Transaction cost economics and organization theory. *Industrial and Corporate Change, 2*(2), 107–156.

Wu, L., Chuang, C. H., & Hsu, C. H. (2014). Information sharing and collaborative behaviors in enabling supply chain performance: A social exchange perspective. *International Journal of Production Economics, 148*, 122–132.

Zaden, I. A. (1965). Fuzzy sets. *Information and Control, 8*, 338–353.

Zajac, E. J., & Olsen, C. P. (1993). From transaction cost to transactional value analysis: Implications for the study of interorganizational strategies. *Journal of Management Studies, 30*(1), 131–145.

Zhu, L., & Cheung, S. O. (2021). Equity gap in construction contracting: Identification and ramifications. *Engineering, Construction and Architectural Management*. https://doi.org/10.1108/ECAM-09-2020-0725

Miss Nan Cao studied construction management and received her MEng degree from the Central South University. Miss Cao is completing her doctoral study in voluntary mediation with the Construction Dispute Resolution Research Unit, City University of Hong Kong. Miss Cao is examining the ways to enhance the value of voluntary mediation as a form of alternative dispute resolution to resolve construction disputes. She is a member Civil Mediation Council and has published conference articles and book chapters in construction mediation.

Ingram Content Group UK Ltd.
Milton Keynes UK
UKHW020624070623
423007UK00002B/28